FROM CLASSICAL MECHANICS TO QUANTUM FIELD THEORY

A TUTORIAL

FROM CLASSICAL MECHANICS TO QUANTUM FIELD THEORY

A TUTORIAL

Manuel Asorey
Universidad de Zaragoza, Spain

Elisa Ercolessi
University of Bologna & INFN-Sezione di Bologna, Italy

Valter Moretti
University of Trento & INFN-TIFPA, Italy

World Scientific

W JERSEY • LONDON • SINGAPORE • BEIJING • SHANGHAI • HONG KONG • TAIPEI • CHENNAI • TOKYO

Published by

World Scientific Publishing Co. Pte. Ltd.
5 Toh Tuck Link, Singapore 596224
USA office: 27 Warren Street, Suite 401-402, Hackensack, NJ 07601
UK office: 57 Shelton Street, Covent Garden, London WC2H 9HE

British Library Cataloguing-in-Publication Data
A catalogue record for this book is available from the British Library.

FROM CLASSICAL MECHANICS TO QUANTUM FIELD THEORY, A TUTORIAL

ISBN 978-981-121-048-8

For any available supplementary material, please visit
https://www.worldscientific.com/worldscibooks/10.1142/11556#t=suppl

Desk Editor: Nur Syarfeena Binte Mohd Fauzi

Typeset by Stallion Press
Email: enquiries@stallionpress.com

Preface

This book grew out of the mini courses delivered at the Fall Workshop on Geometry and Physics, in Granada, Zaragoza and Madrid.

The Fall Workshop on Geometry and Physics takes place in the Iberian Peninsula since 1992. The aim of this workshop is to introduce advanced graduate students, PhD students and young researchers, both in mathematics and physics, with current aspects of mathematics and physics. The International Fall Workshop on Geometry and Physics is normally held over four days around the first week of September and attracts many young participants. Two main speakers each deliver a 4-hour mini course, and the rest of the programme is made up of talks by invited speakers and contributed speakers.

The large number of young participants each year, many of them attending year after year, convinced the Scientific Committee that it would be a good idea to organize correlated mini courses over different years. In particular, it was thought that a reasonable exposition to quantum formalism would allow the young participants to properly benefit from seminars and talks dealing with various aspects of quantum theories along with their impact on modern mathematics and mathematical methods. The first of these courses was given in Granada by Elisa Ercolessi (University of Bologna) on Methods of Quantization and it was followed, in the meeting in Zaragoza, by Valter Moretti (University of Trento) with a course on Advanced Quantum Mechanics, while in the meeting in Madrid Manuel Asorey (University of Zaragoza) conluded the series with an Introduction to Quantum Field Theory. The lecturers did an excellent job, obviously with different degrees of sophistication according to their own taste, to present quantum theory from classical mechanics to quantum field theory. Even though these papers have appeared in print, separately in special issues devoted to the proceedings, we felt that collecting them in a single volume would render better the continuity, the spirit and the aim for which they were delivered.

We would like to thank the speakers for undertaking a revision of their published texts with the aim of constructing cross-referencing and an overall index to improve the unitary character of the book. We hope that other students and readers who did not attend the mini courses will benefit from these presentations.

Giuseppe Marmo,
on behalf of the Scientific Committee

Acknowledgments

The Authors would like to thank the Organizers and the Scientific Committee of the Workshop for the invitation to give the courses and for financial support during the stay.

Contents

Chapter 1

A Short Course on Quantum Mechanics and Methods of Quantization

Elisa Ercolessi

Department of Physics and Astronomy, Università di Bologna and INFN-Sezione di Bologna, via Irnerio 46, 40127, Bologna, Italy.
email: elisa.ercolessi@unibo.it

The first part of this paper aims at introducing a mathematical oriented reader to the realm of Quantum Mechanics (QM) and then at presenting the geometric structures that underline the mathematical formalism of QM which, contrary to what is usually done in Classical Mechanics (CM), are usually not taught in introductory courses. The mathematics related to Hilbert spaces and Differential Geometry are assumed to be known by the reader.

In the second part, we concentrate on some quantization procedures, that are founded on the geometric structures of QM -as we have described them in the first part- and represent the ones that are more operatively used in modern theoretical physics. We will first discuss the so-called "Coherent State Approach" which, mainly complemented by "Feynman Path Integral Technique", is the methods which is most widely used in quantum field theory. Finally, we will describe the "Weyl Quantization Approach" which is at the origin of modern tomographic techniques, originally used in optics and now in quantum information theory.

1.1 Introduction

The XIX century was the apex of Classical Mechanics (CM). The newly born tools of differential and integral calculus and new theoretical general principles (such as variational ones) allowed to put on a rigorous basis what we now call Analytical Mechanics and provided the framework to study all mechanical problems, from the simple case of a single point particle, to planetary motion, to rigid bodies. Also, the study of electric and magnetic forces culminated in the work of J.C. Maxwell, which not only provided a unification of these two originally different phenomena using the concept of electromagnetic field, but also unified electromagnetism with the theory of light via the notion of electromagnetic waves. Thus at the beginning of the XX century, physicists (and mathematical physicists) essentially worked with two paradigms, according to which they could study all known phenomena[1]:

- matter: described by corpuscles, denumerable and localized: position and momentum at a given instant of time are the quantities defining their motion, knowing which one can calculate any other observable, such as energy;
- fields: described by waves, continuous and delocalized: to study their motion one needs the concept of wavelength (or frequency), propagation speed and amplitude of the oscillation.

Since the end of the XIX century, more and more compelling experimental evidences started questioning the great success of CM and its paradigms. The new physics emerged when people began to study the interaction of light with matter and matter itself at a microscopic level: blackbody radiation, photoelectric effect, atomic spectra,... To examine all these facts goes beyond the scope of these lectures and a discussion of them can be found in most introductory books in QM (see e.g. [8; 15]). We recall that only in 1924, De Broglie proposed [9] that, in the same way as electromagnetic waves can be described by discretized corpuscles (i.e. quanta of light, later denominated photons), particles composing matter may be described as a wave, whose wavelength is connected to the momentum of the particle via the famous relation: $\lambda = h/p$. Such a conceptually new description of radiation and matter is at the origin of the so-called *particle-wave duality*, that reigns over the quantum world.

One can see this principle "in action" when studying the so-called Schrödinger and Heisenberg approach to Quantum Mehanics (QM), but its effects can be seen up to more recent conceptual developments such as the definition of a quantum field (both matter and interaction field) and the technique of second quantization. An introduction to the techniques of quantizations in Field Thoery can be found in the third part of this volume.

[1] Of course there are points of contacts between these two approaches. There are situations in which point-like particles originate collective motions that can be interpreted as waves, such as in fluid-dynamics. Also, geometric optics and the corpuscular behavior of light can be obtained as a suitable limit of wave theory.

As a direct proof of De Broglie relation, one can look for wave-light behavior of matter. Indeed one can immediately infer that a beam of particles, say electrons, should exhibits phenomena that are typical of waves, such as diffraction and interference. In 1927, Davisson and Germer performed an experiment in which diffraction of electrons through a nickel crystal (Bragg scattering) was observed. This kind of experiment has been repeated with protons, neutrons, helium atoms, ions, the wave relation for material particles always being verified. Let us notice that the De Broglie relation implies that interference/diffraction effects could be observed not only when considering a beam of particles, but also for single ones. The idea of devising an experiment to look at the interference pattern created by the passage of a single electron through two slits (such as in the classical Young experiment for light) dates back to a proposal of Schrödinger. Feynman uses it to introduce the reader to the fundamental concepts of QM, identifying it as "a phenomenon which is impossible [...] to explain in any classical way, and which has in it the heart of quantum mechanics. In reality, it contains the only mystery [of quantum mechanics]." [18], but warns the readers not to believe that such an experiment could ever be performed. On the contrary, following a series of newly developed electron microscopy techniques and some clever innovation, the experiment was done first in 1972 by Merli *et al.* [28] in Bologna and in 1976 by Tonomura *et al.* [37] in Tokyo[2].

The essence and the meaning of the particle-wave duality principle and its consequences has been discussed inside the physics as well as the philosophy community [11; 12; 13] since its formulation and forces us to recognize that the quantum world has to be described by means of a new physical theory, accompanied by a suitable new mathematics, in which classical paradigms are no longer valid: quantum objects are neither particles nor waves and they have to be described in terms of a new set of principles [25]. Thus, in Subsect. 1.2.1 we will introduce the conceptual and mathematical "postulates" that describe the theory. In particular, we will describe the notions of space of states (a Hilbert space $\mathcal{H}$) observables (self-adjoint operators $\mathcal{O}$), and of evolution (Schrödinger equation). We will also look at some key examples. Most of the work done in such section is at the level of Hilbert spaces, where the linearity principle is enforced. However, the physical content of a state is encoded in a vector of the Hilbert space up to multiplications

[2]The story of this experiment, which was defined as the most beautiful one in physics by the journal "Physics World" after questioning its readers, can be found in the website: http://l-esperimento-piu-bello-della-fisica.bo.imm.cnr.it/.

of a (non-zero) complex number[3]. Thus the physical space of state is not $\mathcal{H}$ itself, by the projective Hilbert space $P\mathcal{H}$ which, despite not being a vector space anymore, has a rich geometric structure that will be investigated in Subsect. 1.2.2. More specifically, we will show how the Hermitean form on $\mathcal{H}$, projects down to $P\mathcal{H}$, letting it to inherit both a symplectic and a Riemannian structure which make it a Kähler manifold.

In the last decades, more and more attention has been dedicated to phenomena and applications in QM for which the geometric structure of the space of states plays a fundamental role. This is the case of of the Aharonov-Bohm effect [1] and other problems connected to adiabatic phases [4; 29] as well as other topics that have recently caught the attention of the researchers, such as entanglement [7; 14] or tomography [26]. Geometric structures are also key ingredients to understand how one can start from a classical system and develop a procedure to "quantize" it.

The discussion about the procedure which one can use to pass from the classical to the quantum description of a given physical system accompanied the birth of QM theory and cannot abstract from a discussion about what the principal structures are that define what a classical or a quantum theory is. Since then, different methods of quantization, that allow to find the quantum counterpart of a classical physical problem, have been proposed and developed, which put emphasis on different aspects of the quantum theory and use either analytical or algebraic or geometrical techniques. The second part of this volume will focus on the logic and algebraic structures of quantum mechanics.

Following the seminal paper of Dirac [10], one may start by assuming that quantum states are represented by wave-functions, i.e. -say in the coordinate representation- by square integrable functions over the classical configuration manifold $Q = \{q \equiv (q_1, \cdots, q_n)\}$, taken usually as $\mathbb{R}^n$: $\mathcal{H} = L^2(Q) = \{\psi(q) : \|\psi(q)\|_2 < \infty\}$. Quantum observables are self-adjoint operators on $\mathcal{H}$, so to have a real spectrum and admitting a spectral decomposition. This is the analogue of what we do in a classical context, in which the space of states is given by the phase space $T^*Q = \{(p,q) \equiv (p_1, \cdots, p_n; q_1, \cdots, q_n)\}$ and observables are real (regular, usually C^∞) functions on it[4]: $f(p,q) \in \mathbb{R}$. Notice that both the space of self-adjoint operators on $\mathcal{H}$ and of real regular functions on T^*Q are vector spaces, actually algebra, on which we might want to assign suitable topologies.

---

[3]This holds for the case of a pure state. In the case of a mixed state, the physical content is encoded in a collection of k linearly independent states, defined up to the action of the unitary group $U(k)$. A pure state is recovered when $k = 1$.

[4]Whenever $T^*Q \neq \mathbb{R}^{2n}$, (p,q) represents coordinates in a local chart.

In "quantizing" a classical system, we would like to assign a set of rules that allows us to univocally pick up a self-adjoint operator $\hat{O}_f$ for each classical observable f. We would like this map to preserve the algebraic structures of these two spaces [2], i.e. to satisfy the following properties:

- the map $f \mapsto \hat{O}_f$ is linear;
- if $f = id_{T^*Q}$ then $\hat{O}_{id} = \mathbb{I}_{\mathcal{H}}$;
- if $g = \Phi \circ f$, with $\Phi : \mathbb{R} \to \mathbb{R}$ for which both $\hat{O}_f$ and $\hat{O}_g$ are well defined, then $\hat{O}_g = \hat{O}_{\Phi \circ f} = \Phi(\hat{O}_f)$.

We would also like this map to include the largest class of functions as possible. In particular, it must be possible to find the operators $\hat{p}_j, \hat{q}_j$ associated to the coordinate functions p_j and q_j $(j = 1, \cdots, n)$ on T^*Q, about which we put an additional requirement. From the simple examples one can develop such as the free particle and the harmonic oscillator, one can argue that Classical Poisson Brackets (CPB):

$$\{p_i, q_j\} = \delta_{ij}\,; \;\; \{q_i, q_j\} = \{p_i, p_j\} = 0\,; \;\; \{q_i, id\} = \{p_i, id\} = 0 \tag{1.1}$$

have to be replaced by the following Canonical Commutation Relations (CCR):

$$[\hat{q}_i, \hat{p}_j] = \imath\hbar\delta_{ij}, \;\; [\hat{q}_i, \hat{q}_j] = [\hat{p}_i, \hat{p}_j] = 0\,; \;\; [\hat{q}_i, \mathbb{I}] = [\hat{p}_i, \mathbb{I}] = 0\,. \tag{1.2}$$

From the first of these commutators with $i = j$, one can see that at least one of (and indeed both) the operators $\hat{p}_i, \hat{q}_i$ has to be unbounded [34]. A theorem by Stone and Von-Neumann [34] then states that, up to unitary equivalence, there is only one irreducible representation of such algebra of observables, called the *Weyl algebra*, which on $\mathcal{H} = L^2(Q)$ reads as:

$$\hat{q}_i\psi(q) = q_i\psi(q)\,, \;\; \hat{p}_i\psi(q) = -\imath\hbar\frac{\partial}{\partial q_i}\psi(q)\,. \tag{1.3}$$

Actually, knowing that in classical mechanics we can reconstruct the Poisson bracket between any two (regular, such as polynomials) functions $f(p,q), g(p,q)$ out of (1.1), we would like to find a map such that if, $f \mapsto \hat{O}_f$ and $g \mapsto \hat{O}_g$, then:

$$\hat{O}_{\{f,g\}} = \imath\hbar[\hat{O}_f, \hat{O}_g]\,. \tag{1.4}$$

This is possible if f, g are both functions of only the q- or the p-coordinates or if they are linear in them. However, we know that this does not hold already if we consider a quadratic function such as the Hamiltonian of the 1D harmonic oscillator. This is because, at the quantum level, we have an ordering problem due to the fact that the operators $\hat{p}, \hat{q}$ do not commute, contrary to the functions p, q.

It is therefore evident that we will not be able to find a correspondence $f \mapsto \hat{O}_f$ which fulfils all (five) requirements[5], at least if we insist on trying to quantize the whole space of observables.

We also remark that one is interested in the development of a technique that is not only suited to describe $T^*Q = \mathbb{R}^{2n}$, but the more general case in which the phase space is a symplectic manifold (Γ, ω).

Geometric quantization is one of the main approaches that has been developed to deal with these questions. The idea behind this approach is to overcome the problems described above by restricting the space of quantizable observables. More precisely, one would like to assign to (Γ, ω) a separable Hilbert space $\mathcal{H}$ and mapping $Q : f \mapsto \hat{O}_f$ from a *sub-Lie algebra* $\emptyset$ (as large as possible) of real-valued functions on Γ into self-adjoint linear operators on $\mathcal{H}$, satisfying:

(1) $Q : f \mapsto \hat{O}_f$ is linear;
(2) $\hat{O}_{id_\Gamma} = \mathbb{I}_{\mathcal{H}}$;
(3) $\hat{O}_{\{f,g\}} = \imath\hbar[\hat{O}_f, \hat{O}_g]$, $\forall f, g \in \emptyset$;
(4) for $\Gamma = \mathbb{R}^{2n}$ and $\omega = \omega_0$, the standard symplectic form, we should recover the operators $\hat{q}_i, \hat{p}_i$ as in (1.3);
(5) the procedure is functorial, in the sense that for any two symplectic manifolds (Γ_1, ω_1), (Γ_2, ω_2) and a symplectic diffeomorphism $\Phi : \Gamma_1 \to \Gamma_2$, the composition with Φ should map $\emptyset_1$ into $\emptyset_2$ and there should exist a unitary operator $U_\Phi : \mathcal{H}_1 \to \mathcal{H}_2$ such that $O^{(1)}_{f \circ \Phi} = U^\dagger_\Phi O^{(2)}_f U_\Phi$, $\forall f \in \emptyset_2$.

The solution to this problem was first given by Kostant and Souriau and goes through two main steps, called *prequantization* and *polarization*. The theory of geometric quantization has become a topic of study both in physics and in mathematics, but is goes beyond the scope of these lectures. We refer the interested reader to ref. [2] for a review and an exhaustive list of references.

Instead, in these lectures we will review some other methods of quantization, which are also founded on the geometric structures of QM and represent the ones that are more operatively used in modern theoretical physics. We will first introduce coherent states and Bargmann-Fock representation, both for bosonic and fermionic systems. Then we will move to discuss Feynman's approach to path integral. We will devote the last section to Weyl-Wigner formalism, to finally introduce the so-called non-commutative *-product and discuss the quantum to classical transitions.

---

[5]This is not only because we insist on Eq. (1.4). In [2], the interested reader may found a detailed discussions and many examples showing all existing inconsistencies among the various assumptions we made.

1.2 Overview of Quantum Mechanics

1.2.1 *Fundamental definitions and examples*

Whenever teaching introductory courses in either classical or quantum physics, one is lead to first answer the following question: what are the minimal conceptual and mathematical structures required for the description of a physical system?

According to modern theoretical set up, we essentially need a few main ingredients that can be identified with:

- a space of states $\mathcal{S}$, a state being an object which is able to encode and describe all degrees of freedom of the system;
- a space of observables $\mathcal{O}$, which can be applied to a state to change it in some specified manner;
- a law of evolution, which fixes (possibly in a unique way) how states and/or observables change as we let the system evolve, eventually under the influence of external and internal forces;
- a pairing $\mu : \mathcal{S} \times \mathcal{O} \to \mathbb{R}$, which produces a real number out of a state and an observable and corresponds to a measurement process.

In CM, $\mathcal{S}$ is the collections of independent coordinates and momenta $\{q_i, p_i\}$ describing the system whereas mathematically $\mathcal{S}$ is a symplectic manifold; $\mathcal{O}$ is given by (usually differentiable) real functions on $\mathcal{S}$; the evolution laws are given in terms of differential equations, which can be derived as a special observable, the Hamiltonian, is specified; while μ is a probability measure and the process of measure corresponds to evaluate the average value of an observable function on the space of states.

Similarly, the so-called "Postulates" of QM give the mathematical framework and explain the conceptual physical interpretation for a system governed by quantum rules. We will try to illustrate them by means of some simple but paradigmatic examples. For a more detailed exposition, we refer the interested reader to QM textbooks [8; 15] or books on Quantum Information and Computation [32].

The interested reader may find in the second part of this volume a discussion of the postulates of QM with emphasis on the algebraic approach, which includes a detailed explanation of the non-boolean logic and of the von Neumann algebra of observables.

1.2.1.1 *The space of states*

Postulate 1. A physical (pure) state of a QM system is represented by a ray in a complex and separable Hilbert space $\mathcal{H}$, i.e. by an equivalence class of unit vectors under the relation: $|\psi_1\rangle \simeq |\psi_2\rangle \Leftrightarrow |\psi_2\rangle = e^{i\alpha}\,|\psi_1\rangle$, $\alpha \in \mathbb{R}$.

We remark that linearity and complex field are needed in order to accommodate all interference phenomena that radiation and matter display in the quantum realm, while a scalar product is necessary to define the notion of probability . The vectors are to be normalized to one since their norm gives the total probability. Also, multiplying by an overall pure phase factor does not change the physical content. Separability is necessary to have a finite or denumerable complete set of orthonormal vectors, according to which we can decompose any state as a linear superposition of such possible alternatives.

In the following, we will use Dirac notation and represent a vector $\psi \in \mathcal{H}$ with the "ket" $|\psi\rangle$, while its dual $\psi^* \in \mathcal{H}^*$ is represented through the "bra" $\langle\psi|$, in such a way that the scalar product $\langle\psi, \phi\rangle$ between any two vectors can be represented (according to Riesz theorem) as the "bracket": $\langle\psi|\phi\rangle$.

A ray in $\mathcal{H}$ can be univocally determined by the projection operator:

$$\rho_\psi \equiv |\psi\rangle \langle\psi| \, , \tag{1.5}$$

where we have supposed (as we will do in the following, if not specified differently) that the vector is normalized. We recall that ρ_ψ is a bounded, positive semi-definite, trace-one operator such that $\rho_\psi^2 = \rho_\psi$. This also means that it is a rank-one projector.

In many experimental and theoretical problems, it is interesting to consider the possibility that a system might be prepared not in a unique state, but in a statistical mixture or mixed state, i.e. a collection of states $\{|\psi_1\rangle, |\psi_2\rangle, \cdots, |\psi_n\rangle\}$ with, respectively, probabilities $p_1, p_2, \cdots, p_n$. Such a state is represented by a so-called density matrix operator, defined as:

$$\rho \equiv \sum_{j=1}^{n} p_j |\psi_j\rangle \langle\psi_j| \, . \tag{1.6}$$

Again, ρ is bounded, positive semi-definite and trace-one, but $\rho^2 \neq \rho$ (for $n > 1$). Notice that now ρ is a rank n operator.

Example 1.2.1. *Explicit realizations.*

The simplest example one can consider is that of a two-level system, able to describe, e.g., the polarization degrees of freedom of a photon or the spin of an electron. In this case $\mathcal{H} = \mathbb{C}^2$.

More generally, one can consider an n-level system, whose space of states is given by $\mathbb{C}^n$.

In the case of infinite dimensions, $\mathcal{H}$ is often realized as the space of square-integrable functions over an open domain $\mathcal{D} \in \mathbb{R}^n$, the latter usually representing the classical configuration space of the system:

$$\mathcal{H} = \left\{ \psi(x) \, : \, \int_{\mathcal{D}} dx \, |\psi(x)|^2 < \infty \right\} \equiv L^2(\mathcal{D}) \, . \tag{1.7}$$

In this case, $\rho(x) \equiv |\psi(x)|^2$ is interpreted as a probability density, $|\psi(x)|^2 \, dx$ representing the probability of finding our system in the volume $[x, x+dx]$ of the classical configuration space. The function $\psi(x)$ is called a wave-function.

1.2.1.2 *Observables*

Postulate 2. The space of observables $\mathcal{O}$ of a QM system is given by the set of all self-adjoint operators on $\mathcal{H}$.

This means that every observable $\hat{O}$ has a real spectrum and admits a spectral decomposition[6]:

$$\hat{O} = \sum_{\lambda} \lambda \hat{P}_{\lambda} \, , \quad \hat{P}_{\lambda} \equiv |\psi_{\lambda}\rangle \langle \psi_{\lambda}| \, , \tag{1.8}$$

where $|\psi_\lambda\rangle$ is the eigenvectors of $\hat{O}$ corresponding to the eigenvalue λ: $\hat{O} \, |\psi_\lambda\rangle = \lambda \, |\psi_\lambda\rangle$. Since $\{|\psi_\lambda\rangle\}_\lambda$ form an orthonormal basis for $\mathcal{H}$, we have the relations:

$$\hat{P}_{\lambda} \hat{P}_{\mu} = \delta_{\lambda\mu} \hat{P}_{\lambda} \, , \tag{1.9}$$

$$\sum_{\lambda} \hat{P}_{\lambda} = \mathbb{I} \, . \tag{1.10}$$

Given a pure state $|\psi\rangle \in \mathcal{H}$, one can consider the mean value (ore expectation value) of an observable $\hat{O}$ over a state defined as:

$$\left\langle \hat{O} \right\rangle \equiv \langle \psi | \, \hat{O} \, |\psi\rangle = \mathrm{Tr} \left[\rho_{\psi} \hat{O} \right] . \tag{1.11}$$

Notice that (1.11) defines a paring between $\mathcal{H}$ and $\mathcal{O}$ (quadratic in the vectors and linear in the observables), which depends on the scalar product, which we will discuss in Subsect. 2.1.4.

One can also consider the variance of $\hat{O}$ over the state $|\psi\rangle$ as given by:

$$\triangle \hat{O} \equiv \left\langle \hat{O}^2 \right\rangle - \left\langle \hat{O} \right\rangle^2 . \tag{1.12}$$

[6] Strictly speaking, in writing expression (1.8) we have assumed the spectrum to be discrete. An analogous formula holds in the case of a continuous spectrum with the sum replaced by the integral over the spectral measure. For simplicity, we will stick to the discrete notation also in the rest of this part.

More generally, one can define the mean value (and variance) of $\hat{O}$ over a mixed state by taking averages over the statistical mixture, i.e. by:

$$\left\langle \hat{O} \right\rangle \equiv \sum_{j=1}^{N} p_j \mathrm{Tr}\left[\rho_{\psi_j} \hat{O}\right] = \mathrm{Tr}\left[\rho \hat{O}\right] . \tag{1.13}$$

Example 1.2.2. *Position and momentum.*
Let us consider $\mathcal{H} = L^2(\mathbb{R}) = \{\psi(x)\}$. On this space we can define two important operators, the position $\hat{x}$ and the momentum $\hat{p}$ respectively given by the multiplication by the coordinate and the derivative with respect to it position:

$$\hat{x}\psi(x) = x\psi(x) , \tag{1.14}$$

$$\hat{p}\psi(x) = -\imath\hbar \frac{d}{dx}\psi(x) , \tag{1.15}$$

which satisfy the so called canonical commutation relation (CCR):

$$[\hat{x}, \hat{p}] \subseteq \imath\hbar\mathbb{I} , \tag{1.16}$$

where the inclusion comes from the fact that, as explained previously, the operators q, p are unbounded. Notice that Eq. (1.16) looks the same as in the classical case, if we replace the Poisson brackets with the Lie commutator. It is said that Eqs. (1.14, 1.15) realize the CCR in the *coordinate representation.*

Relation (1.16) implies that neither $\hat{x}$ nor $\hat{p}$ can be realized as bounded operators: thus one needs to specify their domains $\mathcal{D}_x, \mathcal{D}_p \subset \mathcal{H}$, on which they are (at least essentially) self-adjoint. Usually, one chooses $\mathcal{D}_x = \mathcal{D}_p = \mathcal{S}(\mathbb{R})$, the space of Schwartz functions. The spectrum of both these operators is continuous and equal to $\mathbb{R}$, while the corresponding generalized eigenfunctions are respectively given by:

$$|x_0\rangle = \delta(x - x_0) \ \text{ with eigenvalue } x_0 \in \mathbb{R} , \tag{1.17}$$

$$|p_0\rangle = e^{\imath p_0 x} \qquad \text{with eigenvalue } p_0 \in \mathbb{R} . \tag{1.18}$$

Let us recall that the Fourier transform:

$$\psi(p) \equiv \frac{1}{\sqrt{2\pi}} \int dx\, e^{ipx} \psi(x) \tag{1.19}$$

is a unitary operator on $L^2(\mathbb{R})$. Thus, we may work as well on $\tilde{\mathcal{H}} = \{\psi(p)\}(\sim \mathcal{H})$, where the $\hat{x}$ and the $\hat{p}$ operators are now given by:

$$\hat{x}\psi(p) = -\imath\hbar \frac{d}{dp}\psi(p) , \tag{1.20}$$

$$\hat{p}\psi(p) = p\psi(p) . \tag{1.21}$$

Eqs. (1.20, 1.21) define the so-called *momentum representation* of the CCR.

This construction can be easily generalized to higher dimensions, i.e. to $\mathcal{H} = L^2(\mathbb{R}^n)$, by defining two sets of operators $\hat{x}_j$ and $\hat{p}_j$ ($j = 1, 2, \cdots, n$), such that: $[\hat{x}_j, \hat{p}_k] = \imath\hbar\delta_{jk}\,\mathbb{I}$ and $[\hat{x}_j, \hat{x}_k] = [\hat{p}_j, \hat{p}_k] = 0$.

Example 1.2.3. *Fermionic (or two-level) systems.*
In the finite-dimensional case, all self-adjoint operators are bounded and observables are represented by $n \times n$ Hermitean matrices: $\mathcal{O} = \mathbb{M}_n$. For a two-level system[7], a basis of all Hermitean operators is given by the identity $\mathbb{I}$ and the set of the three Pauli matrices σ_α, ($\alpha = 1, 2, 3$):

$$\mathbb{I} = \begin{pmatrix} 1 & 0 \\ 0 & 1 \end{pmatrix}, \ \sigma_1 = \begin{pmatrix} 0 & 1 \\ 1 & 0 \end{pmatrix}, \ \sigma_2 = \begin{pmatrix} 0 & -\imath \\ \imath & 0 \end{pmatrix}, \ \sigma_3 = \begin{pmatrix} 1 & 0 \\ 0 & -1 \end{pmatrix}. \tag{1.22}$$

It is not difficult to see that the operators $S_\alpha = \hbar\sigma_\alpha/2$ yields the fundamental representation (i.e. spin 1/2) of the $SU(2)$ algebra since:

$$[S_\alpha, S_\beta] = i\hbar\epsilon^{\alpha\beta\gamma} S_\gamma\,. \tag{1.23}$$

The canonical vectors

$$|+\rangle = \begin{pmatrix} 1 \\ 0 \end{pmatrix}, \ |-\rangle = \begin{pmatrix} 0 \\ 1 \end{pmatrix} \tag{1.24}$$

are the eigenstates of the S_3 operator, with eigenvalues $\pm\hbar/2$. It is convenient to define the two ladder operators:

$$\sigma^\pm = \sigma_1 \pm i\sigma_2\,, \tag{1.25}$$

satisfying the algebra commutators:

$$\left[\sigma^+, \sigma^-\right] = \sigma^z\,, \ \left[\sigma^z, \sigma^\pm\right] = 2\sigma^\pm \tag{1.26}$$

and the anti-commutation relations:

$$\left(\sigma^+\right)^2 = \left(\sigma^-\right)^2 = 0\,, \ \{\sigma^+, \sigma^-\} = \mathbb{I}\,. \tag{1.27}$$

From the latter, it follows immediately that $\mathcal{H}$ is generated by the two states: $|0\rangle \equiv |-\rangle$ and $|1\rangle \equiv \sigma^+ |0\rangle = |+\rangle$, while $\sigma^+ |1\rangle = 0$ as well as $\sigma^- |0\rangle = 0$.

The operators $\sigma^\pm$ satisfying (1.27) are called *fermionic creation/annihilation* operators, and $|0\rangle$ is interpreted as the vacuum state.

We can also define the *number operator*: $N \equiv \sigma^+\sigma^-$, for which $|0\rangle, |1\rangle$ are eigenvectors with eigenvalues 0, 1 respectively. This represents an algebraic way of encoding Pauli exclusion principle, which states that two fermions cannot occupy the same state.

---

[7] A two-level system is what is a called a *qubit* in the context of quantum information theory [32].

Example 1.2.4. *Bosonic systems.*
Bosonic creation/annihilation operators are defined as two operators $a, a^\dagger$ on $\mathcal{H}$ such that:

$$[a, a^\dagger] = \mathbb{I}\,. \tag{1.28}$$

Notice that this relation implies that the operator a and its adjoint $a^\dagger$ cannot be bounded, hence $\mathcal{H}$ has to be infinite dimensional. It is useful to define the number operator $N \equiv a^\dagger a$, which is self-adjoint and satisfies: $[N, a] = a,\ [N, a^\dagger] = a^\dagger$.

Then it is an exercise to show [8; 15] that N is such that:

- its spectrum is composed of all integers: $\sigma(N) = \{0, 1, 2, \cdots\}$;
- its eigenvector $|0\rangle$ corresponding to the lowest eigenvalue $\lambda_0 = 0$ is such that: $a\,|0\rangle = 0$;
- its eigenvector $|n\rangle$ corresponding to the $n-th$ eigenvalue $\lambda_n = n$ is given by:

$$|n\rangle = \frac{1}{\sqrt{n!}}(a^\dagger)^n|\psi_0\rangle\,. \tag{1.29}$$

Also, the following relations hold:

$$a^\dagger|\psi_n\rangle = \sqrt{n+1}|\psi_{n+1}\rangle\,, \tag{1.30}$$

$$a|\psi_n\rangle = \sqrt{n}|\psi_{n-1}\rangle\,, \tag{1.31}$$

showing that a $(a^\dagger)$ moves from one eigenstate to the previous (next) one, acting as ladder operators. The complete set of normalized eigenstates $\{|n\rangle\}_{n=0}^{\infty}$ provide a basis for $\mathcal{H}$, which is called Fock basis.

It is interesting to look for the coordinate representation of (1.28). On $\mathcal{H} = L^2(\mathbb{R}) = \{\psi(x)\}$, one has [15][8]:

$$a = \frac{1}{\sqrt{2}}\left(x + \frac{d}{dx}\right), \tag{1.32}$$

$$a^\dagger = \frac{1}{\sqrt{2}}\left(x - \frac{d}{dx}\right), \tag{1.33}$$

while:

$$|0\rangle = C_0\, e^{-x^2/2}\,, \tag{1.34}$$

$$|n\rangle = C_n e^{-x^2/2} H_n(x)\,, \tag{1.35}$$

where $H_n(x)$ is the n-th Hermite polynomial, and C_n normalization constants.

[8]See also sect. 3.2.1 of the third part of this volume.

As a final remark, which will be useful in the following, we notice that, using the coordinate representation:

$$a = \frac{1}{\sqrt{2}} (\hat{x} + i\hat{p}) ,$$
$$a^\dagger = \frac{1}{\sqrt{2}} (\hat{x} - i\hat{p}) . \tag{1.36}$$

As we will see in the next subsection, these operators are necessary to describe the quantum 1D harmonic oscillator.

Example 1.2.5. *Composite systems.*
For completeness, we very briefly review what happens when the system we consider is composed of N independent degrees of freedom/particles[9]. In this case, the Hilbert space of the total system is given by the tensor product of the Hilbert space of each particle: $\mathcal{H} = \otimes_{j=1}^{N} \mathcal{H}_j$. systems States in $\mathcal{H}$ of the form:

$$|\psi\rangle = |\psi\rangle_1 \otimes |\psi_2\rangle \otimes \cdots \otimes |\psi_N\rangle \quad \text{with } |\psi_j\rangle \in \mathcal{H}_j \tag{1.37}$$

are said to be *separable*. A state that cannot be written as so, is called *entangled* . Entanglement is a truly quantum property: it encodes the possibility of knowing some properties of one subsystem by measuring observables on the other part. Consider for example two spin-1/2 particles, A and B, in the singlet state (we omit the symbol of tensor product):

$$|\psi\rangle = \frac{|+\rangle_A |-\rangle_B + |-\rangle_A |+\rangle_B}{\sqrt{2}} . \tag{1.38}$$

In this state, the spins of neither A nor B are defined, but are opposite. When an experimentalist, say Alice, performs a measurement on the system A of the spin along the third component, she finds $+\hbar/2$ or $-\hbar/2$ with 50% probability but Bob, without performing any measurement on its qubit B, can infer that the latter has the opposite value of the spin.

This phenomenon, that Einstein himself defined as "that spooky action at a distance", is at the origin of much work and interesting discussions in the history of development of QM (the EPR paradox, the theory of hidden variables and Bell's inequality, for example) and it is now at the heart of recent applications in the field of quantum information and quantum computation, such as teleportation. The interested reader can look at [32].

As a final remark, let us observe that different states might encode a different level of entanglement, the singlet state given above being an example of a

[9]We are assuming that the particles are distinguishable.

maximally entangled state. We can measure the entanglement of a state by using physical quantities, called entanglement witnesses. For instance, for a pure state described by a density matrix ρ_{AB}, one can define the von Neumann entropy:

$$S(\rho_{AB}) \equiv \text{Tr}_A\left[\rho_A \log \rho_A\right], \tag{1.39}$$

where $\rho_A = \text{Tr}_B\left[\rho_{AB}\right]$ is the partial trace over the system B (of course one can interchange the role of A and B to get the same result). We will not further discuss this topic and refer the interested reader for example to [7], where a discussion of the geometric aspects of entanglement is also presented.

1.2.1.3 *Dynamical evolution*

Postulate 3. The dynamical evolution of a quantum state $|\psi(t)\rangle \in \mathcal{H}$ is specified by a suitable self-adjoint operator $\hat{H}$, called Hamiltonian, and governed by the Schrödinger equation:

$$\imath\hbar\frac{d}{dt}\left|\psi(t)\right\rangle = \hat{H}\left|\psi(t)\right\rangle . \tag{1.40}$$

In the following, we will always assume that $\hat{H}$ is time-independent. In this case, $\hat{H}$ being self-adjoint, the equation can be solved by introducing the (strongly-continuous) one-parameter group of unitary operators (the evolution operator):

$$U(t) = e^{-\frac{\imath t\hat{H}}{\hbar}}, \tag{1.41}$$

since then:

$$\left|\psi\left(t\right)\right\rangle = U(t)\left|\psi\left(0\right)\right\rangle . \tag{1.42}$$

Notice that, since $U(t)$ is unitary, scalar products, hence probabilities, are conserved.

The (possibly generalized) eigenvectors of $\hat{H}$ are the only stationary states :

$$\hat{H}\left|\psi_\lambda\right\rangle = \lambda\left|\psi_\lambda\right\rangle \Leftrightarrow \left|\psi_\lambda(t)\right\rangle = e^{-\frac{\imath E_\lambda}{\hbar}t}\left|\psi_\lambda\right\rangle . \tag{1.43}$$

Using the fact that the eigenstates form an orthonormal basis in $\mathcal{H}$, it is easy to see that the evolution of any state can be written in the form:

$$\left|\psi(t)\right\rangle = \sum_\lambda c_\lambda e^{-\frac{\imath E_\lambda}{\hbar}t}\left|\psi_\lambda\right\rangle \;\text{ if }\; \left|\psi(0)\right\rangle = \sum_\lambda c_\lambda\left|\psi_\lambda\right\rangle . \tag{1.44}$$

Example 1.2.6. *Free particle on* $\mathbb{R}^n$.
This system is easily described in the coordinate representation when $\mathcal{H} = \{\psi(x)\}$, by the Hamiltonian operator:

$$\hat{H}_0 = \frac{\hat{\vec{p}}^2}{2m} = -\frac{\hbar^2}{2m}\vec{\nabla}_x^{\,2}, \tag{1.45}$$

where m is the mass of the particle.

Notice that, as function of the momentum, the Hamiltonian (1.45) has the same expression as in the classical case: to obtain it we just need to replace the function $\vec{p}$ with the operator $\hat{\vec{p}}$. This can be considered as a first rule of quantization.

The corresponding (continuous) eigenvalues and (generalized) eigenfunctions are easily found to be:

$$E_p = \frac{\hbar|\vec{p}\,|^2}{2m}, \quad \psi_p(\vec{x}) = e^{\imath\vec{p}\cdot\vec{x}}. \tag{1.46}$$

Example 1.2.7. *The 1D harmonic oscillator.*
Many interesting dynamical systems are described by a Hamiltonian of the form:

$$H = \frac{\vec{p}^{\,2}}{2m} + V(\vec{x}), \tag{1.47}$$

that gives the energy of a particle in an external potential V. In the previous example, we have seen that, to get the quantum version of a free particle, we need just to replace the momentum $\vec{p}$ with the operator $\hat{\vec{p}}$ in the classical Hamiltonian function. In a similar way, one is led to consider the quantum version of (1.47) to be given by the operator:

$$H = \frac{\hat{\vec{p}}^{\,2}}{2m} + V(\hat{\vec{x}}). \tag{1.48}$$

To see to what extent we can use this approach, let us consider the 1D harmonic oscillator whose classical Hamiltonian is given by

$$H = \frac{\hbar\omega}{2}\left(p^2 + x^2\right). \tag{1.49}$$

Here we work with suitable units in which both x and p are adimensional and $\{x, p\} = 1/\hbar$. Thus we can consider the quantum Hamiltonian:

$$\hat{H} = \frac{\hbar\omega}{2}(\hat{p}^2 + \hat{x}^2), \tag{1.50}$$

with $[\hat{x}, \hat{p}] = \mathbb{I}$. One may argue, however, that (1.49) can be rewritten in three different equivalent ways:

$$H = \frac{\hbar\omega}{2}(p^2 + x^2) = \frac{\hbar\omega}{2}(x + ip)(x - ip) = \frac{\hbar\omega}{2}(x - ip)(x + ip), \tag{1.51}$$

since x, p are commuting functions. At the quantum level, however, this equivalence no longer holds true (see (1.16)) and one would obtain different quantum Hamiltonians, specifically, differing by some constants. This ambiguity in the process of quantization can be resolved by introducing the so-called "symmetrization postulate" [8]: to obtain the quantum Hamiltonian $\hat{H}$ out of the classical one H, any monomial in the variables x, p that classically reads as $x^m p^n$ has to be replaced with the symmetric version obtained by writing the products of m times x-factors and n times p-factors in all possible orders. Applying this rule to our problem, we exactly obtain Eq. (1.50).

Thus, in the coordinate representation, we get:

$$\hat{H} = \frac{\hbar\omega}{2}\left(-\frac{d^2}{dx^2} + x^2 + 1\right). \tag{1.52}$$

Making use of (1.36), the latter can be rewritten as:

$$\hat{H} = \frac{\hbar\omega}{2}\left(a^\dagger a + \frac{1}{2}\right) = \frac{\hbar\omega}{2}\left(N + \frac{1}{2}\right). \tag{1.53}$$

Thus we see that the Hamiltonian of the 1D harmonic oscillator is (up to a constant) the number operator of bosonic type, whose spectral problem has been solved in the previous subsection.

1.2.1.4 *Measurement and probability*

Since the very beginning, the notion of measurement has played an important rôle in the discussion about the interpretation of QM [6], and is central to analyze possible connections with applications, such as in optimization and control problems [32]. However, such a study goes well beyond the scope of these lectures[10]. For completeness, we will give here only the definition of a *positive operator valued measure* (POVM), that we will specialize in one example.

The definition rests on the hypothesis that the outcome of a quantum measure is a random variable X and therefore its possible values form a measure space $(\mathcal{X}, \mu)$. We will denote with $\mathcal{M}$ the set of all measurable subsets of $\mathcal{X}$ and with $\mathcal{B}_{sa}^+$ the set of positive self-adjoint bounded operators on the Hilbert space $\mathcal{H}$.

Postulate 4. *A POVM is specified by a function $F : \mathcal{M} \to \mathcal{B}_{sa}^+$ such that $F(\mathcal{X}) = \mathbb{I}_\mathcal{H}$ and with the property that the probability that a measurement on a (pure or mixed) state ρ yields a result $X \in \mathcal{A}$, for any $\mathcal{A} \in \mathcal{M}$, is given by:*

$$Pr\{X \in \mathcal{A}\} = \mathrm{Tr}\left[\rho F(\mathcal{A})\right]. \tag{1.54}$$

[10] For a thorough discussion of this topic see sect. 2 of the second part of this volume.

This means that the outcome X of a measure on ρ is described by a probability density distribution given by

$$p(X) = \mathrm{Tr}\left[\rho F(X)\right] = \sum_{j=1}^{n} p_j \left\langle \psi_j \right| F(X) \left| \psi_j \right\rangle . \tag{1.55}$$

where to write the last expression we have taken into account Eq. (1.6).

Example 1.2.8. *Projection valued measures.*
This is a particular case of the above definition when F gives projection operators. This emerges, for example, when considering an observable $\hat{O}$: the possible outcomes of a measure are its eigenvalues, so that the space $\mathcal{X}$ is given by its spectrum $\sigma(\hat{O}) \subset \mathbb{R}$, while F is the map that associates to any measurable subset $\mathcal{A} \subset \sigma(\hat{O})$ the projection operator $P_{\mathcal{A}}$ on the corresponding eigenspace. More specifically, assuming for simplicity that $\hat{O}$ has only a discrete non-degenerate spectrum so that it can be decomposed by means of the spectral decomposition (1.8), we have that F is the map that associates to any point $\lambda_\alpha \in \sigma(\hat{O})$ the projector P_α on the corresponding one-dimensional eigenspace. Thus Eq. (1.55) reads:

$$p_\alpha = p(\lambda_\alpha) = \mathrm{Tr}\left[\rho P_\alpha\right] = \left\langle \psi_\alpha \right| \rho \left| \psi_\alpha \right\rangle . \tag{1.56}$$

If considering a pure state $\rho_\phi = \left|\phi\right\rangle \left\langle\phi\right|$, Eq. (1.56) simply becomes: $p_\alpha = |\left\langle \phi | \psi_\alpha \right\rangle|^2$.

Notice that probability densities are defined by using the notion of the trace, which depends on the Hermitean product defined in $\mathcal{H}$, and gives the pairing between states and observables mentioned at the beginning of this Sect.

1.2.2 *Geometric quantum mechanics*

In this Sect. we will describe the geometrical structures that appear in QM. To do so, we will stick to the case of a finite n-level system, so that $\mathcal{H} = \mathbb{C}^n$. The starting observation is that, being a vector space, $\mathcal{H}$ is a manifold and there is a natural identification of the tangent space at any point $\psi \in \mathcal{H}$ with $\mathcal{H}$ itself: $T_\psi \mathcal{H} \approx \mathcal{H}$, so that we have the identification: $T\mathcal{H} \approx \mathcal{H} \times \mathcal{H}$, with $T\mathcal{H}$ the tangent bundle of $\mathcal{H}$. Thus, in the following, vectors of the Hilbert space play a double rôle, as points of the space and as tangent vectors at a given point[11]. Also, to discuss the geometry of the space of states in QM in more detail, we consider the realification $\mathcal{H}_\mathbb{R}$ of $\mathcal{H}$, whose tangent bundle is $T\mathcal{H}_\mathbb{R} \approx \mathcal{H}_\mathbb{R} \times \mathcal{H}_\mathbb{R}$, so to look at $\mathcal{H}$ as to a real manifold.

[11]Here we will write vectors as $\psi, \phi, \cdots$, instead of using the Dirac notation $\left|\psi\right\rangle, \left|\phi\right\rangle, \cdots$.

1.2.2.1 *Vector and tensor fields*

A differentiable map $t \mapsto \psi(t)$, with $\psi(0) = \psi$, gives a curve in $\mathcal{H}$ passing through the point ψ, so that we can define the tangent vector to the curve at $\psi \in \mathcal{H}$ to be given by: $(d\psi(t)/dt)|_{t=0}$. Then a vector field Γ is given once we assign a tangent vector at every point $\psi \in \mathcal{H}$ in a smooth way, i.e. we give a smooth global section of $T\mathcal{H}$:

$$\Gamma : \mathcal{H} \to T\mathcal{H} \, ; \, \psi \mapsto (\psi, \phi) \, , \psi \in \mathcal{H} \, , \, \phi \in T_\psi \mathcal{H} \approx \mathcal{H} \, . \tag{1.57}$$

Also, $\Gamma(\psi)$ denotes the vector field evaluated at the point $\psi \in \mathcal{H}$, with tangent vector $\phi \in \mathcal{H}$ at ψ as given by Eq. (1.57).

Vector fields are derivations on the algebra of functions and we can define the Lie derivative along Γ of a function f as follows:

$$(\mathcal{L}_\Gamma(f))(\psi) = \frac{d}{dt} f(\psi(t))|_{t=0} \, . \tag{1.58}$$

We can introduce local coordinates by choosing an orthonormal basis $\{e_i\}$, so that vectors (and tangent vectors) will be represented by n-tuples of complex numbers $\psi = (\psi^1, \cdots, \psi^n)$ with $\psi^j \equiv \langle e_j | \psi \rangle$. Thus we have[12]:

$$(\mathcal{L}_\Gamma(f))(\psi) = \phi^i(\psi) \frac{\partial f}{\partial \psi^i}(\psi) \, . \tag{1.59}$$

Summation over repeated indexes is understood here and in the following.

In the following, we consider examples of:

(i) constant vector fields, given by $\phi = const.$ in the second argument of Eq. (1.57), and originating the one-parameter group:

$$\mathbb{R} \ni t \mapsto \psi(t) = \psi + t\phi \, ; \tag{1.60}$$

(ii) linear vector fields, with $\phi(\psi)$ being a linear and homogeneous function of ψ: $\phi = A\psi$, for some linear operator A. In this case:

$$\psi(t) = \exp\{tA\} \psi \, . \tag{1.61}$$

Example 1.2.9. *Dilation vector field and linear structure.*
The dilation vector field is defined as:

$$\Delta : \psi \mapsto (\psi, \psi) \, , \tag{1.62}$$

which corresponds to $A = \mathbb{I}$ in ii) above. In this case Eq. (1.61) becomes:

$$\psi(t) = e^t \psi \, . \tag{1.63}$$

---

[12] As ψ_j is complex: $\psi_j = q_j + ip_j$, $q_j, p_j \in \mathbb{R}$, the derivative here has to be understood simply as: $\partial/\partial\psi_j = \partial/\partial q_j - i\partial/\partial p_j$ (see also later on).

Eq. (1.62) gives an identification of $\mathcal{H}$ with the fiber $T_\psi\mathcal{H}$. The latter carries a natural linear structure, and through Δ we can give a tensorial characterization of the linear structure of the base space $\mathcal{H}$.

Thus, we can associate to every linear operator $\mathbb{A}$ the linear vector field:

$$\begin{aligned} \mathbb{X}_\mathbb{A} : \mathcal{H} &\to T\mathcal{H} \\ \psi &\mapsto (\psi, \mathbb{A}\psi) \end{aligned} . \tag{1.64}$$

In local coordinates, we have $\mathbb{X}_\mathbb{A} \equiv A^i{}_j \psi^j \, \partial/\partial\psi^i$, $A^i{}_j$ being the representative matrix of the linear operator.

We notice that, while linear operators form an associative algebra, vector fields do not, yielding instead a Lie algebra. Nevertheless we can recover an associative algebra by defining now the $(1,1)$ tensor:

$$\mathbb{T}_\mathbb{A} \equiv A^i{}_j d\psi^j \otimes \frac{\partial}{\partial\psi^i} . \tag{1.65}$$

It is easy to verify that we can recover the vector field $\mathbb{X}_\mathbb{A}$ from the latter and the dilation field via:

$$\mathbb{X}_\mathbb{A} = \mathbb{T}_\mathbb{A}(\Delta) . \tag{1.66}$$

The geometric structures just introduced allow to interpret the Schrödinger equation (1.40) as a classical evolution equation on the complex vector space $\mathcal{H}$. Indeed, the Hamiltonian H defines the linear vector field[13] Γ_H:

$$\begin{aligned} \Gamma_H : \mathcal{H} &\to T\mathcal{H} \\ \psi &\mapsto (\psi, -(i/\hbar)\, H\psi) \end{aligned} . \tag{1.67}$$

Thus Eq. (1.40) reads as:

$$\mathcal{L}_{\Gamma_H}\psi \equiv \frac{d}{dt}\psi = -\frac{i}{\hbar} H\psi . \tag{1.68}$$

1.2.2.2 *Riemannian and symplectic forms*

Let us denote with

$$\begin{aligned} h : \mathcal{H} \times \mathcal{H} &\to \mathbb{C} \\ (\phi, \psi) &\mapsto h(\phi,\psi) \equiv \langle\phi|\psi\rangle \end{aligned} \tag{1.69}$$

the Hermitean scalar product on $\mathcal{H}$.

[13] We use here the notation Γ_H instead of $\mathbb{X}_H$ to stress that we include an additional factor $(-i/\hbar)$ in its definition.

More properly, h should be seen as a $(0,2)$ (constant) tensor field evaluated at the point $\varphi \in \mathcal{H}$ and with $\phi, \psi \in T_\varphi \mathcal{H}$. Thus we should write, in a more precise way:

$$h(\varphi)\left(\Gamma_\phi(\varphi), \Gamma_\psi(\varphi)\right) = \langle \phi | \psi \rangle . \tag{1.70}$$

Since the r.h.s. of this equation does not depend on φ, from Eq. (1.40), we obtain the chain of equalities:

$$\begin{aligned} 0 &= \mathcal{L}_{\Gamma_H}(h(\phi,\psi)) = (\mathcal{L}_{\Gamma_H} h)(\phi,\psi) + h(\mathcal{L}_{\Gamma_H}\phi,\psi) + h(\phi,\mathcal{L}_{\Gamma_H}\psi) \\ &= (\mathcal{L}_{\Gamma_H} h)(\phi,\psi) + \frac{i}{\hbar}\left\{\langle H\phi|\psi\rangle - \langle \phi|H\psi\rangle\right\}, \end{aligned} \tag{1.71}$$

which implies in turn, as H is self-adjoint, that:

$$\mathcal{L}_{\Gamma_H} h = 0. \tag{1.72}$$

This means that the Hermitean structure h is invariant[14] under the unitary flow of Γ_H. Vice versa, if the Hermitean structure is not invariant then H will not be self-adjoint w.r.t. the given Hermitean structure.

Let us now decompose the Hermitean structure into its real and imaginary parts:

$$h(\cdot,\cdot) = g(\cdot,\cdot) + i\omega(\cdot,\cdot), \tag{1.73}$$

so defining the two $(0,2)$ tensors

$$g(\phi,\psi) \equiv \frac{\langle\phi|\psi\rangle + \langle\psi|\phi\rangle}{2}, \tag{1.74}$$

$$\omega(\phi,\psi) \equiv \frac{\langle\phi|\psi\rangle - \langle\psi|\phi\rangle}{2\imath}. \tag{1.75}$$

Clearly g is symmetric and ω is skew-symmetric, while Eq. (1.72) implies that both tensors are separately invariant under Γ_H, since $\omega(\phi, i\psi) = g(\phi,\psi)$. Also, ω is closed because it is represented by a constant (and unitarily invariant) matrix. In summary, we have proved that ω is a *symplectic form* and g is a (Riemannian) *metric form*.

Let us recall also that $\mathcal{H}$ is endowed with a natural complex structure J, the $(1,1)$ tensor defined by:

$$\begin{aligned} J : \mathcal{H} &\to \mathcal{H} \\ \phi &\to -\imath\phi \end{aligned} \qquad \text{with } J^2 = -\mathbb{I}. \tag{1.76}$$

Since $\omega(J\phi,\psi) = g(\phi,\psi)$, the complex structure J is said to be *compatible* with the pair (g,ω) and the triple (g,ω,J) is called *admissible*. This means that we

[14] This means that Γ_H is a Killing vector field for the Hermitean structure.

can reconstruct the Hermitean structure h when any two of the tensors g, ω, J are given. This is so because:

$$h(\phi, \psi) = \omega(J\phi, \psi) + \imath\omega(\phi, \psi) = g(\phi, \psi) - \imath g(J\phi, \psi). \tag{1.77}$$

Notice also that:

$$\omega(J\phi, J\psi) = \omega(\phi, \psi) \text{ as well as } g(J\phi, J\psi) = g(\phi, \psi). \tag{1.78}$$

We can summarize what has been proven up to now by saying that $\mathcal{H}$ is a Kähler manifold [22], with Hermitean metric h, ω being the fundamental two-form and g the Kähler metric.

As a final remark, we observe that a vector field Γ_H of the form (1.67) is such that:

$$(i_{\Gamma_H}\omega)(\psi) = \omega\left(-\frac{i}{\hbar}H\phi, \psi\right) = \frac{1}{2\hbar}\left[\langle H\phi|\psi\rangle + \langle\psi|H\phi\rangle\right]. \tag{1.79}$$

If we define now the quadratic function

$$f_H(\phi) = \frac{1}{2\hbar}\langle\phi|H\phi\rangle \tag{1.80}$$

and its differential as the one-form

$$df_H(\phi) = \frac{1}{2}\left[\langle\cdot|H\phi\rangle + \langle\phi|H\cdot\rangle\right] = \frac{1}{2}\left[\langle\cdot|H\phi\rangle + \langle H\phi|\cdot\rangle\right] \tag{1.81}$$

(the last passage following from H being self-adjoint), then it is easy to prove that: $(i_{\Gamma_H}\omega)(\psi) = df_H(\phi)(\psi)\,\forall\psi$. Hence we have:

$$i_{\Gamma_H}\omega = df_H, \tag{1.82}$$

i.e. Γ_H is *Hamiltonian* w.r.t. the symplectic structure, with f_H as quadratic Hamiltonian.

Example 1.2.10. *The projective Hilbert space.*
We have already seen that a physical state is not identified with a unique vector in some Hilbert space, but rather with a "ray", i.e. an equivalence class of vectors differing by multiplication through a nonzero complex number: even fixing the normalization, an overall phase remains unobservable. Quotienting with respect to these identifications, we get the following double fibration:

$$\begin{array}{ccc} \mathbb{R}_+ & \hookrightarrow \mathcal{H}_0 = \mathcal{H} - \{\mathbf{0}\} \\ & \downarrow \\ U(1) \hookrightarrow & \mathbb{S}^{2n-1} \\ & \downarrow \\ & P(\mathcal{H}) \end{array} \tag{1.83}$$

whose final result is the projective Hilbert space $P\mathcal{H} \simeq \mathbb{C}P^{n-1}$.

We have also already mentioned that every equivalence class $[|\psi\rangle]$ can be identified with the rank-one projector: $\rho_\psi = |\psi\rangle\langle\psi|$, where, as usual, the vector $|\psi\rangle$ is supposed to be normalized. The space of rank-one projectors is usually denoted as $\mathcal{D}_1^1(\mathcal{H})$ and it is then clear that in this way we can identify it with $P\mathcal{H}$.

More explicitly, points in $\mathbb{CP}^n$ are equivalence classes of vectors $\mathbf{Z} = (Z^0, Z^1, \cdots, Z^n) \in \mathbb{C}^{n+1}$ w.r.t. the equivalence relation $Z \approx \lambda Z$; $\lambda \in \mathbb{C} - \{0\}$. The space $\mathbb{CP}^n$ is endowed with:

(i) the Fubini-Study metric [3], whose pull-back to $\mathbb{C}^{n+1}$ is given by:

$$g_{FS} = \frac{1}{(\mathbf{Z}\cdot\bar{\mathbf{Z}})^2}\left[(\mathbf{Z}\cdot\bar{\mathbf{Z}})d\mathbf{Z}\otimes_S d\bar{\mathbf{Z}} - (d\mathbf{Z}\cdot\bar{\mathbf{Z}})\otimes_S(\mathbf{Z}\cdot d\bar{\mathbf{Z}})\right], \tag{1.84}$$

where $\mathbf{Z}\cdot\bar{\mathbf{Z}} = Z^a\bar{Z}^a$, $d\mathbf{Z}\cdot\bar{\mathbf{Z}} = dZ^a\bar{Z}^a$, $d\mathbf{Z}\otimes_S d\bar{\mathbf{Z}} = dZ^a d\bar{Z}^a + d\bar{Z}^a dZ^a$;

(ii) the compatible Kostant Kirillov Souriau symplectic form [23]:

$$\omega_{FS} = \frac{i}{(\mathbf{Z}\cdot\bar{\mathbf{Z}})^2}\left[(\mathbf{Z}\cdot\bar{\mathbf{Z}})d\mathbf{Z}\wedge d\bar{\mathbf{Z}} - (d\mathbf{Z}\cdot\bar{\mathbf{Z}})\wedge(\mathbf{Z}\cdot d\bar{\mathbf{Z}})\right] = d\theta_{FS}, \tag{1.85}$$

where:

$$\theta_{FS} = \frac{1}{2i}\frac{\overline{\mathbf{Z}}d\mathbf{Z} - \mathbf{Z}d\overline{\mathbf{Z}}}{\mathbf{Z}\cdot\overline{\mathbf{Z}}}. \tag{1.86}$$

This shows that $P\mathcal{H}$ is intrinsically a Kähler manifold. A more detailed discussion of theses structures, together with the ones that will be defined in the next sections, may be found in [14].

We end this example with a remark. The careful reader might have noticed that it appears that the most natural setting for QM is not primarily the Hilbert space itself but rather $P\mathcal{H} = \mathcal{D}_1^1(\mathcal{H})$, which is not a vector space. On the other hand, we know that the superposition rule is the key ingredient to interpret all interference phenomena that yields one of the fundamental building blocks of QM. In [27], it is shown that a superposition of rank-one projectors which yields another rank-one projector is possible, but requires the arbitrary choice of a fiducial projector ρ_0. This procedure is equivalent to the introduction of a connection on the bundle, usually called the Pancharatnam connection [29].

1.2.2.3 *Geometric structures on the Hilbert space*

In the following, points in $\mathcal{H}_\mathbb{R}$ will be denoted by Latin letters $(x = u + \imath v, \cdots)$ while we will use Greek letters $(\psi, \cdots)$ for tangent vectors.

Denoting with X_ψ the constant vector field $X_\psi \equiv (x, \psi)$, we can regard g of Eq. (1.74) and ω of Eq. (1.75) as $(0,2)$ tensors by setting:

$$g\,(x)\,(X_\psi, X_\phi) \equiv g\,(\psi, \phi)\,, \tag{1.87}$$

$$\omega\,(x)\,(X_\psi, X_\phi) \equiv \omega\,(\psi, \phi)\,. \tag{1.88}$$

Also we can promote J as defined in Eq. (1.76) to a $(1,1)$ tensor field by setting:

$$J\,(x)\,(X_\psi) = (x, J\psi)\,, \tag{1.89}$$

making, as we have already noticed before, the tensorial triple (g, J, ω) admissible and $\mathcal{H}_\mathbb{R}$ a linear Kähler manifold [36].

The $(0,2)$-tensors g and ω being non-degenerate, we can consider the invertible maps $\hat{g}, \hat{\omega} : T\mathcal{H}_\mathbb{R} \to T^*\mathcal{H}_\mathbb{R}$ such that, for any $X, Y \in T\mathcal{H}_\mathbb{R}$: $g(X,Y) = \hat{g}(X)(Y)$ and $\omega(X,Y) = \hat{\omega}(X)(Y)$. Then we can define two $(2,0)$ contravariant tensors given respectively by a metric tensor G and a Poisson tensor Λ such that

$$G(\hat{g}(X), \hat{g}(Y)) = g(X,Y)\,, \quad \Lambda(\hat{\omega}(X), \hat{\omega}(Y)) = \omega(X,Y)\,. \tag{1.90}$$

We can now use G and Λ to define a Hermitean product between any two $\alpha, \beta \in \mathcal{H}^*_\mathbb{R}$ (equipped with the dual complex structure J^*):

$$\langle \alpha, \beta \rangle_{\mathcal{H}^*_\mathbb{R}} \equiv G(\alpha, \beta) + i\Lambda(\alpha, \beta)\,. \tag{1.91}$$

To make these structures more explicit, let us introduce an orthonormal basis $\{e_k\}_{k=1,\cdots,n}$ in $\mathcal{H}$ and global coordinates (q^k, p^k) for $k = 1, \cdots, n$ in $\mathcal{H}_\mathbb{R}$ such that $\langle e_k, x \rangle = (q^k + ip^k)(x)$, $\forall x \in \mathcal{H}$. Then, it is easy to see that:

$$g = dq^k \otimes dq^k + dp^k \otimes dp^k\,, \tag{1.92}$$

$$\omega = dq^k \otimes dp^k - dp^k \otimes dq^k\,, \tag{1.93}$$

$$J = dp^k \otimes \frac{\partial}{\partial q^k} - dq^k \otimes \frac{\partial}{\partial p^k}\,, \tag{1.94}$$

so that

$$G = \frac{\partial}{\partial q^k} \otimes \frac{\partial}{\partial q^k} + \frac{\partial}{\partial p^k} \otimes \frac{\partial}{\partial p^k}\,, \tag{1.95}$$

$$\Lambda = \frac{\partial}{\partial q^k} \otimes \frac{\partial}{\partial p^k} - \frac{\partial}{\partial p^k} \otimes \frac{\partial}{\partial q^k}\,. \tag{1.96}$$

As a convenient shorthand notation, we can introduce complex coordinates: $z^k \equiv q^k + ip^k$, $\bar{z}^k \equiv q^k - ip^k$, so that we can write

$$G + i \cdot \Lambda = 4\frac{\partial}{\partial z^k} \otimes \frac{\partial}{\partial \bar{z}^k}\,, \tag{1.97}$$

with $\frac{\partial}{\partial z^k} \equiv \frac{1}{2}\left(\frac{\partial}{\partial q^k} - i\frac{\partial}{\partial p^k}\right)$, $\frac{\partial}{\partial \bar{z}^k} \equiv \frac{1}{2}\left(\frac{\partial}{\partial q^k} + i\frac{\partial}{\partial p^k}\right)$. Also:

$$J = -i\left(dz^k \otimes \frac{\partial}{\partial z^k} - d\bar{z}^k \otimes \frac{\partial}{\partial \bar{z}^k}\right). \tag{1.98}$$

1.2.2.4 *Geometric structures on the space of functions and operators*

The geometric structures just examined allow us to introduce two (non-associative) real brackets on smooth, real-valued functions on $\mathcal{H}_{\mathbb{R}}$:

- the (symmetric) Jordan bracket $\{f,h\}_g \equiv G(df,dh)$;
- the (antisymmetric) Poisson bracket $\{f,h\}_\omega \equiv \Lambda(df,dh)$.

By extending both these brackets to complex functions via complex linearity, we eventually obtain a complex bracket $\{.,.\}_{\mathcal{H}}$ defined as:

$$\{f,h\}_{\mathcal{H}} = \langle df, dh\rangle_{\mathcal{H}^*_{\mathbb{R}}} \equiv \{f,h\}_g + i\{f,h\}_\omega\,. \tag{1.99}$$

Explicitly, in complex coordinates, we can write:

$$\{f,h\}_g = 2\left(\frac{\partial f}{\partial z^k}\frac{\partial h}{\partial \bar{z}^k} + \frac{\partial h}{\partial z^k}\frac{\partial f}{\partial \bar{z}^k}\right),\ \{f,h\}_\omega = \frac{2}{i}\left(\frac{\partial f}{\partial z^k}\frac{\partial h}{\partial \bar{z}^k} - \frac{\partial h}{\partial z^k}\frac{\partial f}{\partial \bar{z}^k}\right). \tag{1.100}$$

In particular, if we associate to any operator $A \in gl(\mathcal{H})$ the quadratic function:

$$f_A(x) = \frac{1}{2}\langle x, Ax\rangle = \frac{1}{2}z^\dagger A z \tag{1.101}$$

(where z is the column vector $(z^1, ..., z^n)$), it follows immediately from Eq. (1.100) that, for any $A, B \in gl(\mathcal{H})$, we have:

$$\{f_A, f_B\}_g = f_{AB+BA}\,, \tag{1.102}$$

$$\{f_A, f_B\}_\omega = f_{\frac{AB-BA}{i}}\,. \tag{1.103}$$

This means that the Jordan bracket of any two quadratic functions f_A and f_B is related to the (commutative) Jordan bracket of A and B, $[A,B]_+$, defined as:

$$[A,B]_+ \equiv AB + BA\,, \tag{1.104}$$

while their Poisson bracket is related to the commutator product (the Lie bracket) $[A,B]_-$ defined as:

$$[A,B]_- \equiv \frac{1}{i}(AB - BA)\,. \tag{1.105}$$

Notice that, if A and B are Hermitean, their Jordan product and their Lie bracket will be Hermitean as well. Hence, the set of Hermitean operators on $\mathcal{H}_{\mathbb{R}}$, equipped with the binary operations (1.104) and (1.105), becomes a *Lie-Jordan algebra*[15] with the associative product:

$$(A, B) \equiv \frac{1}{2}\left([A, B]_+ + \imath\,[A, B]_-\right) = AB\,. \tag{1.106}$$

Going back to quadratic functions, it is not hard to check that:

$$\{f_A, f_B\}_{\mathcal{H}} = 2 f_{AB}\,, \tag{1.107}$$

leading to:

$$\{\{f_A, f_B\}_{\mathcal{H}}, f_C\}_{\mathcal{H}} = \{f_A, \{f_B, f_C\}_{\mathcal{H}}\}_{\mathcal{H}} = 4 f_{ABC}\,,\ \forall A, B, C \in gl(\mathcal{H})\,. \tag{1.108}$$

It is also clear that f_A will be a real function iff A is Hermitean. Thus, the Jordan and Poisson brackets will define a Lie-Jordan algebra structure on the set of real, quadratic functions, and, according to Eq. (1.108), the bracket $\{\cdot,\cdot\}_{\mathcal{H}}$ will be an associative bracket.

We are ready now to define, for any function f, two vector fields, the *gradient* ∇f of f and the *Hamiltonian vector field* X_f defined by:

$$\begin{array}{lcl} g(\nabla f, \cdot) = df & \text{or} & G(\cdot, df) = \nabla f \\ \omega(X_f, \cdot) = df & & \Lambda(\cdot, df) = X_f \end{array}\,, \tag{1.109}$$

that allow to rewrite the Jordan and the Poisson brackets as:

$$\{f, h\}_g = g(\nabla f, \nabla h)\,, \tag{1.110}$$

$$\{f, h\}_\omega = \omega(X_f, X_h)\,. \tag{1.111}$$

Explicitly, in coordinates:

$$\nabla f = \frac{\partial f}{\partial q^k}\frac{\partial}{\partial q^k} + \frac{\partial f}{\partial p^k}\frac{\partial}{\partial p^k} = 2\left(\frac{\partial f}{\partial z^k}\frac{\partial}{\partial \bar{z}^k} + \frac{\partial f}{\partial \bar{z}^k}\frac{\partial}{\partial z^k}\right), \tag{1.112}$$

$$X_f = \frac{\partial f}{\partial p^k}\frac{\partial}{\partial q^k} - \frac{\partial f}{\partial q^k}\frac{\partial}{\partial p^k} = 2i\left(\frac{\partial f}{\partial z^k}\frac{\partial}{\partial \bar{z}^k} - \frac{\partial f}{\partial \bar{z}^k}\frac{\partial}{\partial z^k}\right). \tag{1.113}$$

Also: $J(\nabla f) = X_f$ and $J(X_f) = -\nabla f$.

In particular, if we start from a linear Hermitean operator $A : \mathcal{H} \to \mathcal{H}$ to which we can associate: (i) the quadratic function f_A as in Eq. (1.101) and (ii)

[15] We remark parenthetically that all this extends without modifications to the infinite-dimensional case, if we assume: A, B to be bounded self-adjoint operators on the Hilbert space $\mathcal{H}$.

the vector field $X_A : \mathcal{H} \to T\mathcal{H}$ via $x \longmapsto (x, Ax)$, then we find:

$$\nabla f_A = X_A \,, \tag{1.114}$$

$$X_{f_A} = J(X_A) \,. \tag{1.115}$$

1.3 Methods of Quantization

In the following, we will give an introduction to some quantization techniques which are widely used in the framework of Quantum Field Theory [5; 21]: coherent states, Feynman path integral and finally the Weyl-Wigner map. To do so, we will need some basic knowledge of the Heisenberg-Weyl algebra and group, that we will recall in the next subsection.

1.3.0.1 *The Heisenberg-Weyl Group*

We describe the algebra Heisenberg-Weyl algebra and group in detail just for one degree of freedom, since the generalization to an arbitrary number is obvious.

Starting from the canonical degrees of freedom q, p, we can construct the so-called creation/annihilation operators:

$$a^\dagger = \frac{\hat{q} - \imath\hat{p}}{\sqrt{2\hbar}} \,, \; a = \frac{\hat{q} + \imath\hat{p}}{\sqrt{2\hbar}} \tag{1.116}$$

which, in virtue of the CCR (1.2), satisfy:

$$[a, a^\dagger] = \mathbb{I} \,, \; [a, \mathbb{I}] = [a^\dagger, \mathbb{I}] = 0 \,. \tag{1.117}$$

We say that $\{\hat{q}, \hat{p}, \mathbb{I}\}$ or equivalently $\{a, a^\dagger, \mathbb{I}\}$ generate a Lie algebra, the Heisenberg-Weyl algebra w_1.

Introducing the (imaginary) elements:

$$e_1 = \imath\hat{p}/\sqrt{\hbar} \,, \; e_2 = \imath\hat{q}/\sqrt{\hbar} \,, \; e_3 = \imath\mathbb{I} \,, \tag{1.118}$$

we can equivalently say that the Heisenberg-Weyl algebra w_1 is the real 3-dimensional Lie algebra generated by the set $\{e_1, e_2, e_3\}$ such that:

$$[e_1, e_2] = e_3 \,, \; [e_1, e_3] = [e_2, e_3] = 0 \,. \tag{1.119}$$

A generic element $x \in w_1$ can therefore be written in one of the following form:

$$x = x_1 e_1 + x_2 e_2 + s e_3 \qquad x_1, x_2, s \in \mathbb{R} \tag{1.120}$$

$$= \imath s\mathbb{I} + \imath(P\hat{p} - Q\hat{q})/\hbar \qquad s, P, Q \in \mathbb{R} \tag{1.121}$$

$$= \imath s\mathbb{I} + (\alpha a^\dagger - \alpha^* a) \qquad s \in \mathbb{R} \,, \; \alpha \in \mathbb{C} \,, \tag{1.122}$$

where the relationship between the coefficients of the three expressions can be easily derived by the reader.

The commutator between any two elements of the algebra is easily found to be given by:

$$[x, y] = B(x, y)e_3 \quad \text{with } B(x, y) = x_1 y_2 - x_2 y_1 \,. \tag{1.123}$$

Notice that $B(x, y)$ is the standard symplectic form on the plane of coordinates (x_1, x_2), i.e. (P, Q).

The Heisenberg-Weyl W_1 group is, as usual, obtained by exponentiation. Explicitly, using the complex representation (1.122), one has:

$$g = e^x = e^{\imath s \mathbb{I}} D(\alpha) \quad \text{with} \quad D(\alpha) \equiv e^{\alpha a^\dagger - \alpha^* a} \tag{1.124}$$

and

$$D(\alpha)D(\beta) = e^{\imath \mathrm{Im}\,(\alpha\beta^*)} D(\alpha + \beta) = e^{2\imath \mathrm{Im}\,(\alpha\beta^*)} D(\beta)D(\alpha) \tag{1.125}$$

a formula that can be proved via the Baker-Campbell-Hausdorff formula[16].

The latter expression shows that the operators $e^{\imath s \mathbb{I}} D(\alpha)$ give a representation of the group W_1, fixed by (s, α), with s real and α complex. The group W_1 is nilpotent and its center is given by the subgroup generated by the elements with $(s, 0)$, whose most general representation is fixed by a real number λ via: $T^\lambda((s, 0)) = e^{\imath s \lambda} \mathbb{I}$. Any $\lambda \neq 0$ gives a non-equivalent infinite-dimensional representation of W_1, and any Unitary Irreducible Representation (UIR) is of this kind [33][17].

1.3.1 *Coherent states and Bargmann-Fock representation*

In this section, we will show explicit realizations of the representations of the Weyl group, by means of the so called coherent states, introduced first by Glauber and Sudarshan; the interested reader may find more on this subject and a thorough overview of applications in [24].

1.3.1.1 *Definition and basic properties*

Let us consider a Hilbert $\mathcal{H}$, on which we have defined a couple of bosonic creation/annihilation operators: $[a, a^\dagger] = \mathbb{I}$. As in (1.29), we denote with $|n\rangle$ $(n = 0, 1, 2, ...)$ the orthonormal states of the Fock basis, which are eigenstates of the number operator $N = a^\dagger a$, with eigenvalues n: $N|n\rangle = n|n\rangle$. In the following,

---

[16] This formula reads: $e^{A+B} = e^A e^B e^{-[A,B]/2} = e^B e^A e^{[A,B]/2}$ whenever A, B are such that $[A, B] = c\mathbb{I}$ $(c \in \mathbb{C})$.

[17] In the particular case $\lambda = 0$ the unitary representation is one-dimensional and fixed by a pair of real numbers μ, ν via: $T(g) = e^{\mu x_1 + \nu x_2}$.

we will make use of the following identities which encode the orthonormality and completeness properties of such a basis:

$$\langle n|n'\rangle = \delta_{nn'}, \tag{1.126}$$

$$\sum_{n=0}^{\infty} |n\rangle\langle n| = \mathbb{I}. \tag{1.127}$$

A *coherent state* is, by definition, an eigenstate $|\alpha\rangle$ of the annihilation operator[18] with eigenvalue $\alpha \in \mathbb{C}$:

$$a|\alpha\rangle = \alpha|\alpha\rangle. \tag{1.128}$$

An explicit expression for $|\alpha\rangle$ may be found by expanding it on the Fock basis: $|\alpha\rangle = \sum_{n=0}^{\infty} a_n|n\rangle$. After imposing (1.128), one finds:

$$|\alpha\rangle = \sum_{n=0}^{\infty} \frac{\alpha^n}{\sqrt{n!}}|n\rangle = \sum_{n=0}^{\infty} \frac{\alpha^n}{n!}\left(a^\dagger\right)^n|0\rangle = e^{\alpha a^\dagger}|0\rangle\,,\ \forall\alpha \in \mathbb{C}\,. \tag{1.129}$$

These states are not orthogonal, since:

$$\langle\alpha|\beta\rangle = e^{\alpha^*\beta}\,, \tag{1.130}$$

but form a complete set, as:

$$\mathbb{I} = \int \frac{(d\mathrm{Re}\,\alpha)\,(d\mathrm{Im}\,\alpha)}{\pi} e^{-|\alpha|^2}|\alpha\rangle\langle\alpha|\,. \tag{1.131}$$

Thus, the set: $\{|\tilde{\alpha}\rangle = e^{-|\alpha|^2/2}e^{\alpha a^\dagger}|0\rangle\}_{\alpha\in\mathbb{C}}$ is an overcomplete set of normalized vectors.

We leave the details of the proof of these identities to the reader, by only noticing that to show (1.130), one has to use the Baker-Campbell-Hausdorff formula and the fact that:

$$e^{\alpha^* a}|0\rangle = \sum_{n=0}^{\infty} \frac{(\alpha a)^n}{n!}|0\rangle = |0\rangle.$$

To show (1.131), instead, one can make use of the following chain of identities:

$$\begin{aligned}
\int \frac{(d\mathrm{Re}\,\alpha)\,(d\mathrm{Im}\,\alpha)}{\pi}|\tilde{\alpha}\rangle\langle\tilde{\alpha}| &= \sum_{n=0}^{\infty}\sum_{m=0}^{\infty} \frac{1}{\sqrt{n!}}\frac{1}{\sqrt{m!}}|n\rangle\langle m| \int \frac{(d\mathrm{Re}\,\alpha)\,(d\mathrm{Im}\,\alpha)}{\pi}\alpha^n\left(\alpha^*\right)^m e^{-|\alpha|^2} \\
&= \sum_{n=0}^{\infty}\sum_{m=0}^{\infty} \frac{1}{\sqrt{n!}}\frac{1}{\sqrt{m!}}|n\rangle\langle m|\frac{1}{\pi}\int_0^{2\pi} d\theta \int_0^{\infty} \rho d\rho e^{i(n-m)\theta}\rho^{n+m}e^{-\rho^2} \\
&= \sum_n\sum_m n!\delta_{nm}\frac{1}{\sqrt{n!}}\frac{1}{\sqrt{m!}}|n\rangle\langle m| = \sum_{n=0}^{\infty}|n\rangle\langle n| = \mathbb{I}\,,
\end{aligned}$$

where polar coordinates $\alpha = \rho e^{i\theta}$ have been defined.

[18]One can easily show that the creation operator does not admit eigenvectors with non-zero eigenvalue.

Coherent states satisfy several interesting properties that we now list:

(1) The mean value of creation/annihilation operators on a coherent state is non-zero. More explicitly:

$$\langle\tilde{\alpha}|a|\tilde{\alpha}\rangle = \alpha\,, \ \langle\tilde{\alpha}|a^\dagger|\tilde{\alpha}\rangle = \alpha^*\,. \tag{1.132}$$

(2) A coherent state is also denominated "displaced vacuum" since it can be written as:

$$|\tilde{\alpha}\rangle = D(\alpha)|0\rangle\,, \tag{1.133}$$

where we have introduced the *displacement operator*

$$D(\alpha) \equiv e^{\alpha a^\dagger - \alpha^* a}\,, \tag{1.134}$$

so called because: $D^\dagger(\alpha)aD(\alpha) = a + \alpha$.

The latter expressions can be proved by noticing that:

$$D(\alpha) = e^{\alpha a^\dagger - \alpha^* a} = e^{\frac{|\alpha|^2}{2}} e^{\alpha a^\dagger} e^{-\alpha^* a}\,,$$
$$D(\alpha)^\dagger = e^{-\alpha a^\dagger + \alpha^* a} = e^{-\frac{|\alpha|^2}{2}} e^{\alpha^* a} e^{-\alpha a^\dagger}\,.$$

It is interesting to look at formula (1.133) in the coordinate representation, i.e. when we choose $\mathcal{H} = L^2\left(\mathbb{R} = \{q\},\ d\mu = dq\right)$ and:

$$\hat{q} = q \ \Rightarrow \ e^{\imath p_0 \hat{q}}\psi(q) = e^{\imath p_0 q}\psi(q)\,, \tag{1.135}$$

$$\hat{p} = -\imath\hbar\frac{d}{dq} \ \Rightarrow \ e^{-\imath q_0 \hat{p}}\psi(q) = \psi(q - q_0)\,. \tag{1.136}$$

If we now recall that the vacuum $|0\rangle$ is represented by the function $\psi_0(q) = \frac{1}{\pi^{1/4}}e^{-q^2/2}$, with just some algebra we can show that $|\tilde{\alpha}\rangle$ is represented by:

$$\psi_\alpha(q) = e^{\imath\frac{p_0 q_0}{2}} e^{-\imath q_0 \hat{p}} e^{p_0 \hat{q}}\psi_0(q) = \frac{1}{\pi^{1/4}} e^{-\imath\frac{p_0 q_0}{2}} e^{\imath p_0 q} e^{-(q-q_0)^2/2}\,, \tag{1.137}$$

where we have set $\alpha = (q_0 + ip_0)/\sqrt{2}$. This shows that, like the vacuum, a coherent state is represented by a wave function of gaussian type, with a mean value displaced from $q = 0$ to $q = q_0$.

The reader might have recognized in Eq. (1.133) the operators that define the Heisenberg-Weyl group that we have encountered in the previous section. Indeed, the set of coherent states that we have just defined gives an explicit example of an irreducible representation of W_1.

1.3.1.2 *Physical properties*

Coherent states play an important role in physical problems. They were introduced in the context of optics and possess some very interesting properties that we now briefly present.

- Number of particle distribution.
  Coherent states are given by an infinite superposition of Fock states, hence they have no definite number of particles. However we may easily see that:

$$\begin{aligned}\langle\tilde{\alpha}|N|\tilde{\alpha}\rangle &= e^{-|\alpha|^2}\sum_{nm}\frac{(\alpha^*)^n}{\sqrt{n!}}\frac{(\alpha)^m}{\sqrt{m!}}\langle n|N|m\rangle = e^{-|\alpha|^2}\sum_{n=0}^{\infty}\frac{|\alpha|^{2n}}{n!}n \\ &= e^{-|\alpha|^2}|\alpha|^2\sum_{n=1}^{\infty}\frac{\left(|\alpha|^2\right)^{n-1}}{(n-1)!} = e^{-|\alpha|^2}|\alpha|^2\sum_{m=0}^{\infty}\frac{|\alpha|^{2m}}{m!} = |\alpha|^2 .\end{aligned} \tag{1.138}$$

  The last expression before the final result shows that the probability of finding the value m is given by:

$$p_m = \frac{|\alpha|^{2m}}{m!}e^{-|\alpha|^2} . \tag{1.139}$$

  This means that, in a coherent state, the number of particles obeys a Poisson distribution, with average $|\alpha|^2$, i.e. a distribution that could be obtained form a classical counting of particles which are randomly distributed (with fixed mean).
- Quasi-classical states.
  Let us consider the equations of motion of a classical 1D oscillator: $\dot{q}(t) = \omega p(t)$, $\dot{p}(t) = -\omega q(t)$. Setting $\alpha(t) \equiv [q(t) + \imath p(t)]/\sqrt{2}$, we can summarize them in the complex equation: $\dot{\alpha}(t) = -\imath\omega\alpha(t)$, whose solution is easily found to be:

$$\alpha(t) = \alpha(t=0)\, e^{-\imath\omega t} \equiv \alpha_0\, e^{-\imath\omega t} , \tag{1.140}$$

  with energy

$$E = \frac{\hbar\omega}{2}\left[q(t)^2 + p(t)^2\right] = \hbar\omega|\alpha(t)|^2 = \hbar\omega|\alpha_0|^2 . \tag{1.141}$$

  We notice that formulas (1.116) give the quantum counterpart of what we are doing here at the classical level. To compare the two cases, let us recall that an operator $\hat{A}$ evolves in time as: $\hat{A}(t) = e^{\frac{\imath tH}{\hbar}}\hat{A}e^{-\frac{\imath tH}{\hbar}}$ so that its mean value on a state, $\langle \hat{A} \rangle_t \equiv \langle \psi | \hat{A}(t) | \psi \rangle$ evolves according to the equation:

$$\imath\hbar \frac{d}{dt} \langle \hat{A} \rangle_t = \langle [\hat{A}, \hat{H}] \rangle . \tag{1.142}$$

In particular, it is immediate to see that for $a, a^\dagger$ and $H = \hbar\omega \left(a^\dagger a + 1/2\right)$ one has:

$$\langle a \rangle_t = \langle a \rangle_0 e^{-\imath\omega t} , \quad \langle a^\dagger \rangle_t = \langle a^\dagger \rangle_0 e^{\imath\omega t} , \tag{1.143}$$

$$\langle H \rangle_t = \hbar\omega \left(\langle a^\dagger a \rangle_t + \frac{1}{2} \right) . \tag{1.144}$$

These expectation values coincide with the classical solution (1.140) if:

$$\langle a \rangle_0 = \alpha_0 , \quad \langle a^\dagger \rangle_0 = \alpha_0^* \tag{1.145}$$

and with the classical energy (1.141) if:

$$\langle a^\dagger a \rangle_0 = |\alpha_0|^2 , \tag{1.146}$$

up to a content term which becomes negligible in the (classical) limit $|\alpha_0|^2 \gg \hbar\omega$.
A state that satisfies these conditions is called quasi-classical. It is easy to see that a coherent state $|\alpha\rangle$ does indeed satisfy Eqs. (1.145, 1.146). Moreover, it is not difficult to show the vice versa, i.e. that all quasi-classical states are coherent states.

To show this result, we suppose that $|\psi\rangle$ is a quasi-classical state and consider the operator $b = a - \alpha_0 \mathbb{I}$, so that:

$$\begin{aligned} \| b|\psi\rangle \|^2 &= \langle \psi | b^\dagger b | \psi \rangle = \langle \psi | \left(a^\dagger - \alpha_0^* \right) \left(a - \alpha_0 \right) | \psi \rangle \\ &= |\alpha_0|^2 - \alpha_0 \alpha_0^* - \alpha_0^* \alpha_0 + |\alpha_0|^2 = 0 . \end{aligned}$$

Since the scalar product is non-degenerate, one must have

$$0 = b|\psi\rangle = (a - \alpha_0)|\psi\rangle \Leftrightarrow a|\psi\rangle = \alpha_0 |\psi\rangle ,$$

showing that $|\psi\rangle$ is the coherent state $|\alpha_0\rangle$.

- Minimal uncertainty states.
Coherent states are also called minimal uncertainty states because they saturate Heisenberg uncertainty inequality:

$$\Delta\hat{q}\Delta\hat{p} \geq \frac{\hbar}{2} . \tag{1.147}$$

This can be easily seen, by calculating:

$$\langle \alpha_0 | \hat{q} | \alpha_0 \rangle = \frac{\alpha_0 + \alpha_0^*}{\sqrt{2\hbar}} , \quad \langle \alpha_0 | \hat{p} | \alpha_0 \rangle = \frac{\alpha_0 - \alpha_0^*}{\sqrt{2\hbar} i}$$

and

$$\langle \alpha_0 | \hat{q}^2 | \alpha_0 \rangle = \frac{1 + (\alpha_0 + \alpha_0^*)^2}{2\hbar}, \quad \langle \alpha_0 | \hat{p}^2 | \alpha_0 \rangle \frac{1 - (\alpha_0 - \alpha_0^*)^2}{2\hbar}$$

so that

$$\Delta \hat{q} = \sqrt{\langle \hat{q}^2 \rangle - \langle \hat{q} \rangle^2} = \sqrt{\frac{\hbar}{2}},$$
$$\Delta \hat{p} = \sqrt{\langle \hat{p}^2 \rangle - \langle \hat{p} \rangle^2} = \sqrt{\frac{\hbar}{2}}.$$

Let us notice that, for a coherent state, both $\Delta \hat{q}$ and $\Delta \hat{p}$ are minimal, and equal to $\sqrt{\hbar/2}$. It is possible to define also the so-called "squeezed" states for which:

$$\Delta \hat{q} = \frac{1}{\sqrt{2}} e^{-s} \quad \text{and} \quad \Delta \hat{p} = \frac{1}{\sqrt{2}} e^{+s}, \tag{1.148}$$

with $s \in \mathbf{R}$, which are again of minimal uncertainty. They can be constructed by applying the displacement operator after having applied to the vacuum another operator, called squeezing operator $\delta(s)$:

$$|\psi_s\rangle = D(\alpha)\, \delta(s)\, |0\rangle \quad \text{with} \quad \delta(s) \equiv e^{\frac{s}{2}\left(a^2 - a^{\dagger 2}\right)}. \tag{1.149}$$

In the coordinate representation, a squeezed state is given by the wave-function:

$$\psi_s(q) = \pi^{-\frac{1}{4}} e^{\frac{s}{2}} \exp\left(\imath \frac{q_0 p_0}{2}\right) \exp\left[-\frac{(q - q_0)^2}{2e^{2s}}\right], \tag{1.150}$$

i.e. by a Gaussian distribution centered in $q = q_0$ and variance e^s. Such states are usually obtained in non-linear interaction problems in optics, when one adds to the harmonic oscillator Hamiltonian a term of the kind: $H_{int} = a^2 + (a^\dagger)^2$.

We can summarize what we have seen in this subsection by saying that coherent states are those states (and the only ones) for which the quantum expectation values of the observables satisfy the same dynamical laws as the corresponding classical functions (position, momentum, energy) on phase space and for which the corresponding variances about such classical values get minimized.

1.3.1.3 *Bargmann-Fock representation*

In this subsection, instead of using greek letters $\alpha, \beta, \cdots$, we will denote variables in $\mathbb{C}$ with $z, z', \cdots$ and therefore a coherent states will be represented by the ket $|z\rangle$ and its corresponding eigenvalue with respect to the operator a is denoted with $z \in \mathbb{C}$.

Given any $|\psi\rangle \in \mathcal{H}$, we can use the resolution of the identity (1.131) to write:

$$|\psi\rangle = \left(\int \frac{(d\mathrm{Re}\, z)\,(d\mathrm{Im}\, z)}{\pi} e^{-|z|^2} |z\rangle\langle z|\right) |\psi\rangle = \int \frac{(d\mathrm{Re}\, z)\,(d\mathrm{Im}\, z)}{\pi} e^{-|z|^2} |z\rangle \psi(z^*)\,, \tag{1.151}$$

where $\psi(z^*) \equiv \langle z|\psi\rangle$ is the wave functions associated to the vector ψ in the coherent state basis.

It is not difficult to verify that, for any $|\psi\rangle\,, |\phi\rangle \in \mathcal{H}$:

$$\langle\psi|\phi\rangle = \int \frac{(d\mathrm{Re}\, z)\,(d\mathrm{Im}\, z)}{\pi} e^{-|z|^2} \psi(z^*)^* \phi(z^*)\,, \tag{1.152}$$

which shows that we are working in the Hilbert space $\mathcal{H}_{BF}$ of all anti-holomorphic functions in z such that:

$$\|\psi\|^2_{BF} \equiv \int \frac{(d\mathrm{Re}\, z)\,(d\mathrm{Im}\, z)}{\pi} e^{-|z|^2} |\psi(z^*)|^2 < \infty\,, \tag{1.153}$$

thus obtaining the so-called Bargmann-Fock representation. This Hilbert space might be thought of as the completion of the linear space of polynomials in the variable z^*: $\mathcal{P} = \{P(z^*) = a_0 + a_1 z^* + \cdots + a_n (z^*)^n\}$ with respect to the scalar product defined by the measure:

$$d\mu(z) \equiv \frac{(d\mathrm{Re}\, z)\,(d\mathrm{Im}\, z)}{\pi} e^{-|z|^2}\,. \tag{1.154}$$

In this representation, the vectors of the Fock basis are given by the monomials in z^*. Indeed, from (1.129), one has:

$$\Phi_n(z^*) = \langle z|n\rangle = \sum_{m=0}^{\infty} \frac{(z^*)^m}{\sqrt{m!}} \langle m|n\rangle = \frac{(z^*)^n}{\sqrt{n!}}\,. \tag{1.155}$$

It is interesting to see how the creation/annihilation operators $a^\dagger/a$ are represented on $\mathcal{H}_{BF}$. From the definition of a coherent state, we know that $a|z\rangle = z|z\rangle$, i.e.: $\langle z|a^\dagger = z^*\langle z|$. Therefore we can write:

$$\begin{aligned} a^\dagger|z\rangle &= \sum_{m=0}^{\infty} \frac{(z^*)^m}{\sqrt{m!}} a^\dagger|m\rangle = \sum_{m=0}^{\infty} \frac{(z^*)^m}{\sqrt{m!}} \sqrt{m+1}|m+1\rangle \\ &= \sum_{n=1}^{\infty} \frac{(z^*)^{n-1}}{\sqrt{n!}} n|n\rangle = \frac{\partial}{\partial z} \sum_{n=0}^{\infty} \frac{(z^*)^n}{\sqrt{n!}} |n\rangle\,. \end{aligned}$$

This means that $a^\dagger|z\rangle = \frac{\partial}{\partial z}|z\rangle$ and $\langle z|a = \frac{\partial}{\partial z^*}\langle z|$. In other words, in $\mathcal{H}_{BF}$:

(i) a acts as the derivative with respect to z^*:

$$a : f(z^*) = \langle z|f\rangle \mapsto \langle z|a|f\rangle = \frac{\partial}{\partial z^*}\langle z|f\rangle = \frac{\partial}{\partial z^*} f(z^*)\,; \tag{1.156}$$

(ii) $a^\dagger$ acts as multiplication by z^*:

$$a^\dagger : f(z^*) = \langle z|f\rangle \mapsto \langle z|a^\dagger|f\rangle = z^*\langle z|f\rangle = z^* f(z^*) . \qquad (1.157)$$

We leave to the reader the proof that the operators $\frac{\partial}{\partial z^*}, z^*$ are one the adjoint of the other with respect to the measure (1.154) and that they satisfy: $\left[\frac{\partial}{\partial z^*}, z^*\right] = \mathbb{I}$.

We can now represent in $\mathcal{H}_{BF}$ any operator A by means of an integral representation:

$$(A\psi)(z^*) = \int d\mu(z')A(z^*, z')\psi(z'^*) , \qquad (1.158)$$

with a kernel $A(z^*, z')$ given by:

$$A(z^*, z') = \sum_{m,k=0}^{\infty} A_{mk}\frac{(z')^k}{\sqrt{k!}}\frac{(z^*)^m}{\sqrt{m!}} = \langle z|A|z'\rangle , \; A_{mn} \equiv \langle m|A|n\rangle . \qquad (1.159)$$

To prove these relations, we notice that, for any $|\psi\rangle = \sum_{n=0}^{\infty} c_n|n\rangle \in \mathcal{H}$:

$$\psi(z^*) = \langle z|\psi\rangle = \sum_{n=0}^{\infty} c_n\langle z|n\rangle = \sum_{n=0}^{\infty} c_n\frac{(z^*)^n}{\sqrt{n!}}$$

and

$$(Af)(z^*) = \langle z|A|f\rangle = \sum_{n=0}^{\infty} c_n\langle z|A|n\rangle = \sum_{n,m=0}^{\infty} c_n\langle z|m\rangle\langle m|A|n\rangle$$

$$= \sum_{n,m=0}^{\infty} c_n A_{mn}\frac{(z^*)^m}{\sqrt{m!}} .$$

We can now use the identity

$$\delta_{kn} = \int d\mu(z')\frac{(z')^k}{\sqrt{k!}}\frac{(z'^*)^n}{\sqrt{n!}} ,$$

to write:

$$(Af)(z^*) = \sum_{n,m,k=0}^{\infty} c_n A_{mk}\delta_{kn}\frac{(z^*)^m}{\sqrt{m!}} = \int d\mu(z') \sum_{n,m,k=0}^{\infty} c_n A_{mk}\frac{(z')^k}{\sqrt{k!}}\frac{(z'^*)^n}{\sqrt{n!}}\frac{(z^*)^m}{\sqrt{m!}}$$

$$= \int d\mu(z')\left[\sum_{m,k=0}^{\infty} A_{mk}\frac{(z')^k}{\sqrt{k!}}\frac{(z^*)^m}{\sqrt{m!}}\right]\left(\sum_{n=0}^{\infty} c_n\frac{(z'^*)^n}{\sqrt{n!}}\right)$$

Using such integral representation, it is very simple to calculate the trace of an operator, since we can write:

$$\mathrm{Tr}_{\mathcal{H}}[A] = \sum_{n=0}^{\infty}\langle n|A|n\rangle = \int d\mu(z)\sum_{n=0}^{\infty}\langle n|A|z\rangle\langle z|n\rangle = \int d\mu(z)\sum_{n=0}^{\infty}\langle z|n\rangle\langle n|A|z\rangle$$

$$= \int d\mu(z)e^{-|z|^2}\langle z|\left(\sum_{n=0}^{\infty}|n\rangle\langle n|\right)A|z\rangle = \int d\mu(z)\langle z|A|z\rangle .$$

Thus:

$$\mathrm{Tr}_{\mathcal{H}}[A] = \int d\mu(z) A(z^*, z)\,. \tag{1.160}$$

This formula is very useful in some applications we will encounter in the following.

Example 1.3.1.

(i) The "delta"-operator:

$$\delta(z^* - z_0^*) : \psi(z^*) \mapsto \psi(z_0^*) \tag{1.161}$$

can be written as:

$$\begin{aligned}\psi(z_0^*) = \langle z_0|\psi\rangle = \langle z_0|\left(\int d\mu(z)|z\rangle\langle z|\right)|\psi\rangle = \int d\mu(z)\langle z_0|z\rangle\psi(z^*)\\ = \int d\mu(z) e^{z_0^* z}\psi(z^*)\,,\end{aligned} \tag{1.162}$$

showing that its kernel is given by: $e^{z_0^* z}$.

(ii) The kernel of the annihilation operator a is given by:

$$\langle z|a|z'\rangle = z'\langle z|z'\rangle = z' e^{z^* z'}\,. \tag{1.163}$$

(iii) The kernel of the creation operator $a^\dagger$ is given by:

$$\langle z|a^\dagger|z'\rangle = z^*\langle z|z'\rangle = z^* e^{z^* z'}\,. \tag{1.164}$$

(iv) The kernel of the number operator $a^\dagger a$ is given by:

$$\langle z|a^\dagger a|z'\rangle = z^* z'\langle z|z'\rangle = z^* z' e^{z^* z'}\,. \tag{1.165}$$

(v) More generally, the kernel of any operator of the form[19]:

$$K = \sum_{pq} k_{pq} (a^\dagger)^p a^q \tag{1.166}$$

is simply given by the expression:

$$\langle z|K|z\rangle = \sum_{pq} k_{pq}\langle z|(a^\dagger)^p a^q|z\rangle = \sum_{pq} k_{pq}(z^*)^p (z')^q e^{z^* z'}\,. \tag{1.167}$$

[19] A polynomial in $a, a^\dagger$ in which the creation operators are all on the left of the annihilation ones is said to be in its normal form.

(vi) The operator of the one-parameter group generated by N: $A = \exp[-\tau a^\dagger a]$ is not written in the normal form and to calculate its kernel, we may proceed as follows. Noticing that

$$A_{mk} = \langle m| \exp[-\tau a^\dagger a]|k\rangle = \delta_{mk} \exp[-\tau m] \,, \tag{1.168}$$

one gets

$$\begin{aligned} A(z^*, z') &= \sum_{m,k=0}^{\infty} \delta_{mk} \exp[-\tau m] \frac{(z')^k}{\sqrt{k!}} \frac{(z^*)^m}{\sqrt{m!}} \\ &= \sum_{m=0}^{\infty} \frac{(e^{-\tau} z^* z')^m}{m!} = \exp[e^{-\tau} z^* z'] \,. \end{aligned} \tag{1.169}$$

(vii) These results can be used in the context of Quantum Statistical Mechanics [30] to calculate the, say canonical, partition function of a gas of N distinguishable harmonic oscillators, which is given by $Z_N = (Z_1)^N$, where $Z_1 \equiv \mathrm{Tr}_{\mathcal{H}}[e^{-\beta H}]$, with $H = \hbar\omega(a^\dagger a + 1/2)$ and $\beta = 1/k_B T$ the inverse of the absolute temperature, up to Boltzmann constant k_B. Thus one has:

$$Z_1 = e^{-\beta\hbar\omega/2} \mathrm{Tr}_{\mathcal{H}} \left[e^{-\beta\hbar\omega a^\dagger a}\right] . \tag{1.170}$$

Setting $\tau = \beta\hbar\omega$ in the last example, using (1.160) and changing to polar coordinates, we immediately see that:

$$\begin{aligned} Z &= e^{-\beta\hbar\omega/2} \int \frac{(d\mathrm{Re}\, z)\,(d\mathrm{Im}\, z)}{\pi} e^{-|z|^2} \exp[e^{-\beta\hbar\omega}|z|^2] \\ &= e^{-\beta\hbar\omega/2} \int_0^{2\pi} \frac{d\theta}{\pi} \int_0^{\rho} \rho d\rho e^{-\rho^2(1-e^{-\beta\hbar\omega})} \\ &= \frac{1}{2\sinh(\beta\hbar\omega/2)} \,. \end{aligned} \tag{1.171}$$

1.3.1.4 *Generalized coherent states and comments*

We will introduce here the notion of generalized coherent states [33], applied to the Heisenberg-Weyl group W_1.

Let us take any UIR of W_1, $T(g)$, and denote with $|\psi_0\rangle$ any (non-zero) vector in the (necessarily infinite-dimensional) representation space $\mathcal{H}$. The stability group of $|\psi_0\rangle$ is given only by the center of W_1, i.e. by the elements of the form $T((s,0))$. We define the set of generalized coherent states as:

$$|\alpha\rangle \equiv T(g)\,|\psi_0\rangle = D(\alpha)\,|\psi_0\rangle \,. \tag{1.172}$$

The set of coherent states that we have studied in the previous section corresponds to the choice: $|\psi_0\rangle = |0\rangle$. Similarly to what was done before, one can show that

$\{|\alpha\rangle\}$ form an overcomplete set of states that generate $\mathcal{H}$. More explicitly, we have the following structures:

- law of transformation

$$D(\alpha)\,|\beta\rangle = e^{\imath \mathrm{Im}\,(\alpha\beta^*)}\,|\alpha+\beta\rangle\,; \tag{1.173}$$

- non-orthogonality condition

$$\langle\alpha|\beta\rangle = e^{\alpha^*\beta}; \tag{1.174}$$

- resolution of identity

$$\mathbb{I} = \int \frac{(d\mathrm{Re}\,\alpha)\,(d\mathrm{Im}\,\alpha)}{\pi} e^{-|\alpha|^2}\,|\alpha\rangle\langle\alpha| \equiv \int d\mu_\alpha|\alpha\rangle\langle\alpha|\,. \tag{1.175}$$

From the latter, it is immediate to see that any $|\psi\rangle \in \mathcal{H}$ can be written as:

$$|\psi\rangle = \int d\mu_\alpha|\alpha\rangle\,\langle\alpha|\psi\rangle = \int d\mu_\alpha\,\psi(\alpha)\,, \tag{1.176}$$

where $\psi(\alpha) = \langle\alpha|\psi\rangle$ is called the symbol of the state $|\psi\rangle$. Clearly: $\langle\psi|\psi\rangle = \int d\mu_\alpha|\psi(\alpha)|^2$.

Such a construction can be easily extended to a finite or an infinite number of creation/annihilation operators of *bosonic type*, i.e. to a set of operators $\{(a_i, a_i^\dagger)\}_i$ satisfying the canonical commutation relations:

$$[a_i, a_j] = [a_i^\dagger, a_j^\dagger] = 0\,, \tag{1.177}$$

$$[a_i, a_j^\dagger] = \delta_{ij}\mathbb{I}\,, \tag{1.178}$$

acting on the bosonic Fock Hilbert space $\mathcal{H}_F$ which is generated by the orthonormal basis $|n_1\cdots n_k\cdots\rangle = \frac{1}{\sqrt{\Pi_j n_j!}}(a_1^\dagger)^{n_1}\cdots(a_k^\dagger)^{n_k}\cdots|0\rangle$, with $n_k \in \mathbb{N}$. For instance, this is the framework in which to discuss the quantization of the electromagnet field: in vacuum[20], each component of the latter satisfies d'Alembert equation and hence can be described by using its Fourier modes, each of which behaves independently as a 1D harmonic oscillator. Thus, each of these modes is described quantum mechanically by a couple of creation/annihilation operators, which in all satisfy Eqs. (1.177, 1.178).

Thanks to (1.177), a coherent state can be defined as the common eigenvector $|\phi\rangle \equiv |\phi_1\,\phi_2\,\cdots\rangle$ of all annihilation operators:

$$a_j|\phi\rangle = \phi_j|\phi\rangle \quad \text{with } \phi_j \in \mathbb{C}\,. \tag{1.179}$$

[20]For the definition of Fock space in a more general QFT see sect. 3.5.1 of the third part of this volume.

The solutions of these equations can be easily found to be given by:

$$|\phi\rangle = \prod_j e^{\phi_j a_j^\dagger}|0\rangle = e^{\sum_j \phi_j a_j^\dagger}|0\rangle\,, \tag{1.180}$$

where $a_j|0\rangle = 0$, for all j. It is also not difficult to prove the following relations:

- non-orthogonality condition

$$\langle\phi|\phi'\rangle = e^{\sum_\alpha \phi^*_\alpha \phi'_\alpha}; \tag{1.181}$$

- resolution of identity

$$\mathbb{I} = \int \left(\prod_\alpha \frac{(d\mathrm{Re}\,\phi_\alpha)\,(d\mathrm{Im}\,\phi_\alpha)}{\pi}\right) e^{\sum_\alpha |\phi_\alpha|^2}|\phi\rangle\langle\phi|\,. \tag{1.182}$$

A similar construction is less simple if we aim at discussing a set of finite or infinite number of creation/annihilation operators of *fermionic type*, i.e. a set of operators $\{(a_i, a_i^\dagger)\}_i$ satisfying the canonical commutation relations:

$$\{a_i, a_j\} = \{a_i^\dagger, a_j^\dagger\} = 0\,, \tag{1.183}$$

$$\{a_i, a_j^\dagger\} = \delta_{ij}\mathbb{I}\,, \tag{1.184}$$

acting on the fermionic Fock Hilbert space $\mathcal{H}_F$ which is generated by the orthonormal basis $|n_1 \cdots n_k \cdots\rangle = (a_1^\dagger)^{n_1} \cdots (a_k^\dagger)^{n_k} \cdots |0\rangle$, with $n_k \in \{0, 1\}$.

If we insist to define coherent states as common eigenvectors $|\xi\rangle \equiv |\xi_1\,\xi_2\,\cdots\rangle$ of all annihilation operators:

$$a_j|\xi\rangle = \xi_j|\xi\rangle\,, \tag{1.185}$$

we see that the commutations relations (1.183) now imply:

$$\xi_i\xi_j + \xi_j\xi_i = 0\,. \tag{1.186}$$

This condition admits non-trivial solutions only if we allow the "numbers" ξ_j to be not in $\mathbb{C}$ but in a Grassmann algebra $\mathbb{G}$.

If we allow so, then coherent states are given by:

$$|\xi\rangle \equiv |\xi_1\xi_2\cdots\rangle = e^{\sum_j \xi_j a_j^\dagger}|0\rangle = \prod_j (1 - \xi_j a_j^\dagger)|0\rangle\,, \tag{1.187}$$

which are vectors in the generalized Fock space:

$$\tilde{\mathcal{H}}_F = \left\{|\psi\rangle = \sum_J \chi_J|\phi_J\rangle \;:\; \chi_J \in \mathbb{G}\,,\, |\phi_J\rangle \in \mathcal{H}_F\right\}. \tag{1.188}$$

The set of states $\{|\xi\rangle\}$ satisfy again the relationships:

- non-orthogonality condition

$$\langle\xi|\xi'\rangle = e^{\sum_\alpha \xi^*_\alpha \xi'_\alpha} ; \tag{1.189}$$

- resolution of identity

$$\mathbb{I} = \int \left(\prod_\alpha d\xi^*_\alpha \, d\xi_\alpha\right) e^{\sum_\alpha \xi^*_\alpha \xi_\alpha} |\xi\rangle\langle\xi| \, , \tag{1.190}$$

where the integration over Grassmann variables has to be of course suitably defined. We have not time to explore this subject here and refer the interested reader to the literature [30; 31].

As a final comment, we want to stress that the definition of generalized coherent states can be extended to a more general group G, including all nilpotent and semi-simple Lie-groups, by looking at their UIRs: $T : g \mapsto T(g)$, $T(g)$ being an unitary operator on some Hilbert space $\mathcal{H}$. As before, choosing a reference vector $|\psi_0\rangle$, we construct the set of coherent states as $T(g)\,|\psi_0\rangle$, taking into account that the fiducial vector might have a non-trivial isotropy group G_0. This means that coherent states are in this case labeled by points in the manifold G/G_0. This gives a hint about the fact that coherent states are related to the geometry of symmetric manifolds and the theory of co-adjoint orbits, a topic we cannot deal with here (see [33]).

1.3.2 *Feynman path integral*

What were presented in the previous sections as well as the Weyl-Wigner approach, that will be considered in the next section, essentially rely on Dirac's view of quantization which starts from classical Poisson brackets involving functions of positions and momenta satisfying Hamilton equations of motion. In his doctoral thesis [16], Feynman presented a very different way to discuss a quantum mechanical system starting from a classical one described in terms of a principle of least action, and not necessarily by Hamilton equations of motion. Such an approach was shown to be particularly suited to quantize systems with infinite number of degrees of freedom (field theories), also in a relativistic context[21].

The aim of this section is to give an introduction to Feynman's approach [17] to quantization, by means of the now-called path-integral technique. The idea is not to concentrate on the evolution of states and/or of operators, but to look directly at probabilities. Thus one tries to construct the kernel of the evolution operator:

$$\mathcal{K}(x, x'; t) = \langle x, t|x'\rangle = \langle x|e^{-\frac{\imath tH}{\hbar}}|x'\rangle \, , \tag{1.191}$$

[21] See also sect. 3.8.1 of the third part of this volume.

which gives the probability of finding the particle at point x at time t, given that it was at point x' at time 0 (we are supposing $t > 0$ and invariance under time-translations).

1.3.2.1 *Path integral in the space of coordinates*

Let us start by considering a system with a Hamiltonian of the form:

$$H = T + V\,,\quad T = \frac{p^2}{2m}\,,\ V = V(x) \tag{1.192}$$

describing a particle with mass m moving in a potential $V(x)$.

In the following, we will use the so-called Trotter formula: for any two self-adjoint operators A, B on some Hilbert space $\mathcal{H}$, one can trivially write:

$$e^{\imath t(A+B)} = [e^{\imath\epsilon(A+B)}]^M\,,\quad t \equiv \epsilon M\,. \tag{1.193}$$

Then, using the fact that

$$e^{\imath\epsilon(A+B)} = e^{\imath\epsilon A}e^{\imath\epsilon B} + O(\epsilon^2) \tag{1.194}$$

we get:

$$e^{\imath(A+B)} = \lim_{\epsilon\to 0}\left[e^{\imath\epsilon A}e^{\imath\epsilon B}\right]^M\,, \tag{1.195}$$

where the limit means $\epsilon \to 0$ and $M \to \infty$ so that $M\epsilon = t$ is kept constant. This formula holds in the operator-norm sense if $\mathcal{H}$ is finite-dimensional. In the infinite-dimensional case, the limit has to be understood in the strong sense and applied to vectors in the appropriate domains.

For the Hamiltonian (1.192), supposing $T + V$ to be self-adjoint on a dense domain $\mathcal{D}(T) \cap \mathcal{D}(V)$, we can re-write the kernel in the following form:

$$\begin{aligned}
\langle x|e^{-\frac{\imath tH}{\hbar}}|x'\rangle &= \lim_{\epsilon\to 0}\langle x|\left[e^{-i\epsilon\frac{T}{\hbar}}e^{-\imath\epsilon\frac{V}{\hbar}}\right]^M|x'\rangle \\
&= \lim_{\epsilon\to 0}\int dx_1 \ldots \int dx_{M-1}\langle x|e^{-\imath\epsilon\frac{T}{\hbar}}e^{-\imath\epsilon\frac{V}{\hbar}}|x_{M-1}\rangle \\
&\quad\times\langle x_{M-1}|e^{-\imath\epsilon\frac{T}{\hbar}}e^{-\imath\epsilon\frac{V}{\hbar}}x_{M-2}\rangle\cdots\langle x_1|e^{-\imath\epsilon\frac{T}{\hbar}}e^{-\imath\epsilon\frac{V}{\hbar}}|x'\rangle \\
&= \lim_{\epsilon\to 0}\int dx_1 \ldots \int dx_{M-1}\prod_{n=1}^{M}\langle x_n|e^{-i\epsilon\frac{T}{\hbar}}e^{-\imath\epsilon\frac{V}{\hbar}}|x_{n-1}\rangle\,.
\end{aligned} \tag{1.196}$$

To obtain this expression, in the second line we have inserted M resolutions of the identities in coordinate space ($\mathbb{I} = \int dx_j|x_j\rangle\langle x_j|$) and defined: $x_M = x$, $x_0 = x'$.

We now insert a resolution of the identity in momentum space ($\mathbb{I} = \int dp_{|}p_j\rangle\langle p_j|$) in each factor of the integrand, to get:

$$\langle x_n|e^{-\imath\epsilon\frac{T}{\hbar}}e^{-\imath\epsilon\frac{V}{\hbar}}|x_n - 1\rangle = \int dp_n\langle x_n|e^{-\imath\epsilon\frac{T}{\hbar}}|p_n\rangle\langle p_n|e^{-\imath\epsilon\frac{V}{\hbar}}|x_{n-1}\rangle$$
$$= \int dp_n e^{-\imath\frac{\epsilon p_n^2}{2m\hbar}}\langle x_n|p_n\rangle e^{-\imath\frac{\epsilon V(x_{n-1})}{\hbar}}\langle p_n|x_{n-1}\rangle$$
$$= \frac{1}{2\pi\hbar}\int dp_n e^{-\imath\frac{\epsilon p_n^2}{2m\hbar}}e^{-\imath\frac{\epsilon V(x_{n-1})}{\hbar}}e^{\imath\frac{p_n(x_n - x_{n-1})}{\hbar}}, \quad (1.197)$$

where we have used the fact that:

$$\langle x_j|p_n\rangle = \frac{1}{\sqrt{2\pi\hbar}}e^{\imath\frac{x_j p_n}{\hbar}}. \quad (1.198)$$

Now we insert (1.197) in (1.196), to find:

$$\langle x|e^{-\imath\frac{tH}{\hbar}}|x'\rangle = \lim_{\epsilon\to 0}\int dx_1...dx_{M-1}\frac{dp_1...dp_m}{(2\pi\hbar)^M}\prod_{n=1}^{M}\left[\exp\left\{\imath\frac{p_n(x_n - x_{n-1})}{\hbar}\right.\right.$$
$$\left.\left. - \imath\frac{\epsilon p_n^2}{2m\hbar} - \imath\frac{\epsilon V(x_{n-1})}{\hbar}\right\}\right]. \quad (1.199)$$

Thanks to the form of the Hamiltonian, we see that the integrals in the p_n's are of Gaussian type and thus they can be easily performed, to finally achieve:

$$\langle x|e^{-\imath\frac{tH}{\hbar}}|x'\rangle = \lim_{\epsilon\to 0}\int dx_1...dx_{M-1}\left(\frac{m}{2\pi\imath\hbar\epsilon}\right)^{\frac{M}{2}}$$
$$\times\exp\left\{\frac{\imath\epsilon}{\hbar}\sum_{n=1}^{M}\left[\frac{m}{2}\left(\frac{x_n - x_{n-1}}{\epsilon}\right)^2 - V(x_{n-1})\right]\right\}. \quad (1.200)$$

Let us remark that we have made no approximations to get this formula, which is mathematically sound as long as the problem of domains is taken into account, and the limit is taken after all integrals have been calculated.

Despite the last comment, it is very tempting to bring the limit inside the integrals. Noticing that:

$$\frac{x_n - x_{n-1}}{\epsilon} \to \dot{x}(t'), \quad V(x_{n-1}) \to V(x(t')), \quad \epsilon\sum_{n=1}^{M} \to \int_0^t dt', \quad (1.201)$$

one can write:

$$\epsilon\sum_{n=1}^{M}\left[\frac{m}{2}\left(\frac{x_n - x_{n-1}}{\epsilon}\right)^2 - V(x_{n-1})\right] \to \int_0^t dt'\left[\frac{m\dot{x}^2(t')}{2} - V(x(t'))\right], \quad (1.202)$$

where the last expression gives the classical action S. If we also define:

$$\int_{x(0)=x'}^{x(t)=x} [\mathcal{D}x] = \lim_{\epsilon\to 0} \int dx_1 ... dx_{M-1} \left(\frac{m}{2\pi i\epsilon}\right)^{\frac{M}{2}} , \tag{1.203}$$

we can eventually write Eq. (1.200) as a functional of the classical action S as:

$$\langle x|e^{-\imath\frac{tH}{\hbar}}|x'\rangle = \int_{x(0)=x'}^{x(t)=x} [\mathcal{D}x] e^{\frac{\imath}{\hbar}S} . \tag{1.204}$$

This is the celebrated path integral formula for the kernel, which is telling us that to get the quantum amplitude (1.191) we have to integrate over all possible paths $x(t)$, starting at x' and ending at x, the exponential of the classical action, evaluated on such trajectories.

Let us remark that formula (1.200) is compact and has an elegant interpretation, but it has (at least at this point of the discussion) only a formal meaning, since we have not specified which class of trajectories, i.e. space of functions, we work on and therefore what measure of integration we need. Looking at the way we have got formula (1.204), it is suggestive to interpret $x(t)$ as a trajectory obtained in configuration space by discretizing it with the set of points $\{x_j = x(t = j\epsilon)\}_j$. Thus, for the very way in which it is constructed, we expect it to have no nice properties such as continuity or differentiability. We will say something more about this in the following. For now, we use formula (1.200) in a symbolic way and refer to Eq. (1.200) to do explicit calculations.

Example 1.3.2. *The free particle.*
As a first example, let us choose $V(x) = 0$ and work in 1D, the calculations being easily generalized to arbitrary dimensions. The kernel (1.200) is given by:

$$\langle x|e^{-\imath\frac{tH}{\hbar}}|x'\rangle = \lim_{\epsilon\to 0} \int dx_1 ... dx_{M-1} \left(\frac{m}{2\pi\imath\hbar\epsilon}\right)^{\frac{M}{2}} \exp\left[\frac{\imath m}{2\hbar}\sum_{n=1}^{M}\frac{(x_n - x_{n-1})}{\epsilon}\right]. \tag{1.205}$$

Using the identity:

$$\int dy \exp\left\{\frac{\imath A}{2}\left[\frac{(x-y)^2}{\alpha} + \frac{(y-z)^2}{\alpha'}\right]\right\} = \exp\left\{\frac{\imath A(x-z)^2}{2(\alpha+\alpha')}\right\}\left(\frac{2\pi\imath\alpha\alpha'}{A(\alpha+\alpha')}\right)^{\frac{1}{2}} , \tag{1.206}$$

it is possible to calculate recursively all integrals, to eventually find:

$$\begin{aligned}\langle x|e^{-\imath\frac{tH}{\hbar}}|x'\rangle &= \lim_{\epsilon\to 0}\left(\frac{m}{2\pi\imath\hbar\epsilon}\right)^{\frac{M}{2}}\left(\frac{2\pi\imath\hbar}{m}\right)^{\frac{M-1}{2}}\left(\frac{\epsilon M}{t}\right)^{\frac{1}{2}}\exp\left[\frac{\imath m(x-x')^2}{2\hbar t}\right] \\ &= \sqrt{\frac{m}{2\pi\imath\hbar t}}\exp\left[\frac{\imath m(x-x')^2}{2\hbar t}\right].\end{aligned} \tag{1.207}$$

Recalling that (i) the classical action is given by:

$$S = \int_0^t dt' \frac{m}{2} \left(\frac{dx}{dt} \right)^2 \tag{1.208}$$

and (ii) the solution of the classical equation of motion $d^2x(t)/dt^2 = 0$, satisfying the boundary conditions $x(0) = x', x(t) = x$, yields the classical trajectory:

$$x_{cl}(t') = \frac{t'(x - x')}{t} + x' , \quad \dot{x}_{cl}(t') = \frac{x - x'}{t} , \tag{1.209}$$

one can easily verify that:

$$S_{clas}(x, x') \equiv S|_{x_{cl}} = \int_0^t dt' \frac{m}{2} \left(\frac{x - x'}{t'} \right)^2 . \tag{1.210}$$

Comparing with (1.207), we see that the quantum kernel is given by the exponential of the classical action:

$$\langle x|e^{-\imath \frac{tH}{\hbar}}|x'\rangle = F(t) \exp\left[\frac{\imath}{\hbar} S_{clas}(x, x') \right] , \quad F(t) = \sqrt{\frac{m}{2\pi i t\hbar}} . \tag{1.211}$$

up to a pre-factor $F(t)$ which depends only on time. It is an interesting fact that it can be calculated exactly by also making use of a semi-classical stationary phase approximation [35].

Example 1.3.3. *The 1D harmonic oscillator.*
Similar calculations can be done for the Hamiltonian of the 1D harmonic oscillator, to get:

$$K(x, x'; t) = F_{ho}(t) \exp\left[\frac{\imath}{\hbar} S_{clas}(x, x') \right] , \tag{1.212}$$

with

$$S_{cl}(x, x') = \frac{\omega m}{2 \sin(\omega t)} \left[(x + x')^2 \cos(\omega t) - 2xx' \right] , \tag{1.213}$$

$$F_{ho} = \sqrt{\frac{m\omega}{2\pi\imath\hbar \sin(\omega t)}} . \tag{1.214}$$

As in the previous example, F_{ho} can be calculated exactly by means of the stationary phase approximation [35].

1.3.2.2 *Feynman integral in imaginary time and partition function*

For a quantum system, the (canonical) partition function can be calculated as:

$$\mathcal{Z} \equiv \mathrm{Tr}\left[e^{-\beta H} \right] = \int dx \langle x|e^{-\beta H}|x\rangle . \tag{1.215}$$

To calculate $\mathcal{Z}$, it is possible to proceed as in the previous section by taking into account the following two minor differences[22]:

- in the expression we have $e^{-\beta H}$ instead of $e^{-\frac{\imath}{\hbar}tH}$;
- boundary conditions are now $x(t) = x(0) = x$, on which we have to integrate.

Leaving all details to the reader, we write here only the final result:

$$\mathcal{Z} = \lim_{\epsilon \to 0} \int_{x_0 = x_M} \left(\prod_{n=1}^{M} d^3 x_n \right) \left(\frac{m}{2\pi\hbar\epsilon} \right)^{\frac{3M}{2}} \exp\left\{ -\frac{\epsilon}{\hbar} \sum_{n=1}^{\infty} \left[\frac{(x_n - x_{n-1})^2}{\epsilon^2} + V(x_n) \right] \right\} \tag{1.216}$$

which can be written, formally, as:

$$\mathcal{Z} = \int_{x(0)=x(\beta\hbar)} [\mathcal{D}x(\tau)] \exp\left\{ -\frac{1}{\hbar} \int_0^{\beta\hbar} d\tau' H[x(\tau')] \right\} . \tag{1.217}$$

We remark also that the same result could have been obtained by performing a "Wick-rotation", i.e. the analytic continuation from the real variable t to the imaginary one $\tau = \imath t$[23]. The measure that appears in (1.217) is the same one used in the context of stochastic processes, i.e. Wiener measure. In this case it is possible to give a rigorous mathematical treatment to make sense of the continuous version of Feynman path-integral [20].

1.3.2.3 *Path integral with coherent states*

In this section, we will work out the path integral formulation of QM using coherent states. In the following, we will concentrate on one single degree of freedom, either bosonic or fermionic, but the construction can be generalized to many body systems and quantum field theory, giving one of the most exploited technique in this research area.

Either in the bosonic or in fermionic case, let us denote with $|\phi_i\rangle$ and $\langle\phi_f|$ the initial and final coherent states. Then we write the propagator in the following

[22]See sect. 3.2.1 of the third part of this volume for an analogue definition in the Euclidean approach to QFT.
[23]Indeed this would have changed the expressions of the derivatives appearing in the action S according to: $dx/dt = \imath dx/d\tau$, $(dx/dt)^2 = -(dx/d\tau)^2$.

way:

$$
\begin{aligned}
&\langle\phi_f|e^{-\imath\frac{tH}{\hbar}}|\phi_i\rangle = \lim_{\epsilon\to 0}\langle\phi_f|(e^{-\imath\frac{\epsilon H}{\hbar}})^M|\phi_i\rangle \\
&= \lim_{\epsilon\to 0}\int\left(\prod_{n=1}^{M-1}\frac{d\phi_n^* d\phi_n}{\mathcal{N}}\right)e^{-\sum_{n=1}^{M-1}\phi_n^*\phi_n}\langle\phi_f|e^{-\frac{\imath\epsilon H}{\hbar}}|\phi_{M-1}\rangle \\
&\quad\times\langle\phi_{M-1}|e^{-\frac{\imath\epsilon H}{\hbar}}|\phi_{M-2}\rangle\ldots\langle\phi_1|e^{-\frac{\imath\epsilon H}{\hbar}}|\phi_i\rangle \\
&= \lim_{\epsilon\to 0}\int\left(\prod_{n=1}^{M-1}\frac{d\phi_n^* d\phi_n}{\mathcal{N}}\right)e^{-\sum_{n=1}^{M-1}\phi_n^*\phi_n}\left(\prod_{n=1}^{M}\langle\phi_n|e^{-\imath\frac{\epsilon H}{\hbar}}|\phi_{n-1}\rangle\right) \\
&= \lim_{\epsilon\to 0}\int\left(\prod_{n=1}^{M-1}\frac{d\phi_n^* d\phi_n}{\mathcal{N}}\right)e^{-\sum_{n=1}^{M-1}\phi_n^*\phi_n}e^{\sum_{n=1}^{M}\phi_n^*\phi_{n-1}}e^{-\imath\frac{\epsilon}{\hbar}\sum_{n=1}^{M}H(\phi_n^*,\phi_{n-1})},
\end{aligned}
\tag{1.218}
$$

having set: $\langle\phi_f|\equiv\langle\phi_M|$, $|\phi_i\rangle\equiv|\phi_0\rangle$.

To arrive at this expression we have divided the time interval t in M intervals of length ϵ, inserted $M-1$ resolutions of the identity written in terms of coherent states:

$$
\mathbb{I} = \int\left(\prod_j\frac{d\phi_j^* d\phi_j}{\mathcal{N}}e^{-\phi_j^*\phi_j}|\phi_j\rangle\langle\phi_j|\right),
$$

with

$$
\phi = \begin{cases} z\in\mathbb{C} \\ \xi\in\mathbb{G}\end{cases},\quad \mathcal{N} = \begin{cases} 2\pi i & \text{for bosons} \\ 1 & \text{for fermions}\end{cases}.
$$

The last line of (1.218) follows after assuming that $H = H(a^\dagger, a)$ is written in its normal form so that:

$$
\langle\phi_n|e^{-\imath\frac{\epsilon H}{\hbar}}|\phi_{n-1}\rangle = \langle\phi_n|1-\frac{\imath\epsilon}{\hbar}H(a^\dagger,a)+\cdots|\phi_{n-1}\rangle = e^{\phi_n^*\phi_{n-1}}e^{-\imath\frac{\epsilon}{\hbar}H(\phi_n^*,\phi_{n-1})},
$$

where $H(\phi_n^*, \phi_{n-1})$ has been obtained from $H = H(a^\dagger, a)$ by means of the substitution: $a^\dagger\mapsto\phi_n^*$, $a\mapsto\phi_n$.

As before, we can interpret $\{\phi_n\}$ as a discretized trajectory, so that:

$$
\begin{aligned}
\phi_n^*\frac{\phi_n-\phi_{n-1}}{\epsilon} &\to \phi_n^*(t')\frac{\partial}{\partial t'}\phi(t'), \\
H(\phi_n^*,\phi_{n-1}) &\to H(\phi_n^*(t'),\phi_{n-1}(t')),
\end{aligned}
$$

while

$$
\begin{aligned}
&-\sum_{n=1}^{M-1}\phi_n^*\phi_n+\sum_{n=1}^{M}\phi_n^*\phi_{n-1}-\frac{\imath\epsilon}{\hbar}\sum_{n=1}^{M}H(\phi_n^*,\phi_{n-1}) \\
&\qquad\to\phi^*(t)\phi(t)+\frac{\imath}{\hbar}\int_0^t dt'\left[(i\hbar)\phi^*(t')\frac{\partial\phi(t')}{\partial t'}-H(\phi^*(t'),\phi(t'))\right].
\end{aligned}
$$

Thus, in a formal way we can write:

$$\langle\phi_f|e^{-\imath\frac{tH}{\hbar}}|\phi_i\rangle = \int_{\phi(0)=\phi_i,\phi(t)=\phi_f} [\mathcal{D}\phi^*\mathcal{D}\phi] e^{\phi^*(t)\phi(t)} e^{\frac{\imath}{\hbar}\int_0^t dt\left[i\hbar\phi^*\frac{\partial\phi}{\partial t'}-H(\phi^*,\phi)\right]}, \tag{1.219}$$

where the integral in the exponential represents the so-called Schrödinger Lagrangian of the classical system, which contains a kinetic term which is linear in the first derivatives with respect to time.

We can now proceed to evaluate the (grancanonical) partition function of the system, by recalling that [31]:

$$\mathcal{Z} = \mathrm{Tr}\left[e^{-\beta(\hat{H}-\mu\hat{N})}\right] = \int \frac{d\tilde{\phi}^* d\tilde{\phi}}{\mathcal{N}} e^{-\tilde{\phi}^*\tilde{\phi}} \langle\zeta\tilde{\phi}|e^{-\beta(\hat{H}-\mu\hat{N})}|\tilde{\phi}\rangle, \tag{1.220}$$

with

$$\zeta = \begin{cases} +1 & \text{for bosons} \\ -1 & \text{for fermions} \end{cases}.$$

From (1.219), putting $\tau' = \imath t', \beta = it, \hbar = 1$, we have:

$$\langle\zeta\tilde{\phi}|e^{-\beta(\hat{H}-\mu\hat{N})}|\tilde{\phi}\rangle = \lim_{\epsilon\to 0}\int \prod_{n=1}^{M-1} \frac{d\phi_n^* d\phi_n}{\mathcal{N}} e^{-i\sum_{n=1}^{M-1}\phi_n^*\phi_n} e^{\sum_{n=1}^{M}\phi_n^*\phi_n}$$
$$\times \prod_{n=1}^{M} \exp\left\{-\epsilon[H(\phi_n^*,\phi_{n-1}) - \mu\phi_n^*\phi_{n-1}]\right\}, \tag{1.221}$$

with the boundary conditions:

$$\phi_0 = \tilde{\phi}, \; \phi_M^* = \zeta\tilde{\phi}^* = \zeta\phi_0^*. \tag{1.222}$$

Thus we get (as before $\zeta = \pm 1$ for bosons/fermions):

$$\mathcal{Z} = \lim_{\epsilon\to 0}\int \left(\prod_{n=1}^{M}\frac{d\phi_n^* d\phi_n}{\mathcal{N}}\right) \exp\left\{-\epsilon\sum_{n=1}^{M}\phi_n^*\frac{\phi_n - \phi_{n-1}}{\epsilon}\right\}$$
$$\times \exp\left\{-\epsilon\sum_{n=1}^{M}[H(\phi_n^*,\phi_{n-1}) - \mu\phi_n^*\phi_{n-1}]\right\}. \tag{1.223}$$

Example 1.3.4. *The 1D bosonic/fermionic harmonic oscillator.*
We consider the Hamiltonian

$$H = \Omega a^\dagger a, \tag{1.224}$$

so that $H(\phi_n^*,\phi_{n-1}) = \Omega\phi_n^*\phi_{n-1}$. The argument in the exponential of (1.223) is then given by:

$$\exp\left\{-\sum_{i,j=1}^{M}\phi_i^* M_{ij}\phi_j\right\} \tag{1.225}$$

with the matrix $\mathbb{M} = [M_{ij}]$ of the form:

$$\mathbb{M} = \begin{vmatrix} 1 & 0 & 0 \ldots & \ldots. & -\zeta\Omega_0 \\ -\Omega_0 & 1 & 0 \ldots & \ldots & 0 \\ 0 & -\Omega_0 & 1 \ldots & \ldots & 0 \\ \ldots & \ldots & \ldots \ldots & \ldots & \ldots \\ 0 & 0 & 0 \ldots & -\Omega_0 & 1 \end{vmatrix}, \quad \Omega_0 \equiv 1 - \frac{\beta}{M}(\Omega - \mu)\,. \tag{1.226}$$

The partition function is then given by an integral over (complex or grassmann) variables of gaussian type , which can be performed to get:

$$\mathcal{Z} = \lim_{\epsilon \to 0} (\det \mathbb{M})^{-\zeta}\,, \tag{1.227}$$

where

$$\det \mathbb{M} = 1 + (-)^{M-1}(-\zeta\Omega_0)(-\Omega_0)^{M-1} = 1 - \zeta\Omega_0^M\,. \tag{1.228}$$

Hence

$$\mathcal{Z} = \lim_{\epsilon \to 0} \left[1 - \zeta\left(\frac{\beta(\Omega - \mu)}{M}\right)^M\right]^{-\zeta} = \left[1 - \zeta e^{-\beta(\Omega - \mu)}\right]^{-\zeta}\,. \tag{1.229}$$

1.3.3 *Weyl quantization*

In this section, we will give an introduction to the quantization à la Weyl, the interested reader can find a more exhaustive discussion in [14].

Weyl quantization is interesting from many points of view and has several advantages. Firstly, it has the virtue of overcoming the problem mentioned in the introduction about the fact that the CCR (1.2) between p and q implies that at least one of the two operators must be unbounded. Also, being founded on geometric concepts, it can be generalized to a generic phase space, i.e. a symplectic manifold which is not necessarily a linear space. Then, The Weyl map (and its inverse, the Wigner map) allow for a quantization on the space of functions $f(p, q)$ over the entire phase space, and not on the space of functions on the configuration space or on a suitable maximal Lagrangian subspace, as it is required by geometric quantization. Finally, in this setting, it is transparent to discuss the limit $\hbar \to 0$, which describes the quantum to classical transition.

1.3.3.1 *The Weyl map*

Let us start by considering a (real) vector space $\mathcal{S}$ endowed with a constant symplectic structure ω, so that $\mathcal{S} \approx \mathbb{R}^{2n}$ for some n. We will denote with $\mathcal{U}(\mathcal{H})$ the set of unitary operators on an abstract Hilbert space $\mathcal{H}$.

A *Weyl system* is a map:

$$\begin{aligned} W:\ & \mathcal{S} \to \mathcal{U}(\mathcal{H}) \\ & z \mapsto \widehat{W}(z), \end{aligned} \tag{1.230}$$

with $\widehat{W}(z)\widehat{W}^{\dagger}(z) = \widehat{W}^{\dagger}(z)\widehat{W}(z) = \widehat{\mathbb{I}}$, such that:

(i) W is strongly continuous;
(ii) for any $z, z' \in \mathcal{S}$:

$$\widehat{W}(z+z') = \widehat{W}(z)\widehat{W}(z')\exp\{-\imath\omega(z,z')/2\hbar\}. \tag{1.231}$$

In other words, a *Weyl map* provides a projective (i.e. up to a factor) unitary representation of the vector space $\mathcal{S}$, thought of as the group manifold of the translation group, in the Hilbert space $\mathcal{H}$.

From (1.231), it is easy to derive that:

$$\widehat{W}(0) = \widehat{\mathbb{I}},\ \ \widehat{W}^{\dagger}(z) = \widehat{W}(-z) \tag{1.232}$$

and

$$\widehat{W}(z)\widehat{W}(z') = \widehat{W}(z')\widehat{W}(z)\exp\{\imath\omega(z,z')/\hbar\},\ \forall z, z'. \tag{1.233}$$

Suppose now that $\mathcal{S}$ splits into the direct sum of two Lagrangian subspaces $\mathcal{S} = \mathcal{S}_1 \oplus \mathcal{S}_2$, so that any vector z can be written as $z = (z_1, 0) + (0, z_2)$, $z_1 \in \mathcal{S}_1$, $z_2 \in \mathcal{S}_2$. The restrictions of W to the Lagrangian subspaces:

$$\begin{aligned} U = W|_{\mathcal{S}_1} : \mathcal{S}_1 \to \mathcal{H}, \\ V = W|_{\mathcal{S}_2} : \mathcal{S}_2 \to \mathcal{H} \end{aligned} \tag{1.234}$$

yield faithful abelian representations of the corresponding Lagrangian subspaces:

$$\begin{aligned} \widehat{U}(z_1+z_1') = \widehat{U}(z_1)\widehat{U}(z_1'),\ \ z_1, z_1' \in \mathcal{S}_1 \\ \widehat{V}(z_2+z_2') = \widehat{V}(z_2)\widehat{V}(z_2'),\ \ z_2, z_2' \in \mathcal{S}_2 \end{aligned}, \tag{1.235}$$

which satisfy:

$$\widehat{U}(z_1)\widehat{V}(z_2) = \widehat{V}(z_2)\widehat{U}(z_1)\exp\{\imath\omega((z_1,0),(0,z_2))/\hbar\}. \tag{1.236}$$

Vice versa, it is simple to show that two faithful representations U and V of two transversal Lagrangian subspaces of a symplectic vector space $\mathcal{S}$ satisfying (1.236), yield a Weyl map by setting:

$$z \mapsto \widehat{W}(z) = \widehat{U}(z_1)\widehat{V}(z_2)\exp\{-\imath\omega((z_1,0),(0,z_2))/2\hbar\}. \tag{1.237}$$

We also notice that $\left\{\widehat{W}(\alpha z)\right\}_{\alpha\in\mathbb{R}}$ is a strongly continuous one-parameter group of unitaries since, from (1.231) we have

$$\widehat{W}(\alpha z)\widehat{W}(\beta z) = \widehat{W}((\alpha+\beta)z),\ \ \alpha, \beta \in \mathbb{R}. \tag{1.238}$$

By Stone's theorem [34], there exists an essentially self-adjoint generator $\widehat{G}(z)$:

$$\widehat{W}(\alpha z) = \exp\left\{\imath\alpha\widehat{G}(z)/\hbar\right\}, \tag{1.239}$$

with $\widehat{G}(\alpha z) = \alpha\widehat{G}(z)$. Thus Eq. (1.233) reads:

$$e^{\imath\alpha\widehat{G}(z)/\hbar}e^{\imath\beta\widehat{G}(z')/\hbar} = e^{\imath\alpha\beta\omega(z,z')/\hbar}e^{\imath\alpha\widehat{G}(z)/\hbar}e^{\imath\beta\widehat{G}(z')/\hbar}, \tag{1.240}$$

giving, at the infinitesimal order, the following commutation relation:

$$\left[\widehat{G}(z), \widehat{G}(z')\right] = -\imath\hbar\omega(z, z'). \tag{1.241}$$

Example 1.3.5. *The free particle.*
The simplest case we can consider is given by $\mathcal{S} = \mathbb{R}^2$ with coordinates $z = (q, p)$ and the standard symplectic form $\omega = dq \wedge dp$, so that $\omega((q,p), (q', p')) = qp' - q'p$. In this case a Weyl system satisfies:

$$\widehat{W}((q,p) + (q', p')) = \widehat{W}(q,p)\widehat{W}(q', p')\exp\left\{-\frac{\imath}{2\hbar}(qp' - q'p)\right\}. \tag{1.242}$$

Now if $\mathcal{S}_1 = \{z_1 \equiv (q, 0)\}$ and $\mathcal{S}_2 = \{z_2 \equiv (0, p)\}$, one has:

$$\widehat{W}(q,p) = \widehat{W}((q,0) + (0,p)) = \widehat{W}(q,0)\widehat{W}(0,p)\exp\{-\imath qp/2\hbar\}. \tag{1.243}$$

Defining the infinitesimal generators as:

$$\begin{aligned} \widehat{U}(q) &= \widehat{W}(q,0) \equiv \exp\left\{\imath q\widehat{P}/\hbar\right\} \\ \widehat{V}(p) &= \widehat{W}(0,p) \equiv \exp\left\{\imath p\widehat{Q}/\hbar\right\} \end{aligned}, \tag{1.244}$$

one immediately finds that Eq. (1.241) gives the standard CCR:

$$\left[\widehat{Q}, \widehat{P}\right] = \imath\hbar\mathbb{I}. \tag{1.245}$$

Thus, using the already-mentioned Baker-Campbell-Hausdorff formula, it is straightforward to see that the Weyl system is given by the map:

$$\widehat{W}(q,p) = \exp\left\{\imath\left(q\widehat{P} + p\widehat{Q}\right)/\hbar\right\}. \tag{1.246}$$

To construct a concrete realization of this Weyl system, we can consider wave functions $\psi \in L^2(\mathbb{R}, dx)$ and define the families of operators (1.244) via:

$$\begin{aligned} \left(\widehat{U}(q)\psi\right)(x) &= \psi(x+q) \\ \left(\widehat{V}(p)\psi\right)(x) &= \exp\{\imath px/\hbar\}\psi(x) \end{aligned}, \tag{1.247}$$

which are one-parameter strongly continuous groups of unitary transformations satisfying:

$$\left(\widehat{U}(q)\widehat{V}(p)\psi\right)(x) = \exp\{\imath qp/\hbar\}\left(\widehat{V}(p)\widehat{U}(q)\psi\right)(x). \tag{1.248}$$

Thus, one has:

$$\widehat{W}(q,p) = \widehat{U}(q)\widehat{V}(p)\exp\{-\imath qp/\hbar\} \tag{1.249}$$

or explicitly:

$$\left(\widehat{W}(q,p)\psi\right)(x) = \exp\{\imath p[x+q/2]/\hbar\}\psi(x+q). \tag{1.250}$$

Also, the infinitesimal generators are given, in the appropriate domains, by:

$$\begin{aligned}\left(\widehat{Q}\psi\right)(x) &= x\psi(x)\\ \left(\widehat{P}\psi\right)(x) &= -\imath\hbar\frac{d\psi}{dx}\end{aligned}. \tag{1.251}$$

Finally, a generic matrix element of $\widehat{W}(q,p)$ is given by:

$$\left\langle\phi,\widehat{W}(q,p)\psi\right\rangle = \exp\{\imath qp/2\hbar\}\int_{-\infty}^{+\infty} dx\phi(x)^{*}\exp\{\imath px/\hbar\}\psi(x+q), \tag{1.252}$$

where $\left\langle\phi,\widehat{W}(q,p)\psi\right\rangle$ is square-integrable for all $\phi,\psi\in L^2(\mathbb{R})$, as it can be seen from the fact that:

$$\left\|\left\langle\phi,\widehat{W}(q,p)\psi\right\rangle\right\|^2 = \int\frac{dqdp}{2\pi\hbar}\left|\left\langle\phi,\widehat{W}(q,p)\psi\right\rangle\right|^2 = \|\phi\|^2\|\psi\|^2. \tag{1.253}$$

Explicitly, if we use a generalized basis of plane-waves: $\{|k\rangle = e^{\imath kx}/\sqrt{2\pi}\}$, we get the formula:

$$\left\langle k'|\widehat{W}(q,p)|k\right\rangle = \delta(k-k'+p/\hbar)e^{\imath q(k+k')/2}, \tag{1.254}$$

which will be useful in the following.

The construction outlined in the previous example can be extended to build a concrete realization of a Weyl system in the general case of a symplectic vector space $(\mathcal{S},\omega)$ which decomposes as the direct sum $\mathcal{S}=\mathcal{S}_1\oplus\mathcal{S}_2$ of the two Lagrangian subspaces $\mathcal{S}_1$, $\mathcal{S}_2$. If $U:\mathcal{S}_1\to\mathcal{H}$, $V:\mathcal{S}_2\to\mathcal{H}$ are unitary, irreducible and strongly continuous representations of $\mathcal{S}_1$ and $\mathcal{S}_2$ respectively on a separable Hilbert space $\mathcal{H}$ which satisfy the additional constraint:

$$\widehat{U}(z_1)\widehat{V}(z_2) = \widehat{V}(z_2)\widehat{U}(z_1)\exp\{\imath\omega((z_1,0),(0,z_2))/\hbar\}, \tag{1.255}$$

we can define the Weyl system as:

$$\widehat{W}(z) = \widehat{U}(z_1)\widehat{V}(z_2)\exp\{-\imath\omega((z_1,0),(0,z_2))/2\hbar\}. \tag{1.256}$$

Thanks to the Von Neumann's theorem [34], we can affirm then that there exists a unitary map $T : \mathcal{H} \to L^2(\mathbb{R}^n, d\mu)$ such that:

$$\begin{aligned} \left(T\widehat{U}(q)T^{-1}\psi\right)(x) &= \psi(x+q) \\ \left(T\widehat{V}(p)T^{-1}\psi\right)(x) &= e^{\imath x\cdot p}\psi(x) \end{aligned}, \tag{1.257}$$

where, as before, we have set $z_1 = (q, 0)$, $z_2 = (0, p)$ and denoted $\widehat{U}(z_1)$, $\widehat{V}(z_2)$ as $\widehat{U}(q)$, $\widehat{V}(p)$ respectively. This also shows that all the representations of the Weyl commutation relations are unitarily equivalent to the Schrödinger representation and hence are unitarily equivalent among themselves.

1.3.3.2 *Linear transformations*

We start by observing that a linear transformation $T : \mathcal{S} \to \mathcal{S}$ that preserves the symplectic structure, $\omega(Tz, Tz') = \omega(z, z')$, $\forall z, z' \in \mathcal{S}$, induces a map:

$$\begin{aligned} \widehat{W}_T &: \mathcal{S} \to \mathcal{H} \\ z &\mapsto \widehat{W}_T(z) \equiv \widehat{W}(Tz) \end{aligned}, \tag{1.258}$$

such that:

$$\widehat{W}_T(z+z') = \widehat{W}_T(z)\widehat{W}_T(z')\exp\{-\imath\omega(z,z')/2\hbar\}, \tag{1.259}$$

as it can be easily proved. Hence $\widehat{W}_T$ is also a Weyl system, which, by von Neumann's theorem, it is unitarily equivalent to $\widehat{W}$. Thus, to the map T there is associated an automorphism $\widehat{U}_T \in \mathcal{U}(\mathcal{H})$ such that:

$$\widehat{W}_T(z) = \widehat{U}_T^{\dagger}\left(\widehat{W}(z)\right)\widehat{U}_T. \tag{1.260}$$

Example 1.3.6. *Fourier transform.*
The action of the (unitary) Fourier transform

$$\begin{aligned} &\mathcal{F} : L^2(\mathbb{R}) \to L^2(\mathbb{R}) \\ &\psi(x) \mapsto \widetilde{\psi}(p) = \int_{-\infty}^{\infty} \frac{dx}{\sqrt{2\pi}}\psi(x)e^{-\imath p\cdot x} \end{aligned} \tag{1.261}$$

translates on the Weyl operators as the transformation:

$$\begin{aligned} \widetilde{\left(e^{\imath x\widehat{P}}\psi\right)}(p) = e^{\imath xp}\widetilde{\psi}(p) \quad &\Rightarrow \left(\widehat{P}\widetilde{\psi}\right)(p) = p\widetilde{\psi}(p) \\ \widetilde{\left(e^{\imath p_0\widehat{Q}}\psi\right)}(p) = \widetilde{\psi}(p-p_0) &\Rightarrow \left(\widehat{Q}\widetilde{\psi}\right)(p) = \imath d\widetilde{\psi}(p)/dp \end{aligned}, \tag{1.262}$$

which is such that:

$$\mathcal{F}^{\dagger}\widehat{Q}\mathcal{F} = -\widehat{P}\,,\ \mathcal{F}^{\dagger}\widehat{P}\mathcal{F} = \widehat{Q}\,. \tag{1.263}$$

It is simple to verify that this is indeed the action induced by the liner map on $\mathbb{R}^2$:

$$(q,p) \to (-p,q)\,, \tag{1.264}$$

for which

$$\begin{aligned} \widehat{U}(q) = \widehat{W}((q,0)) &\mapsto \widehat{W}((0,-p)) = \widehat{V}(-p) \\ \widehat{V}(p) = \widehat{W}((0,p)) &\mapsto \widehat{W}((q,0)) = \widehat{U}(q) \end{aligned}\,. \tag{1.265}$$

Suppose now to have a one-parameter group $\{T_\lambda\}_{\lambda\in\mathbb{R}}$ of linear symplectic transformations, which is generated by the linear vector field Γ. Invariance of the symplectic form is encoded in the infinitesimal relation: $\mathcal{L}_\Gamma\omega = 0$, where $\mathcal{L}$ denotes the Lie derivative. This implies that there exists a globally defined function g such that:

$$i_\Gamma\omega = dg\,, \tag{1.266}$$

which, in addition, will be a quadratic function of the coordinates.

According to (1.260), to the family $\{T_\lambda\}$ we can associate a strongly continuous one-parameter group $\{U_\lambda\}_{\lambda\in\mathbb{R}}$ of unitary operators such that:

$$\widehat{W}(z(\lambda)) = \widehat{U}_\lambda^{\dagger}\,\widehat{W}(z)\,\widehat{U}_\lambda\,, \tag{1.267}$$

where $z(\lambda) = T_\lambda(z)$. Now, through Stone's theorem, we can obtain the self-adjoint generator $\widehat{G}$:

$$\widehat{U}_\lambda = \exp\left\{-i\lambda\widehat{G}/\hbar\right\}\,, \tag{1.268}$$

representing the quantum counterpart of the quadratic function g.

Notice that, in this way, we have achieved a way to quantize all quadratic functions.

Let us consider now a general, not necessarily symplectic, linear transformation $T : \mathcal{S} \to \mathcal{S}$. Denoting with ω_0 the standard symplectic form on $\mathcal{S}$ which in a Darboux chart is written as $\omega_0 = dq^j \wedge dp^j$, we define a new symplectic structure ω_T via:

$$\omega_T(z,z') \equiv \omega_0(Tz,Tz')\,. \tag{1.269}$$

We leave to the reader to prove that we obtain in this way a new Weyl system for $(\mathcal{S},\omega_T)$, which is defined by:

$$\widehat{W}_T(z) \equiv \widehat{W}(Tz) \tag{1.270}$$

and is such that:

$$\widehat{W}_T(z+z') = \widehat{W}_T(z)\,\widehat{W}_T(z')\exp\{-\imath\omega_T(z,z')/2\hbar\}. \tag{1.271}$$

Also, as before, one can define the infinitesimal generators:

$$\widehat{W}_T(\lambda z) = \exp\{i\lambda\widehat{G}(z)\}, \tag{1.272}$$

which satisfy the commutation relations:

$$\left[\widehat{G}(z), \widehat{G}(z')\right] = -\imath\hbar\omega_T(z,z'). \tag{1.273}$$

This observation allows us to consider Weyl systems for a vector space with an arbitrary and translationally invariant symplectic structure ω. Indeed, by Darboux theorem, there always exists an invertible linear transformation T that maps ω_0 in ω, $T:(\mathcal{S},\omega)\to(\mathcal{S},\omega_0)$. Denoting then with $W:(\mathcal{S},\omega_0)\to\mathcal{U}(\mathcal{H})$ the Weyl map with respect to the standard symplectic form, we can define a Weyl system for $(\mathcal{S},\omega)$ by setting: $W\circ T = W_T$ or, more explicitly:

$$\widehat{W}_T(z) \equiv \widehat{W}(Tz). \tag{1.274}$$

In physics, a conspicuous example of a one-parameter group of symplectic transformations is provided by the time evolution of a Hamiltonian system, of which we now give some simple examples[24], leaving the details of the calculations to the reader.

Example 1.3.7. *Free particle evolution.*
For a free particle of mass m, the one-parameter group is given by: $(q,p)\to(q+tp/m,p)$ and can be represented by the matrix:

$$\begin{vmatrix} q(t) \\ p(t) \end{vmatrix} = F(t)\begin{vmatrix} q \\ p \end{vmatrix}, \; F(t) = \begin{vmatrix} 1 & t/m \\ 0 & 1 \end{vmatrix}, \; F(t)F(t') = F(t+t'). \tag{1.275}$$

Then, one finds:

$$\begin{aligned} \widehat{W}_t(q,p) &= \widehat{W}(q(t),p(t)) = \exp\left\{(\imath/\hbar)\left[q(t)\widehat{P} + p(t)\widehat{Q}\right]\right\} \\ &\equiv \exp\left\{(\imath/\hbar)\left[q\widehat{P}_t + p\widehat{Q}_t\right]\right\}, \end{aligned} \tag{1.276}$$

with

$$\widehat{P}_t = \widehat{P}, \;\; \widehat{Q}_t = \widehat{Q} + t\widehat{P}/m. \tag{1.277}$$

[24]Other examples may be found in [14].

As usual, we can define a one-parameter family of unitary operators $\left\{\widehat{F}(t)\right\}_{t\in\mathbb{R}}$ such that:

$$\begin{aligned}\exp\left\{\imath p\widehat{Q}_t/\hbar\right\} &= \widehat{F}^{\dagger}(t)\exp\left\{\imath p\widehat{Q}/\hbar\right\}\widehat{F}(t)\\ \exp\left\{\imath q\widehat{P}_t/\hbar\right\} &= \widehat{F}^{\dagger}(t)\exp\left\{\imath q\widehat{P}/\hbar\right\}\widehat{F}(t)\end{aligned} \tag{1.278}$$

which, in terms of the infinitesimal generator $\widehat{H}$, can be written as:

$$\widehat{F}(t) = \exp\left\{-\imath\widehat{H}t/\hbar\right\}. \tag{1.279}$$

Eqs. (1.277) imply the commutation relations:

$$\left[\widehat{P},\widehat{H}\right] = 0\,,\quad \left[\widehat{Q},\widehat{H}\right] = \frac{\imath\hbar}{m}\widehat{P}. \tag{1.280}$$

If we now look for a quantum operator $\widehat{H}$ given by a quadratic function, as it happens for the generators of linear and homogeneous canonical transformations, i.e. by a Hamiltonian of the type:

$$\widehat{H} = a\widehat{P}^2 + b\widehat{Q}^2 + c\left(\widehat{P}\widehat{Q} + \widehat{Q}\widehat{P}\right), \tag{1.281}$$

it is easy to check that the solution of commutation relations (1.280) is given by:

$$\widehat{H} = \frac{\widehat{P}^2}{2m} + \lambda\widehat{\mathbb{I}}, \tag{1.282}$$

where $\widehat{\mathbb{I}}$ is the identity operator and λ any real constant. Thus, apart from this constant term, the quantum operator associated with the time evolution is the standard quantum Hamiltonian for a free particle of mass m.

Example 1.3.8. *Harmonic oscillator evolution.*
From the classical equations of motion

$$\begin{aligned}q(t) &= q\cos\omega t + p\frac{\sin\omega t}{m\omega}\\ p(t) &= p\cos\omega t - qm\omega\sin\omega t\end{aligned}, \tag{1.283}$$

one finds that $F(t)$ is given, in this case, by:

$$F(t) = \left|\begin{array}{cc}\cos\omega t & \dfrac{\sin\omega t}{m\omega}\\ -m\omega\sin\omega t & \cos\omega t\end{array}\right|. \tag{1.284}$$

Proceeding just as in the previous example, we obtain:

$$\widehat{W}_t(q,p) = \exp\left\{(\imath/\hbar)\left[q\widehat{P}_t + p\widehat{Q}_t\right]\right\}, \tag{1.285}$$

with now:

$$\begin{aligned}\widehat{Q}_t &= \widehat{Q}\cos\omega t + \widehat{P}\frac{\sin\omega t}{m\omega} \\ \widehat{P}_t &= \widehat{P}\cos\omega t - \widehat{Q}m\omega\sin\omega t\end{aligned} . \tag{1.286}$$

Defining again $\widehat{F}(t) = \exp\left\{-\imath\hat{H}t/\hbar\right\}$ and working out the commutation relations, one gets:

$$\left[\widehat{Q},\widehat{H}\right] = \frac{\imath\hbar}{m}\widehat{P},\quad \left[\widehat{P},\widehat{H}\right] = -\imath\hbar m\omega^2\widehat{Q}. \tag{1.287}$$

Thus a quadratic Hamiltonian must now have the form:

$$\widehat{H} = \frac{\widehat{P}^2}{2m} + \frac{1}{2}m\omega^2\widehat{Q}^2 + \lambda\widehat{\mathbb{I}}, \tag{1.288}$$

which, again up to an additive multiple of the identity, is the standard quantum Hamiltonian for the harmonic oscillator.

1.3.3.3 *Quantum mechanics on phase space*

For simplicity, in this Sect. we will work in $S \approx \mathbb{R}^2$, since generalizations to $S \approx \mathbb{R}^{2n}$ are easy to work out.

As a preliminary remark, let us observe that, for any $f \in L^2(\mathbb{R}^2)$, we have the identity:

$$\int \frac{d\xi d\eta dq' dp'}{(2\pi\hbar)^2} f(q',p')\, e^{-\imath\omega_0((q',p'),(\xi,\eta))/\hbar} e^{\imath(\xi p+\eta q)/\hbar} = f(q,-p). \tag{1.289}$$

Defining [19] the symplectic Fourier transform $\mathcal{F}_s(f)$:

$$\mathcal{F}_s(f)(\eta,\xi) = \int \frac{dqdp}{2\pi} f(q,p)\, e^{-\imath\omega_0((q,p),(\xi,\eta))}, \tag{1.290}$$

where as usual $\omega_0((q,p),(\xi,\eta)) = q\eta - p\xi$, we can rewrite the above expression as:

$$\int \frac{d\xi d\eta}{2\pi\hbar}\left[\frac{1}{\hbar}\mathcal{F}_s(f)\left(\frac{\eta}{\hbar},\frac{\xi}{\hbar}\right)\right] e^{\imath(\xi p+\eta q)/\hbar} = f(q,-p). \tag{1.291}$$

The Weyl map, which amounts to make, in Eq.(1.291), the replacement:

$$\exp\{\imath(\xi p + \eta q)/\hbar\} \mapsto \exp\left\{\imath\left(\xi\widehat{P} + \eta\widehat{Q}\right)/\hbar\right\} \equiv \widehat{W}(\xi,\eta), \tag{1.292}$$

gives a map Ω from the space of functions $\mathcal{F}\left(\mathbb{R}^2\right)$ to operators $\mathcal{O}p\left(\mathcal{H}\right)$, i.e.:

$$\Omega\left(f\right) \equiv \int \frac{d\xi d\eta}{2\pi\hbar}\left[\frac{1}{\hbar}\mathcal{F}_s\left(f\right)\left(\frac{\eta}{\hbar},\frac{\xi}{\hbar}\right)\right]\widehat{W}\left(\xi,\eta\right). \tag{1.293}$$

It is simple to show that:

(i) $\Omega\left(f\right)$ is at least a symmetric operator.
This follows from the identity: $\left[\mathcal{F}_s\left(f\right)\left(\eta,\xi\right)\right]^* = \mathcal{F}_s\left(f\right)\left(-\eta,-\xi\right)$ which holds when f is real.

(ii) The action on a wave function $\psi(x)$ is explicitly given by:

$$\left(\Omega\left(f\right)\psi\right)\left(x\right) = \int \frac{d\xi d\eta}{2\pi}\mathcal{F}_s\left(f\right)\left(\eta,\xi\right)\exp\left[\imath\eta\left(x+\hbar\xi/2\right)\right]\psi\left(x+\hbar\xi\right), \tag{1.294}$$

as it can be proven by taking into account that:

$$\left(\widehat{W}\left(\xi,\eta\right)\psi\right)\left(x\right) = \exp\left\{\imath\eta\left[x+\xi/2\right]/\hbar\right\}\psi\left(x+\xi\right). \tag{1.295}$$

(iii) The matrix elements of the Weyl operator $\Omega\left(f\right)$ are given by the expression:

$$\langle\phi|\Omega\left(f\right)|\psi\rangle = \int \frac{dxd\xi d\eta}{2\pi}\mathcal{F}_s\left(f\right)\left(\eta,\xi\right)e^{\imath\eta\left(x+\hbar\xi/2\right)}\phi^*\left(x\right)\psi\left(x+\hbar\xi\right), \tag{1.296}$$

as it is found directly from Eq. (1.294). In particular, in a plane-wave basis:

$$\langle k'|\Omega\left(f\right)|k\rangle = \int \frac{d\xi}{2\pi}\mathcal{F}_s\left(f\right)\left(k'-k,\xi\right)\exp\left\{\imath\hbar\xi\left(k+k'\right)/2\right\}. \tag{1.297}$$

As expected, for $f = q$ and $f = p$, Eq. (1.294) yields[25]:

$$\begin{aligned}\left(\Omega\left(q\right)\psi\right)\left(x\right) &= x\psi\left(x\right)\\ \left(\Omega\left(p\right)\psi\right)\left(x\right) &= \imath\hbar\frac{d\psi}{dx}\end{aligned}, \tag{1.298}$$

i.e.

$$\begin{aligned}\Omega\left(q\right) &= \widehat{Q}\\ \Omega\left(p\right) &= -\widehat{P}\end{aligned}. \tag{1.299}$$

[25] Allowing for distribution-valued transforms, the result follows form the identities: $\mathcal{F}_s\left(q\right)\left(\eta,\xi\right) = 2\pi\imath\,\delta'\left(\eta\right)\delta\left(\xi\right)$ and $\mathcal{F}_s\left(p\right)\left(\eta,\xi\right) = -2\pi\imath\,\delta\left(\eta\right)\delta'\left(\xi\right)$.

Using then the identity: $\mathcal{F}_s(q^n p^m)(\eta,\xi) = 2\pi(-1)^m \imath^{n+m}\delta^{(n)}(\eta)\,\delta^{(m)}(\xi)$, one can see that Ω sends any monomial $q^n p^m$ (with n, m integers) into the operator:

$$(\Omega(q^n p^m)\psi)(x) = \left(\imath\frac{d}{d\xi}\right)^m [(x+\hbar\xi/2)^n \psi(x+\hbar\xi)]\,|_{\xi=0}$$
$$= \frac{1}{2^n}\sum_{k=0}^{n}\binom{n}{k} x^k \left(\imath\hbar\frac{d}{dx}\right)^m \left[x^{n-k}\psi(x)\right], \qquad (1.300)$$

i.e.

$$\Omega(q^n p^m) = \Omega(q^n p^m) = \frac{1}{2^n}\sum_{k=0}^{n}\binom{n}{k}[\Omega(q)]^k\cdot[\Omega(p)]^m\cdot[\Omega(q)]^{n-k}. \qquad (1.301)$$

Let us notice that, for $n = m = 1$, one has

$$\Omega(qp) = \Omega(pq) = \frac{1}{2}[\Omega(q)\cdot\Omega(p) + \Omega(p)\cdot\Omega(q)] \qquad (1.302)$$

which gives a justification for the symmetrization procedure we have talked about while discussing the quantization of the harmonic oscillator in Sect. 2, Example 7.

In general, however:

$$\Omega(fg) \neq \frac{1}{2}(\Omega(f)\cdot\Omega(g) + \Omega(g)\cdot\Omega(f)), \qquad (1.303)$$

meaning that the so-called "Weyl symmetrization procedure" holds only in very special cases.

1.3.3.4 *The Wigner map*

In this subsect. we will see that the Weyl map can be inverted, i.e there exists a map, called the *Wigner map*: $\Omega^{-1}: \mathcal{O}p(\mathcal{H}) \to \mathcal{F}(\mathbb{R}^2)$, such that $\Omega^{-1}(\Omega(f)) = f$. It is defined as follows: given any operator $\widehat{O}$ such that $\mathrm{Tr}\left[\widehat{O}\widehat{W}(x,k)\right]$ exists[26], we have:

$$\Omega^{-1}\left(\widehat{O}\right)(q,p) \equiv \int\frac{dxdk}{2\pi\hbar}\exp\{-\imath\omega_0((x,k),(q,p))/\hbar\}\,\mathrm{Tr}\left[\widehat{O}\widehat{W}^\dagger(x,k)\right]. \quad (1.304)$$

In order to prove Eq. (1.304), we need the expression for the trace:

$$\mathrm{Tr}[\widehat{W}(x,k)\,\widehat{W}^\dagger(\xi,\eta)] = \int dhdh'\left\langle h\left|\widehat{W}(x,k)\right|h'\right\rangle\left\langle h'\left|\widehat{W}^\dagger(\xi,\eta)\right|h\right\rangle. \qquad (1.305)$$

[26] As W is a bounded operator, this is true, e.g., if A is trace-class.

Using Eq. (1.254), we have: $\mathrm{Tr}\left[\widehat{W}(x,k)\,\widehat{W}^{\dagger}(\xi,\eta)\right] = 2\pi\hbar\,\delta(x-\xi)\,\delta(k-\eta)$, which, inserted into (1.304), gives:

$$\Omega^{-1}(\Omega(f))(q,p) = \int \frac{d\xi d\eta}{2\pi}\mathcal{F}_s(\eta,\xi)\exp\{-\imath\omega((\xi,\eta),(q,p))\} = f(q,p).$$

It may be useful to have an expression for the Wigner map directly in terms of the matrix elements of the operators, which for plane waves reads [14]:

$$\Omega^{-1}\left(\widehat{O}\right)(q,p) = \int dke^{\imath qk}\left\langle -p/\hbar + k/2|\widehat{O}| - p/\hbar - k/2\right\rangle. \tag{1.306}$$

Also, from the very definition, it is not difficult to prove that

$$\Omega^{-1}\left(\widehat{W}(q',,p')\right)(q,p) = \exp\{\imath\omega_0((q,p),(q',p'))/\hbar\}. \tag{1.307}$$

Introducing now two resolutions of the identity relative to the coordinates, we can write:

$$\Omega^{-1}\left(\widehat{O}\right)(q,p) = \int dkdxdx'e^{\imath qk}\left\langle -p/\hbar + k/2|\,x\right\rangle\langle x\,|\widehat{O}|\,x'\rangle\left\langle x'\,| - p/\hbar - k/2\right\rangle, \tag{1.308}$$

where the integration over k can be explicitly performed, yielding a delta-function. Thus one obtains the celebrated Wigner formula:

$$\Omega^{-1}\left(\widehat{O}\right)(q,p) = \int d\xi e^{\imath p\xi/\hbar}\left\langle q + \xi/2|\widehat{O}\,|q - \xi/2\right\rangle. \tag{1.309}$$

Notice also that the Wigner transform inverts to:

$$\left\langle x|\widehat{O}|x'\right\rangle = \int \frac{dp}{2\pi\hbar}\exp\{-\imath p(x-x')/\hbar\}\,\Omega^{-1}\left(\widehat{O}\right)\left(\frac{x+x'}{2},p\right). \tag{1.310}$$

Example 1.3.9.

(1) If $\widehat{O} = -\widehat{P}$, since $\widehat{P}\,|m\rangle = \hbar m\,|m\rangle$, we have:

$$\left\langle -p/\hbar + k/2|(-\widehat{P})| - p/\hbar - k/2\right\rangle = (p\hbar + k/2)\langle -p/\hbar + k/2| \\ -p/\hbar - k/2\rangle = p\,\delta(k)$$

and we find, as expected:

$$\Omega^{-1}\left((-\widehat{P})\right)(q,p) = p. \tag{1.311}$$

(2) Setting: $\widehat{O} = \widehat{Q}$, we find at once:

$$\Omega^{-1}\left(\widehat{Q}\right)(q,p) = q\,. \tag{1.312}$$

(3) Consider now $\widehat{O} = |\phi\rangle\langle\psi|$, which is the simplest example of a finite-rank operator. Then it is immediate to see that:

$$\Omega^{-1}\left(|\phi\rangle\langle\psi|\right)(q,p) = \int_{-\infty}^{\infty} d\xi e^{\imath p\xi/\hbar}\phi\left(q+\xi/2\right)\psi^*\left(q-\xi/2\right)\,. \tag{1.313}$$

It is also easy to check formula (1.310):

$$\int \frac{dp}{2\pi\hbar}e^{\{-\imath p(x-x')/\hbar\}}\Omega^{-1}\left(\widehat{O}\right)\left(\frac{x+x'}{2},p\right) = \phi(x)\,\psi^*(x') = \langle x|\phi\rangle\langle\psi|x'\rangle\,.$$

(4) We can now proceed to consider a self-adjoint operator with discrete spectrum: $\widehat{O}|\phi_n\rangle = \lambda_n|\phi_n\rangle$, with $\langle\phi_n|\phi_m\rangle = \delta_{nm}$, $\sum_n |\phi_n\rangle\langle\phi_n| = \mathbb{I}$. Then we can write:

$$\Omega^{-1}\left(\widehat{O}\right)(q,p) = \sum_n \lambda_n \int d\xi e^{\imath p\xi/\hbar}\phi_n\left(q+\xi/2\right)\phi_n^*\left(q+\xi/2\right). \tag{1.314}$$

The most interesting consequence of what was seen above is the fact that the Weyl and Wigner maps establish a bijection [19] between Hilbert-Schmidt operators and square-integrable functions on phase space, which is also strongly bicontinuous.

Indeed the following theorem holds[27]:

f will be square-integrable if and only if $\Omega(f)$ is Hilbert-Schmidt. Similarly, $\Omega^{-1}\left(\widehat{A}\right)$ will be square-integrable if and only if $\widehat{A}$ is Hilbert-Schmidt.

We notice that, since $[\mathcal{F}_s(\eta,\xi)]^* = \mathcal{F}_s(-\eta,-\xi)$ and $\widehat{W}^\dagger(\xi,\eta) = \widehat{W}(-\xi,-\eta)$, the Weyl and Wigner maps preserve conjugation:

$$\Omega(f^*) = \Omega(f)^\dagger\,,\ \ \Omega^{-1}\left(\widehat{O}^\dagger\right) = \Omega^{-1}\left(\widehat{O}\right)^*, \tag{1.315}$$

so guaranteeing that f is real iff $\Omega(f)$ is at least a symmetric operator.

Before looking at some examples, we also observe that Eq. (1.310) implies

$$\mathrm{Tr}_x\left[\widehat{O}\right] \equiv \int dx\,\langle x\,|O|\,x\rangle = \int \frac{dqdp}{2\pi\hbar}\Omega^{-1}\left(\widehat{O}\right)(q,p)\,, \tag{1.316}$$

[27] See [14] for a proof.

as well as

$$\int \frac{dqdp}{2\pi\hbar} f(q,p) = \mathrm{Tr}\left[\Omega(f)\right] . \tag{1.317}$$

This allows for a formal definition of a *trace operation* on phase space given by:

$$\mathrm{Tr}\left[f\right] \equiv \int \frac{dqdp}{2\pi\hbar} f(q,p) . \tag{1.318}$$

Example 1.3.10. *The 1D Harmonic oscillator.*
We go back to the Hamiltonian

$$\widehat{H} = \frac{\widehat{P}^2}{2m} + \frac{1}{2}m\omega^2\widehat{Q}^2 , \tag{1.319}$$

which has eigenvalues: $E_n = \hbar\omega(n+1/2)$ $(n \geq 0)$ and eigenfunctions:

$$\psi_n(x) = \left(\frac{m\omega}{\pi\hbar}\right)^{1/4} \frac{1}{\sqrt{2^n n!}} \exp\left(-\zeta^2/2\right) H_n(\zeta) , \tag{1.320}$$

where $\zeta = x\sqrt{m\omega/\hbar}$. We want to evaluate here the Wigner function associated with the so called Boltzmann factor $\widehat{O} = \exp\left(-\beta\widehat{H}\right)$. From:

$$\left\langle x \left| e^{-\beta\widehat{H}} \right| x' \right\rangle = \sum_{n=0}^{\infty} e^{-\beta E_n} \psi_n^*(x)\, \psi_n(x') , \tag{1.321}$$

inserting the explicit form (1.320) of the eigenfunctions and manipulating the expression (we refer to [14] for details), one finds that the matrix element (1.321) can be expressed as:

$$\left\langle x \left| e^{-\beta\widehat{H}} \right| x' \right\rangle = \sqrt{\frac{m\omega}{\pi\hbar}} e^{-(\zeta^2+\zeta'^2)/2} \sqrt{\frac{z}{1-z^2}} \exp\left[\frac{2z\zeta\zeta' - z^2\left(\zeta^2+\zeta'^2\right)}{1-z^2}\right] .$$

This yields the Wigner function:

$$\Omega^{-1}\left(e^{-\beta\widehat{H}}\right)(q,p) = \frac{1}{\cosh(\beta\hbar\omega/2)} \exp\left\{-\tanh(\beta\hbar\omega/2)\left[\frac{m\omega}{\hbar}q^2 + \frac{p^2}{m\hbar\omega}\right]\right\} . \tag{1.322}$$

Finally, using Eq. (1.322), we find with some long but elementary algebra:

$$\mathrm{Tr}\left[\Omega^{-1}\left(e^{-\beta\widehat{H}}\right)\right] = \int \frac{dqdp}{2\pi\hbar} \Omega^{-1}\left(e^{-\beta\widehat{H}}\right) = \frac{1}{2\sinh(\beta\hbar\omega/2)} , \tag{1.323}$$

which is the expected result for the canonical partition function of the 1D harmonic oscillator (see Example 1.3.1).

1.3.3.5 *The Moyal product and the quantum to classical transition*

The Wigner map allows for the definition of a new algebra structure on the space of functions $\mathcal{F}\left(\mathbb{R}^2\right)$, the *Moyal "$*$"-product* that is defined as:

$$f * g \equiv \Omega^{-1}\left(\widehat{\Omega}(f) \cdot \widehat{\Omega}(g)\right). \tag{1.324}$$

This product is *associative* and *distributive* with respect to the sum, but it is *non-local* and *non-commutative* (since in general $\widehat{\Omega}(f) \cdot \widehat{\Omega}(g) \neq \widehat{\Omega}(g) \cdot \widehat{\Omega}(f)$). Explicitly:

$$(f * g)(q,p) = \int \frac{dxdk}{2\pi\hbar} \exp\left\{-\imath\omega_0\left((x,k),(q,p)\right)/\hbar\right\} \operatorname{Tr}\left[\widehat{\Omega}(f) \cdot \widehat{\Omega}(g)\, \widehat{W}^{\dagger}(x,k)\right], \tag{1.325}$$

with

$$\begin{aligned} \operatorname{Tr}\left[\widehat{\Omega}(f) \cdot \widehat{\Omega}(g)\, \widehat{W}^{\dagger}(x,k)\right] = &\int \frac{d\xi d\eta d\xi' d\eta'}{(2\pi)^2} \mathcal{F}_s(f)(\eta,\xi)\, \mathcal{F}_s(g)(\eta',\xi') \\ &\times \operatorname{Tr}\left[\widehat{W}(\hbar\xi,\hbar\eta)\, \widehat{W}(\hbar\xi',\hbar\eta')\, \widehat{W}^{\dagger}(x,k)\right]. \end{aligned} \tag{1.326}$$

Skipping the details of calculations [14], it is possible to show that the last expression can be recast in the form:

$$\begin{aligned} (f * g)(q,p) &= 4\int \frac{dadbdsdt}{(2\pi\hbar)^2} f(a,b)\, g(s,t) \\ &\quad\times \exp\left\{-\frac{2\imath}{\hbar}\left[(a-q)(t-p) + (s-q)(p-b)\right]\right\} \\ &= 4\int \frac{dadbdsdt}{(2\pi\hbar)^2} f(a,b)\, g(s,t) \\ &\quad\times \exp\left\{2\imath\omega_0\left((q-a,p-b),(q-s,p-t)\right)/\hbar\right\}, \end{aligned}$$

explicitly exhibiting the non-locality of the Moyal product.

There are several equivalent ways of re-writing such an expression, such as[28]:

$$\begin{aligned} (f * g)(q,p) &= \sum_{n,m=0}^{\infty} \left(\frac{i\hbar}{2}\right)^{n+m} \frac{(-1)^n}{n!m!} \left\{\frac{\partial^{m+n} f(a,b)}{\partial a^m \partial b^n} \frac{\partial^{m+n} g(a,b)}{\partial a^n \partial b^m}\right\}\Bigg|_{a=q,b=p} \\ &= f(q,p) \exp\left\{\frac{i\hbar}{2}\left[\frac{\overleftarrow{\partial}}{\partial q}\frac{\overrightarrow{\partial}}{\partial p} - \frac{\overleftarrow{\partial}}{\partial p}\frac{\overrightarrow{\partial}}{\partial q}\right]\right\} g(q,p). \end{aligned} \tag{1.327}$$

[28] All the above expressions for the Moyal product apply of course to functions that are regular enough for the right-hand side of the defining equations to make sense. In particular, they will hold when f, g are Schwartz functions.

The latter form exhibits explicitly the Moyal product as a series expansion in powers of $\hbar$. To lowest order:

$$f * g = fg + \frac{i\hbar}{2}\{f, g\} + \mathcal{O}\left(\hbar^2\right), \tag{1.328}$$

where $\{\cdot, \cdot\}$ is the Poisson bracket. Thus, we see that Planck constant $\hbar$ acts as a "deformation parameter" of the usual associative product structure on the algebra of functions, making the product non-commutative. Indeed, it can be seen, e.g., from the expansion of the exponential in Eq.(1.327), that terms proportional to even powers of $\hbar$ are symmetric under the interchange $f \leftrightarrow g$, but terms proportional to odd powers are *anti*symmetric, and this makes the product non-commutative.

Example 1.3.11.

(1) If $f \equiv 1$ *or* $g \equiv 1$, then:

$$(1 * g)(q, p) = g(q, p), (f * 1)(q, p) = f(q, p). \tag{1.329}$$

(2) If $f = q$ and at least $g \in S^{\infty}\left(\mathbb{R}^2\right)$, then:

$$\begin{aligned} (q * g)(q, p) &= 4 \int \frac{dadbdsdt}{(2\pi\hbar)^2} a g(s, t) \\ &\quad \times \exp\left\{\frac{2\imath}{\hbar}\left[(a - q)(t - p) + (s - q)(p - b)\right]\right\} \\ &= 4 \int \frac{dadbdsdt}{(2\pi\hbar)^2} g(s, t)\left(q + \frac{i\hbar}{2}\frac{\partial}{\partial t}\right) \\ &\quad \times \exp\left\{\frac{2\imath}{\hbar}\left[(a - q)(t - p) + (s - q)(p - b)\right]\right\}. \end{aligned}$$

Integrating by parts in the second integral and using the previous result, one gets:

$$(q * g)(q, p) = \left(q + \frac{\imath\hbar}{2}\frac{\partial}{\partial p}\right) g(q, p). \tag{1.330}$$

$$(g * q)(q, p) = \left(q - \frac{\imath\hbar}{2}\frac{\partial}{\partial p}\right) g(q, p). \tag{1.331}$$

(3) In the same way, if $f = p$, we have:

$$(p * g)(q, p) = \left(p - \frac{\imath\hbar}{2}\frac{\partial}{\partial q}\right) g(q, p). \tag{1.332}$$

(4) If $f = q$ and $g = p$ (or vice versa), one obtains:

$$(q * p)(q,p) = qp + \frac{\imath\hbar}{2}; \quad (p * q)(q,p) = qp - \frac{\imath\hbar}{2}. \tag{1.333}$$

(5) Notice that Eq.(1.330) implies:

$$\widehat{\Omega}(q) \cdot \widehat{\Omega}(g) = \widehat{\Omega}(qg) + \frac{\imath\hbar}{2}\widehat{\Omega}\left(\frac{\partial g}{\partial p}\right) \tag{1.334}$$

and similarly for the others.

Using the Moyal product, we can define the *Moyal Bracket* $\{\cdot,\cdot\}_*$ as:

$$\{\cdot,\cdot\}_* : \mathcal{F}\left(\mathbb{R}^2\right) \times \mathcal{F}\left(\mathbb{R}^2\right) \to \mathcal{F}\left(\mathbb{R}^2\right) \tag{1.335}$$

$$\{f,g\}_* \equiv \frac{1}{\imath\hbar}(f * g - g * f) = \{f,g\} + \mathcal{O}\left(\hbar^2\right), \tag{1.336}$$

We notice that the difference between the Moyal and Poisson brackets is $\mathcal{O}(\hbar^2)$, since the difference $f * g - g * f$ contains only odd powers of $\hbar$.

Being defined in terms of an associative product, the Moyal bracket fulfils all the properties of a Poisson bracket (linearity, anti-symmetry and the Jacobi identity), and defines a new Poisson structure on the (non-commutative) algebra of functions endowed with the Moyal product. In particular, just as for the ordinary Poisson brackets, the Jacobi identity implies that $\{f,\cdot\}_*$ is a derivation (with respect to the $*$-product) on the algebra of functions:

$$\{f, g * h\}_* = \{f,g\}_* * h + g * \{f,h\}_*, \tag{1.337}$$

Now $\{f,\cdot\}_*$ is not necessarily a vector field, contrary to what happens with the standard Poisson bracket $\{f,\cdot\}$. This can be seen by explicitly writing down the second term in (1.336) as: $\{f,g\}_* = \{f,g\} + \hbar^2\{f,g\}_2 + \ldots$, to obtain:

$$\{f,g\}_2(q,p) = \frac{1}{24}\left\{\frac{\partial^3 f}{\partial q^3}\frac{\partial^3 g}{\partial p^3} - 3\frac{\partial^3 f}{\partial p\partial q^2}\frac{\partial^3 g}{\partial q\partial p^2} + 3\frac{\partial^3 f}{\partial p^2\partial q}\frac{\partial^3 g}{\partial q\partial q^2} - \frac{\partial^3 f}{\partial p^3}\frac{\partial^3 g}{\partial q^3}\right\},$$

showing that $\{f,g\}_*$ contains also higher-order derivatives. The reason for that is precisely that the Moyal bracket is non-local. It is only when f is at most a quadratic polynomial that $\{f,\cdot\}_*$ becomes a derivation on the usual pointwise product. Indeed, if this is the case, the Moyal and Poisson brackets of f with other functions coincide.

Finally, using the definitions of the Weyl and Wigner maps, we have in general:

$$\{f,g\}_* = \imath\Omega^{-1}\left(\widehat{\Omega}(f) \cdot \widehat{\Omega}(g) - \widehat{\Omega}(g) \cdot \widehat{\Omega}(f)\right)/\hbar, \tag{1.338}$$

i.e.:

$$\left[\widehat{\Omega}(f), \widehat{\Omega}(g)\right] = -\imath\hbar\widehat{\Omega}\left(\{f,g\}_*\right). \tag{1.339}$$

Unless f and/or g are at most quadratic, $\{f, g\}_* \neq \{f, g\}$. Therefore, the commutator of the quantum operators associated with observables on phases space is not (modulo a multiplicative constant) the quantum operator associated with the Poisson bracket. Generically, it becomes so only to lowest order in $\hbar$, reproducing the so called Ehrenfest theorem [15].

Bibliography

[1] Y. Aharonov, D. Bohm, *Significance of electromagnetic potentials in the quantum theory*, Phys.Rev. **115** (1959) 485; *Further Considerations on Electromagnetic Potentials in the Quantum Theory*, *Ibid.* **123** (1961) 1511

[2] S.T. Ali, M. English, *Quantization methods: a guide for physicists and analysts*, Rev. Math. Phys. **17** (2005) 391

[3] I. Bengtsson, K. Życzkovski, *Geometry of quantum states* (Cambridge University Press, 2006)

[4] M.V. Berry, *Quantal phase factors accompanying adiabatic changes*, Proc. Roy. Soc. A **392** (1984) 45

[5] O. Bratteli, D.W. Robinson, *Operator Algebras and Quantum Statistical Mechanics*, Springer-Verlag (1987)

[6] P. Busch, P. Lahti, P. Mittelstaedt, *The Quantum Theory of Measurement*, Springer-Verlag (1996)

[7] J. Clemente-Gallardo, G. Marmo, *Basics of Quantum Mechanics, Geometrization and some Applications to Quantum Information*, Int. Jour. Geometric Methods in Modern Phys. **5** (2008) 989

[8] C. Cohen-Tannoudji, B. Diu, F. Laloë, *Quantum Mechanics*, Wiley-VHC (2006)

[9] L. de Broglie, *Recherches sur la Theorie des Quanta*, PhD Thesis, Paris (1924)

[10] P.A.M. Dirac, *The fundamental equations of quantum mechanics*, Proc. R. Soc. Lond. **A109** (1925) 642

[11] P.A.M. Dirac, *The Principles of Quantum Mechanics*, Oxford University Press (1962)

[12] A. Einstein, *Concerning an Heuristic Point of View Toward the Emission and Transformation of Light*, Annalen der Physik **17**(6) (1905) 132

[13] G.G. Emch, *Mathematical and Conceptual Foundations of* 20-*th Century Physics*, North-Holland (1984)

[14] E. Ercolessi, G. Marmo, G. Morandi, *From the equations of motion to the canonical commutation relations*, La Rivista del Nuovo Cimento **33** (2010) 401

[15] G. Esposito, G. Marmo, G. Miele, G. Sudarshan, *Advanced Concepts in Quantum Mechanics*, Cambridge University Press (2004)

[16] R.P. Feynman, *A New Approach to Quantum Theory* (L.M. Brown ed.), World Scientific (2005)

[17] R.P. Feynman, A.R. Hibbs, *Quantum Mechanics and Path Integrals*, McGraw-Hills (1965)

[18] R.P. Feynman, R.B. Leighton, M. Sands, *The Feynman Lectures on Physics*, Vol. 3., Addison-Wesley (1965)

[19] G.B. Folland, *Harmonic Analysis in Phase Space*, Princeton University Press (1989)

[20] J. Glimm, A. Jaffe, *Quantum physics. A functional integral point of view*, Springer-Verlag (1987)
[21] R. Haag, *Local Quantum Physics. Fields, Particles, Algebras*, Springer-Verlag (1992)
[22] S. Helgason, *Differential Geometry, Lie Groups, and Symmetric Space*, American Mathematical Society (1978)
[23] A.A. Kirillov, *Lectures on the orbit method*, Graduate studies in mathematics, Vol. 64, American Mathematical Society (2004)
[24] J.R. Klauder, B.S. Skagerstam, *Coherent States, Applications in Physics and Mathematical Physics*, World Scientific (1985)
[25] J.M. Lévy-Leblond, F. Balibar, *Quantique: Rudiments*, Dunod (2007)
[26] V.I. Manko, G. Marmo, A. Simoni, F. Ventriglia, *Tomography in abstract Hilbert spaces*, Open Systems and Information Dynamics **13** (3) (2006) 239
[27] V.I. Manko, G. Marmo, E.C.G. Sudarshan, F. Zaccaria, *Differential geometry of density states*, Repts.Math.Phys. **55** (2005) 405
[28] P.G. Merli, F.G. Missiroli, G. Pozzi, *On the statistical aspect of electron interference phenomena*, Am.J.Phys. **44** (1976) 306
[29] G. Morandi, *The Role of Topology in Classical and Quantum Physics*, Springer-Verlag (1992)
[30] G. Morandi, F. Napoli, E. Ercolessi, *Statistical Mechanics. An Intermediate Course*, World Scientific (2001)
[31] J.W. Negele, H. Orland, *Quantum Many-Particle Systems*, Westview Press (2008)
[32] M.A. Nielsen, I.L. Chuang, *Quantum Computation and Quantum Information: 10th Anniversary Edition*, Cambridge (2010)
[33] A. Perelemov, *Generalized Coherent States and its Applications*, Springer-Verlag (1986)
[34] M. Reed, B. Simon, *Methods of Modern Mathematical Physics, Vol. I. Functional Analysis*, Academic Press (1980)
[35] L.S. Schulman, *Techniques and Applications of Path Integrals*, Wiley (1981)
[36] L. Schwartz, *Lectures on Complex Analytic Manifolds*, Narosa, New Dehli (1986)
[37] A. Tonomura, J. Endo, T. Matsuda, T. Kawasaki, H. Esawa, *Demonstration of single-electron buildup of an interference pattern*, Am.J.Phys. **57** (1989) 117

Chapter 2

Mathematical Foundations of Quantum Mechanics: An Advanced Short Course

Valter Moretti

Department of Mathematics of the University of Trento and INFN-TIFPA, via Sommarive 14, I-38122 Povo (Trento), Italy.
email: valter.moretti@unitn.it

Within these lectures, I review the formulation of Quantum Mechanics, and quantum theories in general, from a mathematically advanced viewpoint essentially based on the orthomodular lattice of elementary propositions, discussing some fundamental ideas, mathematical tools and theorems also related to the representation of physical symmetries. The final step consists of an elementary introduction of the so-called (C*-) algebraic formulation of quantum theories.

2.1 Introduction: Summary of Elementary Facts of QM

A concise account of the basic structure of quantum mechanics and *quantization procedures* has already been extensively presented in the first part of this book, with several crucial examples. In the rest of Section 1, we quickly review again some elementary facts and properties, either of physical or mathematical nature, related to Quantum Mechanics, without fully entering into the mathematical details.

Section 2 is instead devoted to present some technical definitions and results of spectral analysis in complex Hilbert spaces, especially the basic elements of spectral theory, including the classic theorem about spectral decomposition of (generally unbounded) selfadjoint operators and the so called measurable functional calculus. A brief presentation of the three most important operator topologies for applications in QM closes Section 2.

Within Section 3, the *corpus* of the lectures, we pass to analyse the mathematical structure of QM from a finer and advanced viewpoint, adopting the framework based on orthomodular lattices' theory. This approach permits one to justify some basic assumptions of QM, like the mathematical nature of the observables represented by selfadjoint operators and the quantum states viewed as trace class operators. QM is essentially a probability measure on the non-Boolean lattice $\mathcal{L}(\mathcal{H})$ of elementary observables. A key tool of that analysis is the theorem by Gleason characterising the notion probability measure on $\mathcal{L}(\mathcal{H})$ in terms of certain trace class operators. We also discuss the structure of the algebra of observables in the presence of superselection rules after having introduced the mathematical notion of von Neumann algebra. The subsequent part of the third section is devoted to present the idea of quantum symmetry, illustrated in terms of Wigner and Kadison theorems. Some basic mathematical facts about groups of quantum symmetries are introduced and discussed, especially in relation with the problem of their unitarisation. Bargmann's condition is stated. The particular case of a strongly continuous one-parameter unitary group will be analysed in detail, mentioning von Neumann's theorem and the celebrated Stone theorem, remarking its use to describe the time evolution of quantum systems. A quantum formulation of Noether theorem ends this part. The last part of Section 3 aims to introduce some elementary results about continuous unitary representations of Lie groups, discussing in particular a theorem by Nelson which proposes sufficient conditions for lifting a (anti)selfadjoint representation of a Lie algebra to a unitary representation of the unique simply connected Lie group associated to that Lie algebra.

The last section closes the paper focussing on elementary ideas and results of the so called algebraic formulation of quantum theories. Many examples and exercises (with solutions) accompany the theoretical text at every step.

2.1.1 *Physical facts about quantum mechanics*

Let us quickly review the most relevant and common features of quantum systems. Next, we will present a first elementary mathematical formulation which will be improved in the rest of the lectures, introducing a suitable mathematical technology.

2.1.1.1 *When a physical system is quantum*

Loosely speaking, Quantum Mechanics is the physics of microscopic world (elementary particles, atoms, molecules). That realm is characterized by a universal physical constant denoted by h and called **Planck constant**. A related constant

– nowadays much more used – is the **reduced Planck constant**, pronounced "h-bar",

$$\hbar : \frac{h}{2\pi} = 1.054571726 \times 10^{-34} J \cdot s \,.$$

The physical dimensions of h (or $\hbar$) are those of an *action*, i.e. *energy* × *time*. A rough check on the appropriateness of a quantum physical description for a given physical system is obtained by comparing the value of some characteristic action of the system with $\hbar$. For a macroscopic pendulum (say, length $\sim 1m$, mass $\sim 1kg$ maximal speed $\sim 1ms^{-1}$), multiplying the period of oscillations and the maximal kinetic energy, we obtain a typical action of $\sim 2Js \gg h$. In this case quantum physics is expected to be largely inappropriate, exactly as we actually know from our experience of every days. Conversely, referring to a hydrogen electron orbiting around its proton, the first *ionization energy* multiplied with the orbital period of the electron (computed using the classical formula with a value of the radius of the order of 1 Å) produces a typical action of the order of h. Here quantum mechanics is necessary.

2.1.1.2 *General properties of quantum systems*

Quantum Mechanics (QM) enjoys a triple of features which seem to be very far from properties of Classical Mechanics (CM). These remarkable general properties concern the physical quantities of physical systems. In QM physical quantities are called *observables*.

(1) Randomness. When we perform a measurement of an observable of a quantum system, the outcomes turn out to be *stochastic*: Performing measurements of the same observable A on completely identical systems prepared in the *same* physical state, one generally obtains different results $a, a', a'' \ldots$.

Referring to the standard interpretation of the formalism of QM (see [2] for a nice up-to-date account on the various interpretations), the randomness of measurement outcomes should not be considered as due to an incomplete knowledge of the state of the system as it happens, for instance, in Classical Statistical Mechanics. Randomness is not *epistemic*, but it is *ontological*. It is a fundamental property of quantum systems.

On the other hand, *QM permits one to compute the* probability distribution *of all the outcomes of a given observable, once the state of the system is known.*

Moreover, it is always possible to prepare a state ψ_a where a certain observable A is *defined* and takes its value a. That is, repeated measurements of A give rise to the same value a with probability 1. (Notice that we can perform simultaneous measurements on identical systems all prepared in the state ψ_a, or we can perform different subsequent measurements on the same system in the state ψ_a. In the second case, these measurements have to be performed very close to each other in time to prevent the state of the system from evolving in view of Schrödinger evolution as said in (3) below.) Such states, where observable take definite values, cannot be prepared for *all* observables simultaneously as discussed in (2) below.

(2) Compatible and Incompatible Observables. The second noticeable feature of QM is the existence of *incompatible observables*. Differently from CM, there are physical quantities which cannot be measured simultaneously. There is no physical instrument capable to do it. If an observable A is *defined* in a given state ψ – i.e. it attains a precise value a with probability 1 in case of a measurement – an observable B *incompatible* with A turns out to be *not defined* in the state ψ – i.e., it generally attains several different values $b, b', b'' \ldots$, *none with probability* 1, in case of measurement. So, if we perform a measurement of B, we generally obtain a spectrum of values described by a probabilistic distribution as preannounced in (1) above.

Incompatibility is a *symmetric* property: A is incompatible with B if and only if B is incompatible with A. However it is not *transitive*.

There are also *compatible observables* which, by definition, can be measured simultaneously. An example is the component x of the position of a particle and the component y of the momentum of that particle, referring to a given inertial reference frame. A popular case of incompatible observables is a pair of *canonically conjugated observables* (see the first part) like the position X and the momentum P of a particle both along the same fixed axis of a reference frame. In this case, there is a lower bound for the product of the standard deviations, resp. ΔX_ψ, ΔP_ψ, of the outcomes of the measurements of these observables in a given state ψ (these measurement has to be performed on different identical systems all prepared in the same state ψ). This lower bound does not depend on the state and is encoded in the celebrated mathematical formula of the *Heisenberg principle* (a theorem in the modern formulations):

$$\Delta X_\psi \Delta P_\psi \geq \hbar/2 \,, \tag{2.1}$$

where Planck constant shows up.

(3) Post Measurement Collapse of the State. In QM, measurements *generally change the state of the system* and produce a post-measurement state from the state on which the measurement is performed. (We are here referring to idealized measurement procedures, since measurement procedures are very often destructive.) If the measured state is ψ, immediately after the measurement of an observable A obtaining the value a among a plethora of possible values $a, a', a'', \ldots$, the state changes to ψ' generally different form ψ. In the new state ψ', the distribution of probabilities of the outcomes of A changes to 1 for the outcome a and 0 for all other possible outcomes. A is therefore *defined* in ψ'.

When we perform repeated and alternated measurements of a pair of incompatible observables, A, B, the outcomes disturb each other: If the first outcome of A is a, after a measurement of B, a subsequent measurement of A produces $a' \neq a$ in general. Conversely, if A and B are compatible, the outcomes of their subsequent measurements do not disturb each other.

In CM there are measurements that, in practice, disturb and change the state of the system. It is however possible to decrease the disturbance arbitrarily, and nullify it in ideal measurements. In QM it is not always possible as for instance witnessed by (2.1).

In QM, there are two types of time evolution of the state of a system. One is the usual one due to the dynamics and encoded in the famous *Schrödinger equation* we shall see shortly. It is nothing but a quantum version of classical *Hamiltonian evolution* as presented in the first part. The other is the sudden change of the state due to measurement procedure of an observable, outlined in (3): The *collapse of the state* (or *wavefunction*) of the system.

The nature of the second type of evolution is still a source of an animated debate in the scientific community of physicists and philosophers of Science. There are many attempts to reduce the collapse of the state to the standard time evolution referring to the quantum evolution of the whole physical system, also including the measurement apparatus and the environment (*de-coherence processes*) [2; 12]. None of these approaches seem to be completely satisfactory up to now.

Remark 2.1.1. *Unless explicitly stated, we henceforth adopt a physical unit system such that* $\hbar = 1$.

2.1.2 *Elementary formalism for the finite dimensional case*

To go on with this introduction, let us add some further technical details to the presented picture to show how practically (1)-(3) have to be mathematically interpreted (reversing the order of (2) and (3) for our convenience). The rest of the paper is devoted to make technically precise, justify and widely develop these ideas from a mathematically more advanced viewpoint than the one of the first part.

To mathematically simplify this introductory discussion, throughout this section, except for Sect 2.1.5, we assume that $\mathcal{H}$ denotes a *finite dimensional* complex vector space equipped with a Hermitian scalar product , henceforth denoted by $\langle \cdot, \cdot \rangle$, where the linear entry is the second one. With $\mathcal{H}$ as above, $L(\mathcal{H})$ will denote the complex algebra of operators $A : \mathcal{H} \to \mathcal{H}$. We remind the reader that, if $A \in L(\mathcal{H})$ with $\mathcal{H}$ finite dimensional, the *adjoint operator*, $A^\dagger \in L(\mathcal{H})$, is the unique linear operator such that

$$\langle A^\dagger x, y \rangle = \langle x, Ay \rangle \quad \text{for all } x, y \in \mathcal{H}\,. \tag{2.2}$$

A is said to be *selfadjoint* if $A = A^\dagger$, so that, in particular

$$\langle Ax, y \rangle = \langle x, Ay \rangle \quad \text{for all } x, y \in \mathcal{H}\,. \tag{2.3}$$

Since $\langle \cdot, \cdot \rangle$ is linear in the second entry and antilinear in the first entry, we immediately have that all eigenvalues of a selfadjoint operator A are real.

Our assumptions on the mathematical description of quantum systems are the following ones in agreement with the discussion in the first part:

- A quantum mechanical system S is always associated to a complex vector space $\mathcal{H}$ (here finite dimensional) equipped with a Hermitian scalar product $\langle \cdot, \cdot \rangle$;
- observables are pictured in terms of *selfadjoint* operators A on $\mathcal{H}$;
- states are equivalence classes of *unit* vectors $\psi \in \mathcal{H}$, where $\psi \sim \psi'$ iff $\psi = e^{ia}\psi'$ for some $a \in \mathbb{R}$.

Remark 2.1.2.

(a) *It is clear that states are therefore one-to-one represented by all of the elements of the complex projective space $P\mathcal{H}$. The states we are considering within this introductory section are called* pure *states. A more general notion of state, already introduced in the first part, will be discussed later.*

(b) *$\mathcal{H}$ is an elementary version of complex Hilbert space since it is automatically complete, it being finite dimensional.*

(c) *Since* $\dim(\mathcal{H}) < +\infty$, *every selfadjoint operator $A \in L(\mathcal{H})$ admits a spectral decomposition*

$$A = \sum_{a \in \sigma(A)} a P_a^{(A)}\,, \tag{2.4}$$

where $\sigma(A)$ is the finite *set of eigenvalues – which must be* real *as A is selfadjoint – and $P_a^{(A)}$ is the orthogonal projector onto the eigenspace associated to a. Notice that $P_a P_{a'} = 0$ if $a \neq a'$ as eigenvectors with different eigenvalue are orthogonal.*

Let us show how the mathematical assumptions (1)-(3) permit us to set the physical properties of quantum systems (1)-(3) into a mathematically nice form.

(1) Randomness: The eigenvalues of an observable A are physically interpreted as the possible values of the outcomes of a measurement of A.

Given a state, represented by the unit vector $\psi \in \mathcal{H}$, the probability to obtain $a \in \sigma(A)$ as an outcome when measuring A is

$$\mu_\psi^{(A)}(a) := ||P_a^{(A)}\psi||^2 .$$

Going along with this interpretation, the expectation value of A ,when the state is represented by ψ, turns out to be

$$\langle A\rangle_\psi := \sum_{a\in\sigma(A)} a\mu_\psi^{(A)}(a) = \langle\psi, A\psi\rangle .$$

So that the identity holds

$$\langle A\rangle_\psi = \langle\psi, A\psi\rangle . \tag{2.5}$$

Finally, the standard deviation ΔA_ψ results to be

$$\Delta A_\psi^2 := \sum_{a\in\sigma(A)} (a - \langle A\rangle_\psi)^2\mu_\psi^{(A)}(a) = \langle\psi, A^2\psi\rangle - \langle\psi, A\psi\rangle^2 . \tag{2.6}$$

Remark 2.1.3.

(a) *Notice that the arbitrary phase affecting the unit vector $\psi \in \mathcal{H}$ ($e^{ia}\psi$ and ψ represent the same quantum state for every $a \in \mathbb{R}$) is armless here.*

(b) *If A is an observable and $f : \mathbb{R} \to \mathbb{R}$ is given, $f(A)$ is interpreted as an observable whose values are $f(a)$ if $a \in \sigma(a)$: Taking (2.4) into account,*

$$f(A) := \sum_{a\in\sigma(A)} f(a)P_a^{(A)} . \tag{2.7}$$

For polynomials $f(x) = \sum_{k=0}^n a_k x^k$, it results $f(A) = \sum_{k=0}^n a_k A^k$ as expected. The selfadjoint operator A^2 can naturally be interpreted this way as the natural observable whose values are a^2 when $a \in \sigma(A)$. For this reason, looking at the last term in (2.6) and taking (2.5) into account,

$$\Delta A_\psi^2 = \langle A^2\rangle_\psi - \langle A\rangle_\psi^2 = \langle(A - \langle A\rangle_\psi I)^2\rangle_\psi = \langle\psi, (A - \langle A\rangle_\psi I)^2\psi\rangle . \tag{2.8}$$

(2) Collapse of the State. If a is the outcome of the (idealized) measurement of A when the state is represented by ψ, the new state immediately after the measurement is represented by the unit vector

$$\psi' := \frac{P_a^{(A)}\psi}{||P_a^{(A)}\psi||} . \tag{2.9}$$

Remark 2.1.4. *Obviously this formula does not make sense if* $\mu_\psi^{(A)}(a) = 0$ *as expected. Moreover the arbitrary phase affecting* ψ *does not lead to troubles, due to the linearity of* $P_a^{(A)}$.

(3) Compatible and Incompatible Observables. Two observables are compatible – i.e. they can be simultaneously measured – if and only if the associated operators *commute*, that is

$$AB - BA = 0\,.$$

Using the fact that $\mathcal{H}$ has finite dimension, one easily proves that the observables A and B are compatible if and only if the associated spectral projectors commute as well

$$P_a^{(A)} P_b^{(B)} = P_b^{(B)} P_a^{(A)} \quad a \in \sigma(A)\,, b \in \sigma(B)\,.$$

In this case,

$$||P_a^{(A)} P_b^{(B)} \psi||^2 = ||P_b^{(B)} P_a^{(A)} \psi||^2$$

has the natural interpretation of the probability to obtain the outcomes a and b for a simultaneous measurement of A and B. If instead A and B are incompatible, it may happen that

$$||P_a^{(A)} P_b^{(B)} \psi||^2 \neq ||P_b^{(B)} P_a^{(A)} \psi||^2\,.$$

Sticking to the case of A and B incompatible, exploiting (2.9),

$$||P_a^{(A)} P_b^{(B)} \psi||^2 = \left|\left| P_a^{(A)} \frac{P_b^{(B)} \psi}{||P_b^{(B)} \psi||} \right|\right|^2 ||P_b^{(B)} \psi||^2 \tag{2.10}$$

has the natural meaning of *the probability of obtaining first* b *and next* a *in a subsequent measurement of* B *and* A.

Remark 2.1.5.

(a) *Notice that, in general, we cannot interchange the rôle of* A *and* B *in (2.10) because, in general,* $P_a^{(A)} P_b^{(B)} \neq P_b^{(B)} P_a^{(A)}$ *if* A *and* B *are incompatible. The measurement procedures "disturb each other" as already said.*

(b) *The interpretation of (2.10) as a probability of subsequent measurements can be given also if* A *and* B *are compatible. In this case, the probability of obtaining first* b *and next* a *in a subsequent measurement of* B *and* A *is identical to the probability of measuring* a *and* b *simultaneously and, in turn, it coincides with the probability of obtaining first* a *and next* b *in a subsequent measurement of* A *and* B.

(c) *A is always compatible with itself. Moreover $P_a^{(A)}P_a^{(A)} = P_a^{(A)}$ just due to the definition of projector. This fact has the immediate consequence that if we obtain a measuring A so that the state immediately after the measurement is represented by $\psi_a = ||P_a^{(A)}\psi||^{-1}\psi$, it will remain ψ_a even after other subsequent measurements of A and the outcome will result to be always a. Versions of this phenomenon, especially regarding the decay of unstable particles, are experimentally confirmed and it is called the* quantum Zeno effect.

Example 2.1.6. An electron admits a triple of observables, S_x, S_y, S_z, known as the components of the *spin*. Very roughly speaking, the spin can be viewed as the angular momentum of the particle referred to a reference frame always at rest with the centre of the particle and carrying its axes parallel to the ones of the reference frame of the laboratory, where the electron moves. In view of its peculiar properties, the spin cannot actually have a complete classical corresponding and thus that interpretation is untenable. For instance, one cannot "stop" the spin of a particle or change the constant value of $S^2 = S_x^2 + S_y^2 + S_z^2$: It is a given property of the particle like the mass. The electron spin is described within an *internal* Hilbert space $\mathcal{H}_s$, which has dimension 2. Identifying $\mathcal{H}_s$ with $\mathbb{C}^2$, the three spin observables are defined in terms of the three Hermitian matrices (occasionally re-introducing the constant $\hbar$)

$$S_x = \frac{\hbar}{2}\sigma_x\,, \qquad S_y = \frac{\hbar}{2}\sigma_y\,, \qquad S_z = \frac{\hbar}{2}\sigma_z\,, \tag{2.11}$$

where we have introduced the well known *Pauli matrices*,

$$\sigma_x = \begin{bmatrix} 0 & 1 \\ 1 & 0 \end{bmatrix}, \quad \sigma_y = \begin{bmatrix} 0 & -i \\ i & 0 \end{bmatrix}, \quad \sigma_z = \begin{bmatrix} 1 & 0 \\ 0 & -1 \end{bmatrix}. \tag{2.12}$$

Notice that $[S_a, S_b] \neq 0$ if $a \neq b$ so that the components of the spin are incompatible observables. In fact one has

$$[S_x, S_y] = i\hbar S_z$$

and this identity holds also cyclically permuting the three indices. These commutation relations are the same as for the observables L_x,L_y,L_z describing the angular momentum referred to the laboratory system which have classical corresponding (we shall return on these observables in example 2.3.77). So, differently from CM, the observables describing the components of the angular momentum are incompatible, they cannot be measured simultaneously. However the failure of the compatibility is related to the appearance of $\hbar$ on the right-hand side of

$$[L_x, L_y] = i\hbar L_z\,.$$

That number is extremely small if compared with macroscopic scales. This is the ultimate reason why the incompatibility of L_x and L_z is negligible for macroscopic systems.

Direct inspection proves that $\sigma(S_a) = \{\pm\hbar/2\}$. Similarly $\sigma(L_a) = \{n\hbar | n \in \mathbb{Z}\}$. Therefore, differently from CM, the values of angular momentum components form a discrete set of reals in QM. Again notice that the difference of two closest values is extremely small if compared with typical values of the angular momentum of macroscopic systems. This is the practical reason why this discreteness disappears at macroscopic level.

Just a few words about the time evolution and composite systems already discussed in the first part are however necessary now, a wider discussion on the time evolution will take place later in this paper.

2.1.3 *Time evolution*

Among the class of observables of a quantum system described in a given inertial reference frame, an observable H called the (quantum) *Hamiltonian* plays a fundamental rôle. We are assuming here that the system interacts with a stationary environment. The one-parameter group of unitary operators associated to H (exploiting (2.7) to explain the notation)

$$U_t := e^{-itH} := \sum_{h \in \sigma(H)} e^{-ith} P_h^{(H)} , \quad t \in \mathbb{R} \tag{2.13}$$

describes the *time evolution of quantum states* as follows. If the state at time $t = 0$ is represented by the unit vector $\psi \in \mathcal{H}$, the state at the generic time t is represented by the vector

$$\psi_t = U_t \psi .$$

Remark 2.1.7. *Notice that ψ_t has norm* 1 *as necessary to describe states, since U_t is norm preserving it being unitary.*

Taking (2.13) into account, this identity is equivalent to

$$i\frac{d\psi_t}{dt} = H\psi_t . \tag{2.14}$$

Equation (2.14) is nothing but a form of the celebrated *Schrödinger equation.* If the environment is not stationary, a more complicated description can be given where H is replaced by a class of Hamiltonian (selfadjoint) operators parametrized in time, $H(t)$, with $t \in \mathbb{R}$. This time dependence accounts for the time evolution of the external system interacting with our quantum system. In that case, it is simply assumed that the time evolution of states is again described by the equation above where H is replaced by $H(t)$:

$$i\frac{d\psi_t}{dt} = H(t)\psi_t \,.$$

Again, this equation permits one to define a two-parameter *groupoid* of unitary operators $U(t_2, t_1)$, where $t_2, t_1 \in \mathbb{R}$, such that

$$\psi_{t_2} = U(t_2, t_1)\psi_{t_1}\,, \quad t_2, t_1 \in \mathbb{R}\,.$$

The groupoid structure arises from the following identities: $U(t,t) = I$ and $U(t_3, t_2)U(t_2, t_1) = U(t_3, t_2)$ and $U(t_2, t_1)^{-1} = U(t_2, t_1)^{\dagger} = U(t_1, t_2)$.

Remark 2.1.8. *In our elementary case where $\mathcal{H}$ is finite dimensional,* Dyson's formula *holds with the simple hypothesis that the map $\mathbb{R} \ni t \mapsto H_t \in L(\mathcal{H})$ is continuous (adopting any topology compatible with the vector space structure of $L(\mathcal{H})$) [5]*

$$U(t_2, t_1) = \sum_{n=0}^{+\infty} \frac{(-i)^n}{n!} \int_{t_1}^{t_2} \cdots \int_{t_1}^{t_2} T[H(\tau_1)\cdots H(\tau_n)]\, d\tau_1 \cdots d\tau_n \,.$$

Above, we define $T[H(\tau_1)\cdots H(\tau_n)] = H(\tau_{\pi(1)})\cdots H(\tau_\pi(n))$, where the bijective function $\pi : \{1, \ldots, n\} \to \{1, \ldots, n\}$ is any permutation with $\tau_{\pi(1)} \geq \cdots \geq \tau_{\pi(n)}$.

2.1.4 *Composite systems*

If a quantum system S is made of two parts, S_1 and S_2, respectively described in the Hilbert spaces $\mathcal{H}_1$ and $\mathcal{H}_2$, it is assumed that the whole system is described in the space $\mathcal{H}_1 \otimes \mathcal{H}_2$ equipped with the unique Hermitian scalar product $\langle \cdot, \cdot \rangle$ such that $\langle \psi_1 \otimes \psi_2, \phi_1 \otimes \phi_2 \rangle = \langle \psi_1, \phi_1 \rangle_1 \langle \psi_2, \phi_2 \rangle_2$ (in the infinite dimensional case $\mathcal{H}_1 \otimes \mathcal{H}_2$ is the Hilbert completion of the afore-mentioned algebraic tensor product). If $\mathcal{H}_1 \otimes \mathcal{H}_2$ is the space of a composite system S as before and A_1 represents an observable for the part S_1, it is naturally identified with the selfadjoint operator $A_1 \otimes I_2$ defined in $\mathcal{H}_1 \otimes \mathcal{H}_2$. A similar statement holds when swapping 1 and 2. Notice that $\sigma(A_1 \otimes I_2) = \sigma(A_1)$ as one easily proves. (The result survives the extension to the infinite dimensional case.)

Remark 2.1.9.

(a) *Composite systems are in particular systems made of many (either identical or not) particles. If we have a pair of particles respectively described in the Hilbert space $\mathcal{H}_1$ and $\mathcal{H}_2$, the full system is described in $\mathcal{H}_1 \otimes \mathcal{H}_2$. Notice that the dimension of the final space is the* product *of the dimension of the component spaces. In CM, the system would instead be described in a space of phases which is the Cartesian* *product of the two spaces of phases. In that case, the dimension would be the* sum, *rather than the product, of the dimensions of the component spaces.*

(b) $\mathcal{H}_1 \otimes \mathcal{H}_2$ *contains the so-called* entangled states. *They are states represented by vectors* not *factorized as* $\psi_1 \otimes \psi_2$, *but they are* linear combinations *of such vectors. Suppose the whole state is represented by the entangled state*

$$\Psi = \frac{1}{\sqrt{2}} \left(\psi_a \otimes \phi + \psi_{a'} \otimes \phi' \right),$$

where $A_1 \psi_a = a \psi_a$ *and* $A_1 \psi_{a'} = a' \psi_{a'}$ *for a certain observable* A_1 *of the part* S_1 *of the total system. Performing a measurement of* A_1 *on* S_1*, due to the collapse of state phenomenon, we automatically act one the whole state and on the part describing* S_2*. As a matter of fact, up to normalization, the state of the full system after the measurement of* A_1 *will be* $\psi_a \otimes \phi$ *if the outcome of* A_1 *is* a*, or it will be* $\psi_{a'} \otimes \phi'$ *if the outcome of* A_1 *is* a'*. It could happen that the two measurement apparatuses, respectively measuring* S_1 *and* S_2*, are localized very far in the physical space. Therefore acting on* S_1 *by measuring* A_1*, we "instantaneously" produce a change of* S_2 *which can be seen performing mesurements on it, even if the measurement apparatus of* S_2 *is very far from the one of* S_1*. This seems to contradict the fundamental relativistic postulate, the* locality *postulate, that there is a maximal speed, the one of light, for propagating physical information. After the famous analysis of Bell, improving the original one by Einstein, Podolsky and Rosen, the phenomenon has been experimentally observed. Locality is truly violated, but in a such subtle way which does not allows superluminal propagation of physical information. Non-locality of QM is nowadays widely accepted as a real and fundamental feature of Nature [1; 2; 6].*

Example 2.1.10. An electron also possesses an *electric charge.* That is another *internal* quantum observable, Q, with two values $\pm e$, where $e < 0$ is the elementary electrical charge of an electron. So there are two types of electrons. *Proper electrons,* whose internal state of charge is an eigenvector of Q with eigenvalue $-e$ and *positrons,* whose internal state of charge is a eigenvector of Q with eigenvalue e. The simplest version of the internal Hilbert space of the electrical charge is therefore $\mathcal{H}_c$ which[1], again, is isomorphic to $\mathbb{C}^2$. With this representation $Q = e\sigma_3$. The full Hilbert space of an electron must contain a factor $\mathcal{H}_s \otimes \mathcal{H}_c$. Obviously this is by no means sufficient to describe an electron, since we must introduce at least the observables describing the position of the particle in the physical space at rest with a reference (inertial) frame.

[1] As we shall say later, in view of a *superselection rule* not all normalized vectors of $\mathcal{H}_c$ represent (pure) states.

2.1.5 *A first look to the infinite dimensional case, CCR and quantization procedures*

All the described formalism, barring technicalities we shall examine in the rest of the paper, holds also for quantum systems whose complex vector space of the states is *infinite* dimensional.

To extend the ideas treated in Sect. 2.1.2 to the general case, dropping the hypothesis that $\mathcal{H}$ is finite dimensional, it seems to be natural to assume that $\mathcal{H}$ is complete with respect to the norm associated to $\langle \cdot, \cdot \rangle$. In particular, completeness assures the existence of spectral decompositions, generalizing (2.4) for instance when referring to *compact* selfadjoint operators (e.g., see [5]). In other words, $\mathcal{H}$ is a *complex Hilbert space.*

The most elementary example of a quantum system described in an infinite dimensional Hilbert space is a quantum particle whose position is along the axis $\mathbb{R}$. In this case, as seen in the first part of this book, the Hilbert space is $\mathcal{H} := L^2(\mathbb{R}, dx)$, dx denoting the standard Lebesgue measure on $\mathbb{R}$. States are still represented by elements of $P\mathcal{H}$, namely equivalence classes $[\psi]$ of measurable functions $\psi : \mathbb{R} \to \mathbb{C}$ with unit norm, $||[\psi]|| = \int_{\mathbb{R}} |\psi(x)|^2 dx = 1$.

Remark 2.1.11. *We therefore have here* two *quotient procedures. ψ and ψ' define the same element $[\psi]$ of $L^2(\mathbb{R}, dx)$ iff $\psi(x) - \psi'(x) \neq 0$ on a zero Lebesgue measure set. Two unit vectors $[\psi]$ and $[\phi]$ define the same state if $[\psi] = e^{ia}[\phi]$ for some $a \in \mathbb{R}$.*

Notation 2.1.12. In the rest of the paper, we adopt the standard convention of many textbooks on functional analysis denoting by ψ, instead of $[\psi]$, the elements of spaces L^2 and tacitly identifying pair of functions which are different on a zero measure set.

The functions ψ defining (up to zero-measure set and phases) states, are called *wavefunctions.* There is a pair of fundamental observables describing our quantum particle moving in $\mathbb{R}$. One is the *position observable.* The corresponding selfadjoint operator, denoted by X, is defined as follows

$$(X\psi)(x) := x\psi(x)\,, \quad x \in \mathbb{R}\,, \quad \psi \in L^2(\mathbb{R}, dx)\,.$$

The other observable is the one associated to the momentum and indicated by P. Restoring $\hbar$ for the occasion, the *momentum operator* is

$$(P\psi)(x) := -i\hbar\frac{d\psi(x)}{dx}\,, \quad x \in \mathbb{R}\,, \quad \psi \in L^2(\mathbb{R}, dx)\,.$$

We immediately face several mathematical problems with these, actually quite naive, definitions. Let us first focus on X. First of all, generally $X\psi \notin L^2(\mathbb{R}, dx)$

even if $\psi \in L^2(\mathbb{R}, dx)$. To fix the problem, we can simply restrict the domain of X to the linear subspace of $L^2(\mathbb{R}, dx)$

$$D(X) := \left\{ \psi \in L^2(\mathbb{R}, dx) \;\middle|\; \int_{\mathbb{R}} |x\psi(x)|^2 dx < +\infty \right\}. \tag{2.15}$$

Though it holds

$$\langle X\psi, \phi\rangle = \langle \psi, X\phi\rangle \quad \text{for all } \psi, \phi \in D(X)\,, \tag{2.16}$$

we cannot say that X is selfadjoint simply because we have not yet given the definition of adjoint operator of an operator defined in a non-maximal domain in an infinite dimensional Hilbert space. In this general case, the identity (2.2) does not define a (unique) operator $X^\dagger$ without further technical requirements. We just say here, to comfort the reader, that X is truly selfadjoint with respect to a general definition we shall give in the next section, when its domain is (2.15).

Like (2.3) in the finite dimensional case, the identity (2.16) implies that all eigenvalues of X must be real if any. Unfortunately, for every fixed $x_0 \in \mathbb{R}$ there is no $\psi \in L^2(\mathbb{R}, dx)$ with $X\psi = x_0\psi$ and $\psi \neq 0$. (A function ψ satisfying $X\psi = x_0\psi$ must also satisfy $\psi(x) = 0$ if $x \neq x_0$, due to the definition of X. Hence $\psi = 0$, as an element of $L^2(\mathbb{R}, dx)$ just because $\{x_0\}$ has zero Lebesgue measure!) All that seems to prevent the existence of a spectral decomposition of X like the one in (2.4), since X does not admit eigenvectors in $L^2(\mathbb{R}, dx)$ (and *a fortiori* in $D(X)$). The definition of P suffers from similar troubles. The domain of P cannot be the whole $L^2(\mathbb{R}, dx)$ but should be restricted to a subset of (weakly) differentiable functions with derivative in $L^2(\mathbb{R}, dx)$. The simplest definition is

$$D(P) := \left\{ \psi \in L^2(\mathbb{R}, dx) \;\middle|\; \exists\, \text{w-}\frac{d\psi(x)}{dx}\,,\; \int_{\mathbb{R}} \left| \text{w-}\frac{d\psi(x)}{dx} \right|^2 dx < +\infty \right\}. \tag{2.17}$$

Above w-$\frac{d\psi(x)}{dx}$ denotes the *weak derivative* of ψ[2]. As a matter of fact $D(P)$ coincides with the *Sobolev space* $H^1(\mathbb{R})$.

Again, without a precise definition of adjoint operator in an infinite dimensional Hilbert space (with non-maximal domain) we cannot say anything more precise about the selfadjointness of P with that domain. We say however that P turns out to be selfadjoint with respect to the general definition we shall give in the next section provided its domain is (2.17).

[2] $f : \mathbb{R} \to \mathbb{C}$, defined up to zero-measure set, is the weak derivative of $g \in L^2(\mathbb{R}, dx)$ if it holds $\int_{\mathbb{R}} g \frac{dh}{dx} dx = -\int_{\mathbb{R}} f h dx$ for every $h \in C_0^\infty(\mathbb{R})$. If g is differentiable, its standard derivative coincide with the weak one.

From the definition of the domain of P and passing to the Fourier-Plancherel transform, onc finds again (it is not so easy to see it)

$$\langle P\psi, \phi \rangle = \langle \psi, P\phi \rangle \quad \text{for all } \psi, \phi \in D(P)\,, \tag{2.18}$$

so that, eigenvalues are real if exist. However P does not admit eigenvectors. The naive eigenvectors with eigenvalue $p \in \mathbb{R}$ are functions proportional to the map $\mathbb{R} \ni x \mapsto e^{ipx/\hbar}$, which does not belong to $L^2(\mathbb{R}, dx)$ nor $D(P)$. We will tackle all these issues in the next section in a very general fashion.

We observe that the space of *Schwartz functions*, $\mathcal{S}(\mathbb{R})$[3] satisfies

$$\mathcal{S}(\mathbb{R}) \subset D(X) \cap D(P)$$

and furthermore $\mathcal{S}(\mathbb{R})$ is dense in $L^2(\mathbb{R}, dx)$ and *invariant* under X and P: $X(\mathcal{S}(\mathbb{R})) \subset \mathcal{S}(\mathbb{R})$ and $P(\mathcal{S}(\mathbb{R})) \subset \mathcal{S}(\mathbb{R})$.

Remark 2.1.13. *Though we shall not pursue this approach within these notes, we stress that X admits a set of eigenvectors if we extend the domain of X to the space $\mathcal{S}'(\mathbb{R})$ of* Schwartz distributions *in a standard way: If $T \in \mathcal{S}'(\mathbb{R})$,*

$$(X(T), f) := (T, X(f)) \quad \textit{for every } f \in \mathcal{S}(\mathbb{R})\,.$$

With this extension, the eigenvectors in $\mathcal{S}'(\mathbb{R})$ of X with eigenvalues $x_0 \in \mathbb{R}$ are the distributions $c\delta(x - x_0)$ (see the first part of this book) This class of eigenvectors can be exploited to build a spectral decomposition of X similar to that in (2.4). Similarly, P admits eigenvectors in $\mathcal{S}'(\mathbb{R})$ with the same procedure. They are just the above exponential functions. Again, this class of eigenvectors can be used to construct a spectral decomposition of P like the one in (2.4). The idea of this procedure can be traced back to Dirac [3] and, in fact, something like ten years later it gave rise to the rigorous theory of distributions *by L. Schwartz. The modern formulation of this approach to construct spectral decompositions of selfadjoint operators was developed by Gelfand in terms of the so called* rigged Hilbert spaces *[4].*

Referring to a quantum particle moving in $\mathbb{R}^n$, whose Hilbert space is $L^2(\mathbb{R}^n, dx^n)$, one can introduce observables X_k and P_k representing position and momentum with respect to the k-th axis, $k = 1, 2, \ldots, n$. These operators, which are defined analogously to the case $n = 1$, have domains smaller than the full Hilbert space. We do not write the form of these domain (where the operators turn out to be properly selfadjoint referring to the general definition we shall state in the next

[3] $\mathcal{S}(\mathbb{R}^n)$ is the vector space of the C^∞ complex valued functions on $\mathbb{R}^n$ which, together with their derivatives of all orders in every set of coordinate, decay faster than every negative integer power of $|x|$ for $|x| \to +\infty$.

section). We just mention the fact that all these operators admit $\mathcal{S}(\mathbb{R}^n)$ as common invariant subspace included in their domains. Thereon

$$(X_k\psi)(x) = x_k\psi(x)\,, \qquad (P_k\psi)(x) = -i\hbar\frac{\partial\psi(x)}{\partial x_k}\,, \quad \psi \in \mathcal{S}(\mathbb{R}^n) \tag{2.19}$$

and so

$$\langle X_k\psi, \phi\rangle = \langle\psi, X_k\phi\rangle\,, \quad \langle P_k\psi, \phi\rangle = \langle\psi, P_k\phi\rangle \quad \text{for all } \psi, \phi \in \mathcal{S}(\mathbb{R}^n)\,, \tag{2.20}$$

By direct inspection, one easily proves that the *canonical commutation relations* (CCR) hold when all the operators in the subsequent formulas are supposed to be restricted to $\mathcal{S}(\mathbb{R}^n)$

$$[X_h, P_k] = i\hbar\delta_{hk}I\,, \quad [X_h, X_k] = 0\,, \quad [P_h, P_k] = 0\,. \tag{2.21}$$

We have introduced the *commutator* $[A, B] := AB - BA$ of the operators A and B generally with different domains, defined on a subspace where both compositions AB and BA makes sense, $\mathcal{S}(\mathbb{R}^n)$ in the considered case. Assuming that (2.5) and (2.8) are still valid for X_k and P_k referring to $\psi \in \mathcal{S}(\mathbb{R}^n)$, (2.21) easily leads to the *Heisenberg uncertainty relations*,

$$\Delta X_{k\psi}\Delta P_{k\psi} \geq \frac{\hbar}{2}\,, \quad \text{for } \psi \in \mathcal{S}(\mathbb{R}^n)\,, \quad ||\psi|| = 1\,. \tag{2.22}$$

Exercise 2.1.14. *Prove inequality (2.22) assuming (2.5) and (2.8).*

Solution. Using (2.5), (2.8) and the Cauchy-Schwarz inequality, one easily finds (we omit the index $_k$ for simplicity),

$$\Delta X_\psi \Delta P_\psi = ||X'\psi||||P'\psi|| \geq |\langle X'\psi, P'\psi\rangle|\,.$$

where $X' := X - \langle X\rangle_\psi I$ and $P' := X - \langle X\rangle_\psi I$. Next, notice that

$$|\langle X'\psi, P'\psi\rangle| \geq |Im\langle X'\psi, P'\psi\rangle| = \frac{1}{2}|\langle X'\psi, P'\psi\rangle - \langle P'\psi, X'\psi\rangle|$$

Taking advantage from (2.20) and the definitions of X' and P' and exploiting (2.21),

$$\begin{aligned}|\langle X'\psi, P'\psi\rangle - \langle P'\psi, X'\psi\rangle| &= |\langle\psi, (X'P' - P'X')\psi\rangle| \\ &= |\langle\psi, (XP - PX)\psi\rangle| = \hbar|\langle\psi, \psi\rangle|\,.\end{aligned}$$

Since $\langle\psi,\psi\rangle = ||\psi||^2 = 1$ by hypotheses, (2.22) is proved. Obviously the open problem is to justify the validity of (2.5) and (2.8) also in the infinite dimensional case.

Another philosophically important consequence of the CCR (2.21) is that they resemble the *classical canonical commutation relations* of the Hamiltonian variables q^h and p_k, referring to the standard *Poisson brackets* $\{\cdot,\cdot\}_P$,

$$\{q^h, p_k\}_P = \delta^h_k\,, \quad \{q^h, q^k\}_P = 0\,, \quad \{p_h, p_k\}_P = 0\,. \tag{2.23}$$

as soon as one identifies $(i\hbar)^{-1}[\cdot,\cdot]$ with $\{\cdot,\cdot\}_P$. This fact, initially noticed by Dirac [3], leads to the idea of "quantization" of a classical Hamiltonian theory (see the first part of this book).

One starts from a classical system described on a symplectic manifold (Γ,ω), for instance $\Gamma = \mathbb{R}^{2n}$ equipped with the standard symplectic form as ω and considers the (real) Lie algebra $(C^\infty(\Gamma,\mathbb{R}), \{\cdot,\cdot\}_P)$. To "quantize" the system one looks for a map associating classical observables $f \in C^\infty(\Gamma,\mathbb{R})$ to quantum observables $\widehat{O}_f$, i.e. selfadjoint operators restricted[4] to a common invariant domain $\mathcal{S}$ of a certain Hilbert space $\mathcal{H}$. (In case $\Gamma = T^*Q$, $\mathcal{H}$ can be chosen as $L^2(Q, d\mu)$ where μ is some natural measure.) The map $f \mapsto \widehat{O}_f$ is expected to satisfy a set of constraints. The most important are listed here:

(1) $\mathbb{R}$-linearity;
(2) $\widehat{O}_{id} = I|_{\mathcal{S}}$;
(3) $\widehat{O}_{\{f,g\}_P} = -i\hbar[\widehat{O}_f, \widehat{O}_g]$;
(4) If (Γ,ω) is $\mathbb{R}^{2n}$ equipped with the standard symplectic form, they must hold $\widehat{O}_{x_k} = X_k|_{\mathcal{S}}$ and $\widehat{O}_{p_k} = P_k|_{\mathcal{S}}$, $k = 1, 2, \ldots, n$.

The penultimate requirement says that the map $f \mapsto \widehat{O}_f$ transforms the real Lie algebra $(C^\infty(\Gamma,\mathbb{R}), \{\cdot,\cdot\}_P)$ into a real Lie algebra of operators whose Lie bracket is $i\hbar[\widehat{O}_f, \widehat{O}_g]$. A map fulfilling these constraints, in particular the third one, is possible if f, g are both functions of only the q or the p coordinates separately or if they are linear in them. But it is false already if we consider elementary physical systems as some discussed in the first part of the book. The ultimate reason of this obstructions is due to the fact that the operators P_k, X_k do not commute,

[4]The restriction should be such that it admits a unique selfadjoint extension. A sufficient requirement on $\mathcal{S}$ is that every $\widehat{O}_f$ is *essentially selfadjoint* thereon, notion we shall discuss in the next section.

contrary to the functions p_k, q^k which do. The problem can be solved, in the paradigm of the so-called *Geometric Quantization* (see the first part of the book), replacing $(C^\infty(\Gamma, \mathbb{R}), \{\cdot,\cdot\}_P)$ with a sub-Lie algebra (as large as possible). There are other remarkable procedures of "quantization" in the literature, we shall not insist on them any further here (see the first part of the book).

Example 2.1.15.
(1) The full Hilbert space of an electron is therefore given by the tensor product $L^2(\mathbb{R}^3, d^3x) \otimes \mathcal{H}_s \otimes \mathcal{H}_c$.
(2) Consider a particle in $3D$ with mass m, whose potential energy is a bounded-below real function $U \in C^\infty(\mathbb{R}^3)$ with polynomial growth. Classically, its Hamiltonian function reads

$$h := \sum_{k=1}^{3} \frac{p_k^2}{2m} + U(x) .$$

A brute force quantization procedure in $L^2(\mathbb{R}^3, d^3x)$ consists of replacing every classical object with corresponding operators. It may make sense at most when there are no ordering ambiguities in translating functions like p^2x, since classically $p^2x = pxp = xp^2$, but these identities are false at quantum level. In our case, these problems do not arise so that

$$H := \sum_{k=1}^{3} \frac{P_k^2}{2m} + U , \tag{2.24}$$

where $(U\psi)(x) := U(x)\psi(x)$, could be accepted as first quantum model of the Hamiltonian function of our system. The written operator is at least defined on $\mathcal{S}(\mathbb{R}^3)$, where it satisfies $\langle H\psi, \phi\rangle = \langle \psi, H\phi\rangle$. The existence of selfadjoint extensions is a delicate issue [5] we shall not address here. Taking (2.19) into account, always on $\mathcal{S}(\mathbb{R}^3)$, one immediately finds

$$H := -\frac{\hbar^2}{2m}\Delta + U ,$$

where Δ is the standard Laplace operator in $\mathbb{R}^3$. If we assume that the equation describing the evolution of the quantum system is again[5] (2.14), in our case we

[5] A factor $\hbar$ has to be added in front of the left-hand side of (2.14) if we deal with a unit system where $\hbar \neq 1$.

find the known form of the Schrödinger equation,

$$i\hbar\frac{d\psi_t}{dt} = -\frac{\hbar^2}{2m}\Delta\psi_t + U\psi_t\,,$$

when $\psi_\tau \in \mathcal{S}(\mathbb{R}^3)$ for τ varying in a neighborhood of t (this requirement may be relaxed). Actually the meaning of the derivative on the left-hand side should be specified. We only say here that it is computed with respect to the natural topology of $L^2(\mathbb{R}^3, d^3x)$.

2.2 Observables in Infinite Dimensional Hilbert Spaces: Spectral Theory

The main goal of this section is to present a suitable mathematical theory, sufficient to extend to the infinite dimensional case the mathematical formalism of QM introduced in the previous section. As seen in Sect. 2.1.5, the main issue concerns the fact that, in the infinite dimensional case, there are operators representing observables which do not have proper eigenvalues and eigenvectors, like X and P. So, naive expansions as (2.4) cannot be literally extended to the general case. These expansions, together with the interpretation of the eigenvalues as values attained by the observable associated with a selfadjoint operator, play a crucial rôle in the mathematical interpretation of the quantum phenomenology introduced in Sect. 2.1.1 and mathematically discussed in Sect. 2.1.2. In particular, we need a precise definition of selfadjoint operator and a result regarding a spectral decomposition in the infinite dimensional case. These tools are basic elements of the so called *spectral theory in Hilbert spaces*, literally invented by von Neumann in his famous book [7] to give a rigorous form to Quantum Mechanics and successively developed by various authors towards many different directions of pure and applied mathematics. The same notion of abstract Hilbert space, as nowadays known, was born in the second chapter of that book, joining and elaborating previous mathematical constructions by Hilbert and Riesz. The remaining part of this section is devoted to introduce the reader to some basic elements of that formalism. Reference books are, e.g., [8; 5; 6; 9; 10].

2.2.1 *Hilbert spaces*

A **Hermitian scalar product** over the complex Hilbert space $\mathcal{H}$ is a map

$$\langle\cdot,\cdot\rangle : \mathcal{H}\times\mathcal{H}\to\mathbb{C}$$

such that, for $a, b\in\mathbb{C}$ and $x, y, z\in\mathcal{H}$,

(i) $\langle x, y\rangle = \langle y, x\rangle^*$;
(ii) $\langle x, ay + bz\rangle = a\langle x, y\rangle + b\langle x, z\rangle$;
(iii) $\langle x, x\rangle \geq 0$ with $x = 0$ if $\langle x, x\rangle = 0$.

The space $\mathcal{H}$ is said to be a (complex) **Hilbert space** if it is complete with respect to the natural norm $||x|| := \sqrt{\langle x, x\rangle}$, $x \in \mathcal{H}$.

In particular, just in view of positivity of the scalar product and regardless the completeness property, the **Cauchy-Schwarz inequality** holds

$$|\langle x, y\rangle| \leq ||x|| \, ||y|| \,, \quad x, y \in \mathcal{H} \,.$$

Another elementary purely algebraic fact is the **polarization identity** concerning the Hermitian scalar product (here, $\mathcal{H}$ is not necessarily complete)

$$\begin{aligned} 4\langle x, y\rangle &= ||x + y||^2 - ||x - y||^2 - i||x + iy||^2 \\ &\quad + i||x - iy||^2 \quad \text{for of } x, y \in \mathcal{H}, \end{aligned} \tag{2.25}$$

which immediately implies the following elementary result.

Proposition 2.2.1. *If $\mathcal{H}$ is a complex vector space with Hermintian scalar product $\langle\ ,\ \rangle$, a linear map $L : \mathcal{H} \to \mathcal{H}$ which is an isometry – $||Lx|| = ||x||$ if $x \in \mathcal{H}$ – also preserves the scalar product – $\langle Lx, Ly\rangle = \langle x, y\rangle$ for $x, y \in \mathcal{H}$.*

The converse proposition is obviously true.

Similar to the above identity, we have another useful identity, for every operator $A : \mathcal{H} \to \mathcal{H}$:

$$\begin{aligned} 4\langle x, Ay\rangle &= \langle x + y, A(x + y)\rangle - \langle x - y, A(x - y)\rangle - i\langle x + iy, A(x + iy)\rangle \\ &\quad + i\langle x - iy, A(x - iy)\rangle \quad \text{for of } x, y \in \mathcal{H} \,. \end{aligned} \tag{2.26}$$

That in particular proves that *if $\langle x, Ax\rangle = 0$ for all $x \in \mathcal{H}$, then $A = 0$.*

(Observe that this fact is not generally true if dealing with real vector spaces equipped with a real symmetric scalar product.)

We henceforth assume that the reader be familiar with the basic theory of normed, Banach and Hilbert spaces and notions like *orthogonal of a set*, *Hilbertian basis* (also called *complete orthonormal systems*) and that their properties and use be well known [8; 5]. We summarize here some basic results concerning *orthogonal sets* and *Hilbertian bases.*

Notation 2.2.2. If $M \subset \mathcal{H}$, $M^\perp := \{y \in \mathcal{H} \mid \langle y, x\rangle = 0 \quad \forall x \in M\}$ denotes the **orthogonal** of M.

Evidently $M^\perp$ is a closed subspace of $\mathcal{H}$. $^\perp$ enjoys several nice properties quite easy to prove (e.g., see [8; 5]), in particular,

$$\overline{\text{span}(M)} = (M^\perp)^\perp \quad \text{and} \quad \mathcal{H} = \overline{\text{span}(M)} \oplus M^\perp \tag{2.27}$$

where the bar denotes the topological closure and $\oplus$ the direct orthogonal sum.

Definition 2.2.3. *A Hilbertian basis N of a Hilbert space $\mathcal{H}$ is a set of* **orthonormal vectors**, *i.e., $||u|| = 1$ and $\langle u, v\rangle = 0$ for $u, v \in N$ with $u \neq v$, such that if $s \in \mathcal{H}$ satisfies $\langle s, u\rangle = 0$ for every $u \in \mathcal{H}$ then $s = 0$.*

Hilbertian bases always exist as a consequence of Zorn's lemma. Notice that, as consequence of (2.27), the following proposition is valid

Proposition 2.2.4. *$N \subset \mathcal{H}$ is a Hilbertian basis of the Hilbert space $\mathcal{H}$ if and only if $\overline{span(N)} = \mathcal{H}$.*

Furthermore, a generalized version of *Pythagorean theorem* holds true.

Proposition 2.2.5. *$N \subset \mathcal{H}$ is a Hilbertian basis of the Hilbert space $\mathcal{H}$ if and only if*

$$||x||^2 = \sum_{u\in N} |\langle u, x\rangle|^2 \quad \textit{for every } x \in \mathcal{H}\,.$$

The sum appearing in the above decomposition of x is understood as the *supremum* of the sums $\sum_{u\in F} |\langle u, x\rangle|^2$ for every finite set $F \subset N$. As a consequence (e.g., see [8; 5]), only a set at most countable of elements $|\langle u, x\rangle|^2$ do not vanish and the sum is therefore interpreted as a standard series that can be re-ordered preserving the sum because it absolutely converges. It turns out that all Hilbertian bases of $\mathcal{H}$ have the same cardinality and $\mathcal{H}$ is separable (it admits a dense counteble subset) if and only if $\mathcal{H}$ has an either finite our countable Hilbertian basis. If $M \subset \mathcal{H}$ is a set of orthonormal vectors, the waeker so-called **Bessel inequality** holds

$$||x||^2 \geq \sum_{u\in M} |\langle u, x\rangle|^2 \quad \text{for every } x \in \mathcal{H}\,,$$

so that Hilbertian bases are exactly orthonormal sets saturating the inequality.

Finally, if $N \subset \mathcal{H}$ is an Hilbertian basis, the decompositions hold for every $x, y \in \mathcal{H}$

$$x = \sum_{u \in N} \langle u, x \rangle u \,, \quad \langle x, y \rangle = \sum_{u \in N} \langle x, u \rangle \langle u, y \rangle \,.$$

In view of the already mentioned fact that only a finite or countable set of elements $\langle u_n, x \rangle$ do not vanish, the first sum is actually a finite sum or at most series $\lim_{m \to +\infty} \sum_{n=0}^{m} \langle u_n, x \rangle u_n$ computed with respect to the norm of $\mathcal{H}$, where the order used to label the element u_n does not matter as the series absolutely converges. The second sum is similarly absolutely convergent so that, again, it can be re-ordered arbitrarily.

We remind the reader of the validity of an elementary though fundamental tecnical result (e.g., see [8; 5]):

Theorem 2.2.6 (Riesz' lemma). *Let $\mathcal{H}$ be a complex Hilbert space. $\phi : \mathcal{H} \to \mathbb{C}$ is linear and continuous if and only if it has the form $\phi = \langle x, \ \rangle$ for some $x \in \mathcal{H}$. The vector x is uniquely determined by ϕ.*

2.2.2 *Types of operators*

Our goal is to present some basic results of *spectral analysis*, useful in QM.

From now on, an **operator** A **in** $\mathcal{H}$ always means a *linear* map $A : D(A) \to \mathcal{H}$, whose **domain**, $D(A) \subset \mathcal{H}$, is a *subspace* of $\mathcal{H}$. In particular, I always denotes the **identity operator** defined on the *whole* space ($D(I) = \mathcal{H}$)

$$I : \mathcal{H} \ni x \mapsto x \in \mathcal{H} \,.$$

If A is an operator in $\mathcal{H}$, $Ran(A) := \{Ax \,|\, x \in D(A)\}$ is the **range** (also known as **image**) of A.

Definition 2.2.7. *If A and B are operators in $\mathcal{H}$,*

$$A \subset B \text{ means that } D(A) \subset D(B) \text{ and } B|_{D(A)} = A \,,$$

where $|_S$ is the standard "restriction to S" symbol. We also adopt usual conventions regarding **standard domains** *of combinations of operators A, B:*

(i) *$D(AB) := \{x \in D(B) \,|\, Bx \in D(A)\}$ is the domain of AB;*
(ii) *$D(A + B) := D(A) \cap D(B)$ is the domain of $A + B$;*
(iii) *$D(\alpha A) = D(A)$ for $\alpha \neq 0$ is the domain of αA.*

With these definitons, it is easy to prove that

(1) $(A + B) + C = A + (B + C)$;
(2) $A(BC) = (AB)C$;

(3) $A(B+C) = AB + BC$;
(4) $(B+C)A \supset BA + CA$;
(5) $A \subset B$ and $B \subset C$ imply $A \subset C$;
(6) $A \subset B$ and $B \subset A$ imply $A = B$;
(7) $AB \subset BA$ implies $A(D(B)) \subset D(B)$ if $D(A) = \mathcal{H}$;
(8) $AB = BA$ implies $D(B) = A^{-1}(D(B))$ if $D(A) = \mathcal{H}$ (so, $A(D(B)) = D(B)$ is A is surjective).

To go on, we define some abstract algebraic structures naturally arising in the space of operators on a Hilbert space.

Definition 2.2.8. *Let $\mathfrak{A}$ be an associative complex algebra $\mathfrak{A}$.*

(1) *$\mathfrak{A}$ is a* **Banach algebra** *if it is a Banach space such that $||ab|| \leq ||a||\,||b||$ for $a, b \in \mathfrak{A}$. A* **unital** *Banach algebra is a Banach algebra with unit multiplicative element $\mathbb{1}$, satisfying $||\mathbb{1}|| = 1$.*
(2) *$\mathfrak{A}$ is a (unital)* ***-algebra** *if it is an (unital) algebra equipped with an anti linear map $\mathfrak{A} \ni a \mapsto a^* \in \mathfrak{A}$, called* **involution**, *such that $(a^*)^* = a$ and $(ab)^* = b^*a^*$ for $a, b \in \mathfrak{A}$.*
(3) *$\mathfrak{A}$ is a (unital)* **C^*-algebra** *if it is a (unital) Banach algebra $\mathfrak{A}$ which is also a *-algebra and $||a^*a|| = ||a||^2$ for $a \in \mathfrak{A}$. A* ***-homomorphism** *from the *-algebra $\mathcal{A}$ to the the *-algebra $\mathcal{B}$ is an algebra homomorphism preserving the involutions (and the unities if both present). A bijective *-homomorphism is called* ***-isomorphism**.

Exercise 2.2.9. *Prove that $\mathbb{1}^* = \mathbb{1}$ in a unital *-algebra and that $||a^*|| = ||a||$ if $a \in \mathfrak{A}$ when $\mathfrak{A}$ is a C^*-algebra.*

Solution. From $\mathbb{1}a = a\mathbb{1} = a$ and the definition of *, we immediately have $a^*\mathbb{1}^* = \mathbb{1}^*a^* = a^*$. Since $(b^*)^* = b$, we have found that $b\mathbb{1}^* = \mathbb{1}^*b = b$ for every $b \in \mathfrak{A}$. Uniqueness of the unit implies $\mathbb{1}^* = \mathbb{1}$. Regarding the second property, $||a||^2 = ||a^*a|| \leq ||a^*||\,||a||$ so that $||a|| \leq ||a^*||$. Everywhere replacing a for a^* and using $(a^*)^*$, we also obtain $||a^*|| \leq ||a||$, so that $||a^*|| = ||a||$.

We remind the reader that a linear map $A : X \to Y$, where X and Y are normed complex vector spaces with resp. norms $||\cdot||_X$ and $||\cdot||_Y$, is said to be **bounded** if

$$||Ax||_Y \leq b||x||_X \quad \text{for some } b \in [0, +\infty) \text{ and all } x \in X. \tag{2.28}$$

As is well known [8; 5], it turns out that the following proposition holds.

Proposition 2.2.10. *An operator $A : X \to Y$, with X, Y normed spaces, is continuous if and only if it is bounded.*

Proof. It is evident that A bounded is continuous because, for $x, x' \in X$, $||Ax - Ax'||_Y \leq b||x - x'||_X$. Conversely, if A is continuous then it is continuous for $x = 0$, so $||Ax||_y \leq \epsilon$ for $\epsilon > 0$ if $||x||_X < \delta$ for $\delta > 0$ sufficiently small. If $||x|| = \delta/2$ we therefore have $||Ax||_Y < \epsilon$ and thus, dividing by $\delta/2$, we also find $||Ax'||_Y < 2\epsilon/\delta$, where $||x'||_X = 1$. Multiplying for $\lambda > 0$, $||A\lambda x'||_Y < 2\lambda\epsilon/\delta$ which can be re-written $||Ax||_Y < 2\frac{\epsilon}{\delta}||x||$ for every $x \in X$, proving that A is bounded. □

For bounded operators, it is possible to define the **operator norm**,

$$||A|| := \sup_{0 \neq x \in X} \frac{||Ax||_Y}{||x||_X} \qquad \left(= \sup_{x \in X,\, ||x||_X = 1} ||Ax||_Y \right).$$

It is easy to prove that this is a true norm on the complex vector spaces $\mathfrak{B}(X, Y)$ of bounded operators $T : X \to Y$, with X, Y complex normed spaces.

An important elementary technical result is stated in the following proposition.

Proposition 2.2.11. *Let $A : S \to Y$ be a bounded operator defined on the subspace $S \subset X$, where X, Y ar enormed spaces with Y complete. If S is dense in X, then A can be continously extended to a bounded opertaor $A_1 : X \to Y$ with the same norm as A and this extension is unique.*

Proof. Uniqueness is obvious from continuity: if $S \ni x_n \to x \in X$ and A_1, A'_1 are continuous extensions, $A_1 x - A'_1 x = \lim_{n\to+\infty} A_1 x_n - A'_1 x_n = \lim_{n\to+\infty} 0 = 0$. Let us construct a linear contiuous extension. If $x \in X$ tere is a sequence $S \ni x_n \to x \in X$ since S is dense. $\{Ax_n\}_{n\in\mathbb{N}}$ is Cauchy because $\{x_n\}_{n\in\mathbb{N}}$ is and $||Ax_n - Ax_m||_Y \leq ||A||||x_n - x_m||_X$. So the limit $A_1 x := \lim_{n\to+\infty} Ax_n$ exists because Y is complete. The limit does not depend on the sequence: if $S \ni x'_n \to x$, then $||Ax_n - Ax'_n|| \leq ||A||||x_n - x'_n|| \to 0$, so A_1 is well defined. It is immediately proven that A_1 is linear form linearity of A and therefore A_1 is an operator extending A on the whole X. By construction $||A_1 x||_Y = \lim_{n\to+\infty} ||Ax_n||_Y \leq \lim_{n\to+\infty} ||A||||x_n||_X \leq ||A||||x||_X$, so $||A_1|| \leq ||A||$, in particular A_1 is bounded. On the other hand,

$$\begin{aligned} ||A_1|| &= \sup\{||A_1 x||\, ||x||^{-1} \mid x \in X \setminus \{0\}\} \geq \sup\{||A_1 x||\, ||x||^{-1} \mid x \in S \setminus \{0\}\} \\ &= \sup\{||Ax||\, ||x||^{-1} \mid x \in X \setminus \{0\}\} = ||A||\,, \end{aligned}$$

so that $||A_1|| \geq ||A||$ and therefore $||A_1|| = ||A||$. □

From now on, $\mathfrak{B}(\mathcal{H}) := \mathcal{B}(\mathcal{H}, \mathcal{H})$ denotes the set of bounded operators $A : \mathcal{H} \to \mathcal{H}$ over the complex Hilbert space $\mathcal{H}$. This set is a complex Hilbert space defining the linear combination of operators $\alpha A + \beta B \in \mathfrak{B}(\mathsf{H})$ for $\alpha, \beta \in \mathbb{C}$ and $A, B \in \mathfrak{B}(\mathsf{H})$ by $(\alpha A + \beta B)x := \alpha Ax + \beta Bx$ for every $x \in \mathcal{H}$.

$\mathfrak{B}(\mathcal{H})$ acquires the structure of a *unital Banach algebra*. The complex vector space structure is the standard one of operators, the associative algebra product is the composition of operators with unit given by I, and the norm being the above defined operator norm,

$$||A|| := \sup_{0 \neq x \in \mathcal{H}} \frac{||Ax||}{||x||} .$$

This definition of $||A||$ can be given also for an operator $A : D(A) \to \mathcal{H}$, if A is bounded and $D(A) \subset \mathcal{H}$ but $D(A) \neq \mathcal{H}$. It immediately arises that

$$||Ax|| \leq ||A|| \, ||x|| \quad \text{if } x \in D(A) .$$

As we already know, $|| \cdot ||$ is a norm over $\mathfrak{B}(\mathcal{H})$. Furthermore it satisfies

$$||AB|| \leq ||A|| \, ||B|| \quad A, B \in \mathfrak{B}(\mathcal{H}) .$$

It is also evident that $||I|| = 1$. Actually $\mathfrak{B}(\mathcal{H})$ is a Banach space so that: $\mathfrak{B}(\mathcal{H})$ *is a unital Banach algebra*. In fact, a fundamental result is the following theorem.

Theorem 2.2.12. *If $\mathcal{H}$ is a Hilbert space, $\mathfrak{B}(\mathcal{H})$ is a Banach space with respect to the norm of operators.*

Proof. The only non-trivial property is completeness of $\mathfrak{B}(\mathcal{H})$. Let us prove it. Consider a Cauchy sequence $\{T_n\}_{n \in \mathbb{N}} \subset \mathfrak{B}(\mathcal{H})$. We want to prove that there exists $T \in \mathfrak{B}(\mathcal{H})$ with $||T - T_n|| \to 0$ as $n \to +\infty$. Let us define $Tx = \lim_{n \to +\infty} Tx$ for every $x \in \mathcal{H}$. The limit exists becouse $\{T_n x\}_{n \in \mathbb{N}}$ is Cauchy from $||T_n x - T_m x|| \leq ||T_n - T_m|| \, ||x||$. Linearity of T is easy to prove from linearity of every T_n. Next observe that $||Tx - T_m x|| = || \lim_n T_n x - T_m x|| = \lim_n ||T_n x - T_m x|| \leq \epsilon ||x||$ is m is sufficiently large. Assuming that $T \in \mathfrak{B}(\mathcal{H})$, the found inequality, dividing by $||x||$ and taking the sup over x with $||x|| \neq 0$ proves that $||T - T_m|| \leq \epsilon$ and thus $||T - T_m|| \to 0$ for $m \to +\infty$ as wanted. This concludes the proof because $T \in \mathfrak{B}(\mathcal{H})$: $||Tx|| \leq ||Tx - T_m x|| + ||T_m x|| \leq \epsilon ||x|| + ||T_m|| ||x||$ and thus $||T|| \leq (\epsilon + ||T_m||) < +\infty$. □

Remark 2.2.13. *The result, with the same proof, is valid for the above defined complex vectorspace $\mathfrak{B}(X, Y)$, provided the normed space Y is $|| \cdot ||_Y$-complete. In* *particular the* **topological dual of** X $X' = \mathfrak{B}(X, \mathbb{C})$ *with* X *complex normed space, is always complete since* $\mathbb{C}$ *is complete.*

$\mathfrak{B}(\mathcal{H})$ is more strongly a unital C^*-algebra, if we introduce the notion of adjoint of an operator. To this end, we have the following general definition concerning also unbounded operators defined on non-maximal domains.

Definition 2.2.14. *Let* A *be a densely defined operator in the complex Hilbert space* $\mathcal{H}$. *Define the subspace of* $\mathcal{H}$,

$$D(A^\dagger) := \{y \in \mathcal{H} \mid \exists z_y \in \mathcal{H} \text{ s.t. } \langle y, Ax\rangle = \langle z_y, x\rangle \ \forall x \in D(A)\}\,.$$

The linear map $A^\dagger : D(A^\dagger) \ni y \mapsto z_y$ *is called the* **adjoint** *operator of* A.

Remark 2.2.15.

(a) *Above,* z_y *is uniquely determined by* y, *since* $D(A)$ *is dense. If both* z_y, z'_y *satisfy* $\langle y, Ax\rangle = \langle z_y, x\rangle$ *and* $\langle y, Ax\rangle = \langle z'_y, x\rangle$, *then* $\langle z_y - z'_y, x\rangle = 0$ *for every* $x \in D(A)$. *Taking a sequence* $D(A) \ni x_n \to z_y - z'_y$, *we conclude that* $||z_y - z'_y|| = 0$. *Thus* $z_y = z'_y$. *The fact that* $y \mapsto z_y$ *is linear can immediately be checked.*

(b) *By construction, we immediately have that*

$$\langle A^\dagger y, x\rangle = \langle y, Ax\rangle \quad \text{for } x \in D(A) \text{ and } y \in D(A^\dagger)$$

and also

$$\langle x, A^\dagger y\rangle = \langle Ax, y\rangle \quad \text{for } x \in D(A) \text{ and } y \in D(A^\dagger)\,,$$

if taking the complex conjugation of the former identity.

Exercise 2.2.16. Prove that $D(A^\dagger)$ can equivalently be defined as the set (subspace) of $y \in \mathcal{H}$ such that the linear functional $D(A) \ni x \mapsto \langle y, Ax\rangle$ is continuous.

Solution. It is a simple application of Riesz' lemma, after having uniquely extended $D(A) \ni x \mapsto \langle y, Ax\rangle$ to a continuous linear functional defined on $\overline{D(A)} = \mathcal{H}$ by continuity.

Remark 2.2.17.

(a) *If* A *is densely defined and* $A^\dagger$ *is also densely defined then* $A \subset (A^\dagger)^\dagger$. *The proof immediately follows form the definition of adjoint.*

(b) *If* A *is densely defined and* $A \subset B$ *then* $B^\dagger \subset A^\dagger$. *The proof immediately follows the definition of adjoint.*

(c) *If $A \in \mathfrak{B}(\mathcal{H})$ then $A^\dagger \in \mathfrak{B}(\mathcal{H})$ and $(A^\dagger)^\dagger = A$. Moreover*

$$||A^\dagger||^2 = ||A||^2 = ||A^\dagger A|| = ||AA^\dagger|| \,.$$

(d) *Directly from given definition of adjoint one has, for densely defined operators A, B on $\mathcal{H}$,*

$$A^\dagger + B^\dagger \subset (A+B)^\dagger \quad \textit{and} \quad A^\dagger B^\dagger \subset (BA)^\dagger \,.$$

Furthermore

$$A^\dagger + B^\dagger = (A+B)^\dagger \quad \textit{and} \quad A^\dagger B^\dagger = (BA)^\dagger \,, \tag{2.29}$$

whenever $B \in \mathfrak{B}(\mathcal{H})$ and A is densely defined.

(e) *From (c) and the last statement in (d) in particular, it is clear that $\mathfrak{B}(\mathcal{H})$ is a unital C^*-algebra with involution $\mathfrak{B}(\mathcal{H}) \ni A \mapsto A^\dagger \in \mathfrak{B}(\mathcal{H})$.*

Definition 2.2.18 (*-representation). *If $\mathfrak{A}$ is a (unital) *-algebra and $\mathcal{H}$ a Hilbert space, a* ***-representation** *on $\mathcal{H}$ is a *-homomorphism $\pi : \mathfrak{A} \to \mathfrak{B}(\mathcal{H})$ referring to the natural (unital) *-algebra structure of $\mathfrak{B}(\mathcal{H})$.*

Exercise 2.2.19. *Prove that $A^\dagger \in \mathfrak{B}(\mathcal{H})$ if $A \in \mathfrak{B}(\mathcal{H})$ and that, in this case $(A^\dagger)^\dagger = A$, $||A|| = ||A^\dagger||$ and $||A^\dagger A|| = ||AA^\dagger|| = ||A||^2$.*

Solution. If $A \in \mathfrak{B}(\mathcal{H})$, for every $y \in \mathcal{H}$, the linear map $\mathcal{H} \ni x \mapsto \langle y, Ax\rangle$ is continuous ($|\langle y, Ax\rangle| \leq ||y||\,||Ax|| \leq ||y||\,||A||\,||x||$) therefore Theorem 2.2.6 proves that there exists a unique $z_{y,A} \in \mathcal{H}$ with $\langle y, Ax\rangle = \langle z_{y,A}, x\rangle$ for all $x, y \in \mathcal{H}$. The map $\mathcal{H} \ni y \mapsto z_{y,A}$ is linear as consequence of the said uniqueness and the antilinearity of the left entry of scalar product. The map $\mathcal{H} \ni y \mapsto z_{y,A}$ fits the definition of $A^\dagger$, so it coincides with $A^\dagger$ and $D(A^\dagger) = \mathcal{H}$. Since $\langle A^\dagger x, y\rangle = \langle x, Ay\rangle$ for $x, y \in \mathcal{H}$ implies (taking the complex conjugation) $\langle y, A^\dagger x\rangle = \langle Ay, x\rangle$ for $x, y \in \mathcal{H}$, we have $(A^\dagger)^\dagger = A$. To prove that $A^\dagger$ is bounded, observe that $||A^\dagger x||^2 = \langle A^\dagger x|A^\dagger x\rangle = \langle x|AA^\dagger x\rangle \leq ||x||\,||A||\,||A^\dagger x||$ so that $||A^\dagger x|| \leq ||A||\,||x||$ and $||A^\dagger|| \leq ||A||$. Using $(A^\dagger)^\dagger = A$ we have $||A^\dagger|| = ||A||$. Regarding the last identity, it is evidently enough to prove that $||A^\dagger A|| = ||A||^2$. First of all, $||A^\dagger A|| \leq ||A^\dagger||\,||A|| = ||A||^2$, so that $||A^\dagger A|| \leq ||A||^2$. On the other hand, $||A||^2 = (\sup_{||x||=1} ||Ax||)^2 = \sup_{||x||=1} ||Ax||^2 = \sup_{||x||=1}\langle Ax|Ax\rangle = \sup_{||x||=1}\langle x|A^\dagger Ax\rangle \leq \sup_{||x||=1} ||x||||A^\dagger Ax|| = \sup_{||x||=1} ||A^\dagger Ax|| = ||A^\dagger A||$. We have found that $||A^\dagger A|| \leq ||A||^2 \leq ||A^\dagger A||$ so that $||A^\dagger A|| = ||A||^2$.

Definition 2.2.20. *Let A be an operator in the complex Hilbert space $\mathcal{H}$.*

(1) *A is said to be* **closed** *if the* **graph** *of A, that is the set*

$$G(A) := \{(x, Ax) \subset \mathcal{H} \times \mathcal{H} \mid x \in D(A)\},$$

is closed in the product topology of $\mathcal{H} \times \mathcal{H}$.

(2) *A is* **closable** *if it admits extensions in terms of closed operators. This is equivalent to saying that the closure of the graph of A is the graph of an operator, denoted by $\overline{A}$, and called the* **closure** *of A.*

(3) *If A is closable, a subspace $S \subset D(A)$ is called* **core** *for A if $\overline{A|_S} = \overline{A}$.*

Referring to (2), we observe that, given an operator A, we can always define the closure of the graph $\overline{G(A)}$ in $\mathcal{H} \times \mathcal{H}$. In general, this closure is not the graph of an operator because there could exist sequences $D(A) \ni x_n \to x$ and $D(A) \ni x'_n \to x$ such that $Tx_n \to y$ and $Tx_n \to y'$ with $y \neq y'$. However, both pairs (x, y) and (x, y') belong to $\overline{G(A)}$. If this case does not take place – and this is equivalent to condition (a) below when making use of linearity – $\overline{G(A)}$ is the graph of an operator, denoted by $\overline{A}$, that is closed by definition. Therefore A admits closed operatorial extensions: at least $\overline{A}$. If, conversely, A admits extensions in terms of closed operators, the intersection of the (closed) graphs of all these extensions $\overline{G(A)}$ is still closed and it is still the graph of an operator again, which must councide with $\overline{A}$ by definition.

Remark 2.2.21.

(a) *Directly from the definition and using linearity, A is closable if and only if there are no sequences of elements $x_n \in D(A)$ such that $x_n \to 0$ and $Ax_n \to y$ with $y \neq 0$ as $n \to +\infty$. Since $\overline{G(A)}$ is the union of $G(A)$ and its accumulation points in $\mathcal{H} \times \mathcal{H}$ and, if A is closable, it is also the graph of the operator $\overline{A}$, we conclude that*

(i) *$D(\overline{A})$ is made of the elements $x \in \mathcal{H}$ such that $x_n \to x$ and $Ax_n \to y_x$ for some sequences $\{x_n\}_{n\in\mathbb{N}} \subset D(A)$ and some $y_x \in D(A)$ and*

(ii) *$\overline{A}x = y_x$.*

(b) *As a consequence of (a) one has that, if A is closable, then $aA + bI$ is closable and $\overline{aA + bI} = a\overline{A} + bI$ for every $a, b \in \mathbb{C}$.* **N.B.** This result generally fails if replacing I for some closable operator B.

(c) *Directly from the definition, A is closed if and only if $D(A) \ni x_n \to x \in \mathcal{H}$ and $Ax_n \to y \in \mathcal{H}$ imply both $x \in D(A)$ and $y = Ax$.*

A useful porposition is the following.

Proposition 2.2.22. *Consider an operator $A : D(A) \to \mathcal{H}$ with $D(A)$ dense in the Hilbert space $\mathcal{H}$. The following facts hold.*

(a) *$A^\dagger$ is closed.*
(b) *A is closable if and only if $D(A^\dagger)$ is dense and, in this case, $\overline{A} = (A^\dagger)^\dagger$.*

Proof. Define the Hermitian scalar product $((x,y)|(x'y')) := \langle x, x'\rangle + \langle y, y'\rangle$ on the vector space defined as the standard direct sum of vector spaces, we henceforth denote by $\mathcal{H} \oplus \mathcal{H}$. It is immediately proven that $\mathcal{H} \oplus \mathcal{H}$ becomes a Hilbert space when equipped with that scalar product. Next define the operator

$$\tau : \mathcal{H} \oplus \mathcal{H} \ni (x,y) \mapsto (-y,x) \in \mathcal{H} \oplus \mathcal{H}\,.$$

It is easy to check that $\tau \in \mathfrak{B}(\mathcal{H} \oplus \mathcal{H})$ and also (referring the adjoint to the space $\mathcal{H} \oplus \mathcal{H}$),

$$\tau^\dagger = \tau^{-1} = -\tau\,. \tag{2.30}$$

Finally, by direct computation, one sees that τ and $^\perp$ (referred to $\mathcal{H} \oplus \mathcal{H}$ with the said scalar product) commute

$$\tau(F^\perp) = (\tau(F))^\perp \quad \text{if } F \subset \mathcal{H} \oplus \mathcal{H}\,. \tag{2.31}$$

Let us pass to prove (a). The following noticeable identity holds true, for an operator $A : D(A) \to \mathcal{H}$ with $D(A)$ dense in $\mathcal{H}$ (so that $A^\dagger$ is defined)

$$G(A^\dagger) = \tau(G(A))^\perp\,. \tag{2.32}$$

Since the right hand side is closed (it being the orthogonal space of a set), the graph of $A^\dagger$ is closed and $A^\dagger$ is therefore closed by definition. To prove (2.32), observe that in view of the definition of τ,

$$\tau(G(A))^\perp = \{(y,z) \in \mathcal{H} \oplus \mathcal{H} \mid ((y,z)|(-Ax,x)) = 0\,, \forall x \in D(A)\}$$

that is

$$\tau(G(A))^\perp = \{(y,z) \in \mathcal{H} \oplus \mathcal{H} \mid \langle y, Ax\rangle = \langle z, x\rangle\,, \forall x \in D(A)\}$$

Since $A^\dagger$ exists, the pairs $(y,z) \in \tau(G(A))^\perp$ can be written $(y, A^\dagger y)$ according to the definition of $A^\dagger$. Therefore $\tau(G(A))^\perp = G(A^\dagger)$ proving (a).

(b) From the properties of $^\perp$ we immediately have $\overline{G(D(A))} = (G(A)^\perp)^\perp$. Since τ and $^\perp$ commute by (2.31), and $\tau\tau = -I$ (2.30),

$$\overline{G(D(A))} = -\tau \circ \tau((G(A)^\perp)^\perp) = -\tau(\tau(G(A))^\perp)^\perp = \tau(\tau(G(A))^\perp)^\perp = \tau(G(A^\dagger))^\perp,$$

where we have omitted the minus sign since it appears in front of a subspace that, by definition, is closed with respect multiplication by scalars and we make use of (2.32). Now suppose that $D(A^\dagger)$ is dense so that $(A^\dagger)^\dagger$ exists. In this case, taking advantage of (2.32) again, we have $\overline{G(A)} = G((A^\dagger)^\dagger)$. Notice that the right-hand side is the graph of an operator so we have obtained that, if $D(A^\dagger)$ is dense, then A is closable. In this case, by definition of closure of an operator, we also have $\overline{A} = (A^\dagger)^\dagger$.

Vice versa, suppose that A is closable, so the operator $\overline{A}$ exists and $G(\overline{A}) = \overline{G(A)}$. In this case $\tau(G(A^\dagger))^\perp = \overline{G(A)}$ is the graph of an operator and thus its graph cannot include pairs $(0, y)$ with $y \neq 0$ by linearity. In other words, if $(0, y) \in \tau(G(A^\dagger))^\perp$ then $y = 0$. This is the same as saying that $((0, y)|(-A^\dagger x, x)) = 0$ for all $x \in D(A^\dagger)$ implies $y = 0$. Summing up, $\langle y|x\rangle = 0$ for all $x \in D(A^\dagger)$ implies $y = 0$. Since $\mathcal{H} = D(A^\dagger)^\perp \oplus (D(A^\dagger)^\perp)^\perp = D(A^\dagger)^\perp \oplus \overline{D(A^\dagger)}$, we conclude that $\overline{D(A^\dagger)} = \mathcal{H}$, that is $D(A^\dagger)$ is dense. □

An immediate corollary follows.

Corollary 2.2.23. *Let $A : D(A) \to \mathcal{H}$ be an operator in the Hilbert space $\mathcal{H}$. If both $D(A)$ and $D(A^\dagger)$ are densely defined then*

$$A^\dagger = \overline{A}^\dagger = \overline{A^\dagger} = (((A^\dagger)^\dagger)^\dagger .$$

The Hilbert space version of the **closed graph theorem** holds (e.g., see [5]).

Theorem 2.2.24 (Closed graph theorem). *Let $A : \mathcal{H} \to \mathcal{H}$ be an operator, $\mathcal{H}$ being a complex Hilbert space. A is closed if and only if $A \in \mathfrak{B}(\mathcal{H})$.*

Exercise 2.2.25. *Prove that, if $B \in \mathfrak{B}(\mathcal{H})$ and A is a closed operator in $\mathcal{H}$ such that $Ran(B) \subset D(A)$, then $AB \in \mathfrak{B}(\mathcal{H})$.*

Solution. AB is well defined by hypothesis and $D(AB) = \mathcal{H}$. Exploiting (c) in remark 2.2.21 and continuity of B, one easily finds that AB is closed as well. Theorem 2.2.24 finally proves that $AB \in \mathfrak{B}(\mathcal{H})$.

Definition 2.2.26. *An operator A in the complex Hilbert space $\mathcal{H}$ is said to be*

(0) **Hermitean** *if $\langle Ax, y\rangle = \langle x, Ay\rangle$ for $x, y \in D(A)$;*
(1) **symmetric** *if it is densely defined and Hermitian, which is equivalent to say that $A \subset A^\dagger$;*
(2) **selfadjoint** *if it is symmetric and $A = A^\dagger$;*
(3) **essentially selfadjoint** *if it is symmetric and $(A^\dagger)^\dagger = A^\dagger$;*

(4) **unitary** *if* $A^\dagger A = AA^\dagger = I$;
(5) **normal** *if it is closed, densely defined and* $AA^\dagger = A^\dagger A$.

Remark 2.2.27.

(a) *If A is unitary then $A, A^\dagger \in \mathfrak{B}(\mathcal{H})$. Furthermore $A : \mathcal{H} \to \mathcal{H}$ is unitary if and only if it is surjective and norm preserving. (See the exercises 2.2.31 below). These operators are nothing but the* **automorphisms** *of the given Hilbert space. Considering two Hilbert spaces $\mathcal{H}, \mathcal{H}'$* **isomorphisms** *are linear maps $T : \mathcal{H} \to \mathcal{H}'$ which are isometric and surjecitve. Notice that T also preserve the scalar products in view of Proposition 2.2.1.*

(b) *A selfadjoint operator A does not admit proper symmetric extensions and essentially selfadjoint operators admit only one self-adjoiunt extension. (See Proposition 2.2.28 below).*

(c) *A symmetric operator A is always closable because $A \subset A^\dagger$ and $A^\dagger$ is closed (Proposition 2.2.22), moreover for that operator the following conditions are equivalent in view of Proposition 2.2.22 and Corollary 2.2.23, as the reader immediately proves:*

(i) $(A^\dagger)^\dagger = A^\dagger$ *(A is essentially selfadjoint);*
(ii) $\overline{A} = A^\dagger$;
(iii) $\overline{A} = (\overline{A})^\dagger$.

(d) *Unitary and selfadjoint operators are cases of normal operators.*

The pair of elementary results of (essentially) selfadjoint operators stated in (b) are worth to be proved.

Proposition 2.2.28. *Let $A : D(A) \to \mathcal{H}$ be a densely defined operator in the Hilbert space $\mathcal{H}$. The following facts are true.*

(a) *If A is selfadjoint, then A does not admit proper symmetric extensions.*
(b) *If A is essentially selfadjoint, then A admits a unique selfadjoint extension, and that this extension is $A^\dagger$.*

Proof. **(a)** Let B be a symmetric extension of A. $A \subset B$ then $B^\dagger \subset A^\dagger$ for (b) in remark 2.2.17. As $A = A^\dagger$ we have $B^\dagger \subset A \subset B$. Since $B \subset B^\dagger$, we conclude that $A = B$.

(b) Let B be a selfadjoint extension of the essentially selfadjoint operator A, so that $A \subset B$. Therefore $A^\dagger \supset B^\dagger = B$ and $(A^\dagger)^\dagger \subset B^\dagger = B$. Since A is essentially selfadjoint, we have found $A^\dagger \subset B$. Here $A^\dagger$ is selfadjoint and B is symmetric because selfadjoint. (b) implies $A^\dagger = B$. That is, every selfadjoint extension of A coincides with $A^\dagger$. □

Another elementary though important result, helping understand why in QM observables are very often described by selfadjoint operators which are unbounded and defined in proper subspaces, is the following proposition (see (c) in remark 2.2.60).

Theorem 2.2.29 (Hellinger-Toepliz theorem). *Let A be a selfadjoint operator in the complex Hilbert space $\mathcal{H}$. A is bounded if and only if $D(A) = \mathcal{H}$ (thus $A \in \mathfrak{B}(\mathcal{H})$).*

Proof. Assume $D(A) = \mathcal{H}$. As $A = A^\dagger$ we have $D(A^\dagger) = \mathcal{H}$. Since $A^\dagger$ is closed, Theorem 2.2.24 implies the $A^\dagger (= A)$ is bounded. Conversely, if $A = A^\dagger$ is bounded, since $D(A)$ is dense, we can continuously extend it to a bounded operator $A_1 : \mathcal{H} \to \mathcal{H}$. That extension, by continuity, trivially satisfies $\langle A_1 x|y\rangle = \langle x|A_1 y\rangle$ for all $x, y \in \mathcal{H}$ thus A_1 is symmetric. Since $A^\dagger = A \subset A_1 \subset A_1^\dagger$, (a) in Proposition 2.2.28 implies $A = A_1$. □

Let us pass to focus on unitary operators. The relevance of unitary operators is evident from the following proposition where it is proven that they preserve the nature of operators with respect to the Hermitian conjugation.

Proposition 2.2.30. *Let $U : \mathcal{H} \to \mathcal{H}$ be a unitary operator in the complex Hilbert space $\mathcal{H}$ and A another operator in $\mathcal{H}$. Prove that $UAU^\dagger$ with domain $U(D(A))$ (resp. $U^\dagger AU$ with domain $U^\dagger(D(A))$) is symmetric, selfadjoint, essentially selfadjoint, unitary, normal if A is respectively symmetric, selfadjoint, essentially selfadjoint, unitary, normal.*

Proof. Since $U^\dagger$ is unitary when U is and $(U^\dagger)^\dagger = U$, it is enough to establish the thesis for $UAU^\dagger$. First of all, notice that $D(UAU^\dagger) = U(D(A))$ is dense if $D(A)$ is dense since U is bijective and isometric and $U(D(A)) = \mathcal{H}$ if $D(A) = \mathcal{H}$ because U is bijective. By direct inspection, applying the definition of adjoint operator, one sees that $(UAU^\dagger)^\dagger = UA^\dagger U^\dagger$ and $D((UAU^\dagger)^\dagger) = U(D(A^\dagger))$. Now, if A is symmetric $A \subset A^\dagger$ which implies $UAU^\dagger \subset UA^\dagger U^\dagger = (UAU^\dagger)^\dagger$ so that $UAU^\dagger$ is symmetric as well. If A is selfadjoint $A = A^\dagger$ which implies $UAU^\dagger = UA^\dagger U^\dagger = (UAU^\dagger)^\dagger$ so that $UAU^\dagger$ is selfadjoint as well. If A is essentially selfadjoint it is symmetric and $(A^\dagger)^\dagger = A^\dagger$, so that $UAU^\dagger$ is symmetric and $U(A^\dagger)^\dagger U^\dagger = UA^\dagger U^\dagger$ that is $(UA^\dagger U^\dagger)^\dagger = UA^\dagger U^\dagger$ which means $((UAU^\dagger)^\dagger)^\dagger = (UAU^\dagger)^\dagger$ so that $UA^\dagger U^\dagger$ is essentially selfadjoint. If A is unitary, we have $A^\dagger A = AA^\dagger = I$ so that $UA^\dagger AU^\dagger = UAA^\dagger U^\dagger = UU^\dagger$ which, since

$U^\dagger U = I = UU^\dagger$, is equivalent to $UA^\dagger U^\dagger UAU^\dagger = UAU^\dagger UA^\dagger U^\dagger = U^\dagger U = I$, that is $(UA^\dagger U^\dagger)UAU^\dagger = (UAU^\dagger)UA^\dagger U^\dagger = I$ and thus $UAU^\dagger$ is unitary as well. If A is normal, $UAU^\dagger$ is normal too, with the same reasoning as in the unitary case. $\square$

Exercise 2.2.31.

(1) *Prove that if A is unitary then $A, A^\dagger \in \mathfrak{B}(\mathcal{H})$.*

Solution. It holds $D(A) = D(A^\dagger) = D(I) = \mathcal{H}$ and $||Ax||^2 = \langle Ax, Ax\rangle = \langle x, A^\dagger Ax\rangle = ||x||^2$ if $x \in \mathcal{H}$, so that $||A|| = 1$. Due to (c) in remark 2.2.17, $A^\dagger \in \mathfrak{B}(\mathcal{H})$.

(2) *Prove that $A : \mathcal{H} \to \mathcal{H}$ is unitary if and only if is surjective and norm preserving.*

Solution. If A is unitary ((3) Def 2.2.26), it is evidently bijective, moreover as $D(A^\dagger) = \mathcal{H}$, $||Ax||^2 = \langle Ax, Ax\rangle = \langle x, A^\dagger Ax\rangle = \langle x, x\rangle = ||x||^2$, so A is isometric. If $A : \mathcal{H} \to \mathcal{H}$ is isometric its norm is 1 and thus $A \in \mathfrak{B}(\mathcal{H})$. Therefore $A^\dagger \in \mathfrak{B}(\mathcal{H})$. The condition $||Ax||^2 = ||x||^2$ can be re-written $\langle Ax, Ax\rangle = \langle x, A^\dagger Ax\rangle = \langle x, x\rangle$ and thus $\langle x, (A^\dagger A - I)x\rangle = 0$ for $x \in \mathcal{H}$. Using $x = y \pm z$ and $x = y \pm iz$, the found indentity implies $\langle z, (A^\dagger A - I)y\rangle = 0$ for all $y, z \in \mathcal{H}$. Taking $z = (A^\dagger A - I)y$, we finally have $||(A^\dagger A - I)y|| = 0$ for all $y \in \mathcal{H}$ and thus $A^\dagger A = I$. In particular, A is injective as it admits the left inverse $A^\dagger$. Since A is also surjective, it is bijective and thus its left inverse $(A^\dagger)$ is also a right inverse, that is $AA^\dagger = I$.

(3) *Prove that, if $A : \mathcal{H} \to \mathcal{H}$ satisfies $\langle x, Ax\rangle \in \mathbb{R}$ for all $x \in \mathcal{H}$ (and in particular if $A \geq 0$, which means $\langle x, Ax\rangle \geq 0$ for all $x \in \mathcal{H}$), then $A^\dagger = A$ and $A \in \mathfrak{B}(\mathcal{H})$.*

Solution. We have $\langle x, Ax\rangle = \langle x, Ax\rangle^* = \langle Ax, x\rangle = \langle x, A^\dagger x\rangle$ where, as $D(A) = \mathcal{H}$, the adjoint $A^\dagger$ is well defined everywhere on $\mathcal{H}$. Thus $\langle x, (A - A^\dagger)x\rangle = 0$ for every $x \in \mathcal{H}$. Using $x = y \pm z$ and $x = y \pm iz$, we obtain $\langle y, (A - A^\dagger)z\rangle = 0$ for all $y, z \in \mathcal{H}$. Choosing $y = (A - A^\dagger)z$, we conclude that $A = A^\dagger$. Theorem 2.2.29 concludes the proof.

Example 2.2.32. The **Fourier transform**, $\mathcal{F} : \mathcal{S}(\mathbb{R}^n) \to \mathcal{S}(\mathbb{R}^n)$, defined as[6]

$$(\mathcal{F}f)(k) := \frac{1}{(2\pi)^{n/2}} \int_{\mathbb{R}^n} e^{-ik\cdot x} f(x) d^n x \tag{2.33}$$

($k \cdot x$ being the standard $\mathbb{R}^n$ scalar product of k and x) is a bijective linear map with inverse $\mathcal{F}_- : \mathcal{S}(\mathbb{R}^n) \to \mathcal{S}(\mathbb{R}^n)$,

$$(\mathcal{F}_- g)(x) := \frac{1}{(2\pi)^{n/2}} \int_{\mathbb{R}^n} e^{ik\cdot x} g(k) d^n k\,. \tag{2.34}$$

[6]In QM, adopting units with $\hbar \neq 1$, $k \cdot x$ has to be replaced for $\frac{k\cdot x}{\hbar}$ and $(2\pi)^{n/2}$ for $(2\pi\hbar)^{n/2}$.

Both $\mathcal{F}$ and $\mathcal{F}_-$ preserve the scalar product

$$\langle \mathcal{F}f, \mathcal{F}g\rangle = \langle f, g\rangle\,, \quad \langle \mathcal{F}_- f, \mathcal{F}_- g\rangle = \langle f, g\rangle, \quad \forall f, g \in \mathcal{S}(\mathbb{R}^n)$$

and thus they also preserve the norm of $L^2(\mathbb{R}^n, d^n x)$, in particular, $||\mathcal{F}|| = ||\mathcal{F}_-|| = 1$. As a consequence of Proposition 2.2.11, using the fact that $\mathcal{S}(\mathbb{R}^n)$ is dense in $L^2(\mathbb{R}^n, d^n x)$, one easily proves that $\mathcal{F}$ and $\mathcal{F}_-$ uniquely continuously extend to unitary operators, respectively, $\hat{\mathcal{F}} : L^2(\mathbb{R}^n, d^n x) \to L^2(\mathbb{R}^n, d^n k)$ and $\hat{\mathcal{F}}_- : L^2(\mathbb{R}^n, d^n k) \to L^2(\mathbb{R}^n, d^n x)$ such that $\hat{\mathcal{F}}^\dagger = \hat{\mathcal{F}}^{-1} = \hat{\mathcal{F}}_-$. The map $\hat{\mathcal{F}}$ is the **Fourier-Plancherel** (unitary) **operator**.

2.2.3 *Criteria for (essential) selfadjointness*

Let us briefly introduce, without proofs (see [5]), some commonly used mathematical tecnology to study (essential) selfadjointness of symmetric operators. If A is a densely defined symmetric operator in the complex Hilbert space $\mathcal{H}$, define the **deficiency indices**, $n_\pm := dim\mathcal{H}_\pm$ (cardinal numbers in general) where $\mathcal{H}_\pm$ are the (closed) subspaces of the solutions of $(A^\dagger \pm iI)x_\pm = 0$ [8; 5; 9].

Proposition 2.2.33. *If A is a densely defined symmetric operator in the complex Hilbert space $\mathcal{H}$ the following holds.*

(a) *A is essentially selfadjoint (thus it admits an unique selfadjoint extension) if $n_\pm = 0$, that is $\mathcal{H}_\pm = \{0\}$.*
(b) *A admits selfadjoint extensions if and only if $n_+ = n_-$ and these extension are labelled by means of n_+ parameters.*

Remark 2.2.34. *If A is symmetric, an easy sufficient condition, due to von Neumann, for $n_+ = n_-$ is that $CA \subset AC$ where $C : \mathcal{H} \to \mathcal{H}$ is a* **conjugation** *that is an isometric surjective* antilinear[7] *map with $CC = I$.*

Indeed, using the definition of $A^\dagger$ and $D(A^\dagger)$ and observing that (from the polarization identity 2.26) $\langle Cy|Cx\rangle = \overline{\langle y|x\rangle}$, the condition

$$CA \subset AC \quad \textit{implies the condition} \quad CA^\dagger \subset A^\dagger C\,.$$

Therefore $A^\dagger x = \pm ix$ if and only if $A^\dagger Cx = C(\pm ix) = \mp iCx$. Since C preserves normality and norm of vectors, we conclude that $n_+ = n_-$.

Taking C as the standard conjugation of functions in $L^2(\mathbb{R}^n, d^n x)$, this result proves in particular that all operators in QM of the Schördinger *form as (2.24) admit selfadjoint extensions when defined on dense domains.*

[7] In other words $C(\alpha x + \beta y) = \overline{\alpha}Cx + \overline{\beta}Cy$ if $\alpha, \beta \in \mathbb{C}$ and $x, y \in \mathcal{H}$.

Exercise 2.2.35. *Prove that a symmetric operator that admits a unique selfadjoint extension is necessarily essentially selfadjoint.*

Solution. By (b) of Proposition 2.2.33, $n_+ = n_-$. If $n_\pm \neq 0$ there are many selfadjoint extension. The only possibility for the uniqueness of the selfadjoint extension is $n_\pm = 0$. (a) of Proposition 2.2.33 implies that A is essentially selfadjoint.

A very useful criterion to establish the essentially selfadjointness of a symmetric operator is due to Nelson. It relies upon an important definition.

Definition 2.2.36. Let A be an operator in the complex Hilbert space $\mathcal{H}$. If $\psi \in \cap_{n\in\mathbb{N}} D(A^n)$ satisfies

$$\sum_{n=0}^{+\infty} \frac{t^n}{n!} ||A^n\psi|| < +\infty \quad \text{for some } t > 0\,,$$

then ψ is said to be an **analytic vector** of A.

We can state Nelson's criterion here [5].

Theorem 2.2.37 (Nelson's essentially selfadjointness criterium). *Let A be a symmetric operator in the complex Hilbert space $\mathcal{H}$, A is essentially selfadjoint if $D(A)$ contains a dense set D of analytic vectors (or – which is equivalent – a set D of analytic vectors whose finite span dense in $\mathcal{H}$).*

The above equivalence is due to the fact that a finite linear combination of analytic vector is an analytic vector as well, the proof being elementary. We have the following evident corollary.

Corollary 2.2.38. *If A is a symmetric operator admitting a Hilbertian basis of eigenvectors in $D(A)$, then A is essentially selfadjoint.*

Example 2.2.39.
(1) For $m \in \{1, 2, \ldots, n\}$, consider the operators X'_m and X''_m in $L^2(\mathbb{R}^n, d^n x)$ with dense domains $D(X'_m) = C_0^\infty(\mathbb{R}^n; \mathbb{C})$, $D(X''_m) = \mathcal{S}(\mathbb{R}^n)$ for $x \in \mathbb{R}^n$ and, for ψ, ϕ in the respective domains,

$$(X'_m\psi)(x) := x_m\psi(x)\,, \quad (X''_m\phi)(x) := x_m\phi(x)\,,$$

where $x_m$is the m-th component of $x \in \mathbb{R}^n$. Both operators are symmetric but not selfadjoint. They admit selfadjoint extensions because they commute with the standard complex conjugation of functions (see remark 2.2.34). It is furthermore possible to prove that both operators are essentially selfadjoint as follows. First define the k-**axis position operator** X_m in $L^2(\mathbb{R}^n, d^n x)$ with domain

$$D(X_m) := \left\{ \psi \in L^2(\mathbb{R}^n, d^n x) \,\middle|\, \int_{\mathbb{R}^n} |x_m \psi(x)|^2 d^k n \right\}$$

and

$$(X_m \psi)(x) := x_m \psi(x)\,, \quad x \in \mathbb{R}^n\,. \tag{2.35}$$

Just by applying the definition of adjoint, one sees that $X_m^\dagger = X_m$ so that X_m is selfdjoint [5]. Again applying the definition of adjoint, one sees (see below) that ${X'_m}^\dagger = {X''_m}^\dagger = X_m$ where we know that the last one is selfadjoint. By definition, X'_m and X''_m are therefore essentially selfadjoint. By (b) in proposition 2.2.28 X'_m and X''_m admit a unique selfadjoint extension which must coincide with X_m itself. We conclude that $C_0^\infty(\mathbb{R}^n; \mathbb{C})$ and $\mathcal{S}(\mathbb{R}^n)$ are *cores* (Def. 2.2.20) for the m-axis position operator.

Let us prove that for ${X'_m}^\dagger = X_m$, the proof for ${X''_m}^\dagger$ is identitcal. By direct inspection, one easily sees that ${X'_m}^\dagger \subset X_m$. Let us prove the converse inclusion. We have that $\phi \in D({X'_m}^\dagger)$ if and only if there exists $\eta_\phi \in L^2(\mathbb{R}^n, d^n x)$ such that $\int \overline{\phi(x)} x_m \psi(x) dx = \int \overline{\eta_\phi(x)} \psi(x) dx$, that is $\int \overline{\phi(x) x_m - \eta_\phi(x)} \psi(x) dx = 0$, for every $\psi \in C_0^\infty(\mathbb{R}^n)$. Fix a compact $K \subset \mathbb{R}^n$, obviously $K \ni x \mapsto \phi(x) x_m - \eta_\phi(x)$ is $L^2(K, dx)$. Since we can $L^2(K)$-approximate that function with a sequence of $\psi_n \in C_0^\infty(\mathbb{R}^n)$ such that $supp(\psi_n) \subset K$, we conclude that $K \ni x \mapsto \phi(x) x_m - \eta_\phi(x)$ is zero a.e.. Since K was arbitrary, we conclude that $\mathbb{R}^n \ni x \mapsto \phi(x) x_m = \eta_\phi(x)$ a.e. In particular, both ϕ and $\mathbb{R}^n \ni x \mapsto x_m \phi(x)$ are $L^2(\mathbb{R}^n, dx)$ (the latter because it is a.e. identical to $\eta_\phi \in L^2(\mathbb{R}^n, dx)$), namely, $D({X'_m}^\dagger) \ni \phi$ implies $\phi \in D(X_m)$. This proves that $D({X'_m}^\dagger) \subset D(X_m)$ and consequently ${X'_m}^\dagger \subset X_m$ as wanted.

(2) For $m \in \{1, 2, \ldots, n\}$, the k-**axis momentum operator**, P_m, is obtained from the position operator using the Fourier-Plancherel unitary operator $\hat{\mathcal{F}}$ introduced in example 2.2.32.

$$D(P_m) := \left\{ \psi \in L^2(\mathbb{R}^n, d^n x) \,\middle|\, \int_{\mathbb{R}^n} |k_m (\hat{\mathcal{F}}\psi)(k)|^2 d^n k \right\}$$

and

$$(P_m \psi)(x) := (\hat{\mathcal{F}}^\dagger K_m \hat{\mathcal{F}} \psi)(x)\,, \quad x \in \mathbb{R}^n\,. \tag{2.36}$$

Above K_m is the m-axis *position operator* just written for functions (in $L^2(\mathbb{R}^n, d^n k)$) whose variable, for pure convenience, is denoted by k instead of x. Since K_m is selfadjoint, P_m is selfadjoint as well, as established in Proposition 2.2.30 as a consequence of the fact that $\hat{\mathcal{F}}$ is unitary.

It is possible to give a more explicit form to P_m if restricting its domain. Taking $\psi \in C_0^\infty(\mathbb{R}^n; \mathbb{C}) \subset \mathcal{S}(\mathbb{R}^n)$ or directly $\psi \in \mathcal{S}(\mathbb{R}^n)$, $\hat{\mathcal{F}}$ reduces to the standard integral

Fourier transform (2.33) with inverse (2.34). Using these integral expressions we easily obtain

$$(P_m\psi)(x) = (\hat{\mathcal{F}}^\dagger K_m \hat{\mathcal{F}}\psi)(x) = -i\frac{\partial}{\partial x_m}\psi(x) \tag{2.37}$$

because in $\mathcal{S}(\mathbb{R}^n)$, which is invariant under the Fourier (and inverse Fourier) integral transformation,

$$\int_{\mathbb{R}^n} e^{ik\cdot x}k_m(\mathcal{F}\psi)(k)d^n k = -i\frac{\partial}{\partial x_m}\int_{\mathbb{R}^n} e^{-ik\cdot x}(\mathcal{F}\psi)(k)d^n k\,.$$

This way leads us to consider the operators P'_m and P''_m in $L^2(\mathbb{R}^n, d^n x)$ with

$$D(P'_m) = C_0^\infty(\mathbb{R}^n;\mathbb{C})\,, \quad D(P''_m) = \mathcal{S}(\mathbb{R}^n)$$

and, for $x \in \mathbb{R}^n$ and ψ, ϕ in the respective domains,

$$(P'_m\psi)(x) := -i\frac{\partial}{\partial x_m}\psi(x)\,, \quad (P''_m\phi)(x) := -i\frac{\partial}{\partial x_m}\phi(x)\,.$$

Both operators are symmetric as one can easily prove by integrating by parts, but not selfadjoint. They admit selfadjoint extensions because they commute with the conjugation $(C\psi)(x) = \overline{\psi(-x)}$ (see remark 2.2.34). It is furthermore possible to prove that both operators are essentially selfadjoint by direct use of Proposition 2.2.33 [5]. However we already know that P''_m is essentially selfadjoint as it coincides with the essentially selfadjoint operator $\hat{\mathcal{F}}^\dagger K''_m\hat{\mathcal{F}}$ beacause $\mathcal{S}(\mathbb{R}^n)$ is invariant under $\hat{\mathcal{F}}$.

The unique selfadjoint extension of both operators turns out to be P_m. We conclude that $C_0^\infty(\mathbb{R}^n;\mathbb{C})$ and $\mathcal{S}(\mathbb{R}^n)$ are *cores* for the m-axis momentum operator.

With the given definitions of selfadjoint operators, X_k and P_k, $\mathcal{S}(\mathbb{R}^n)$ turn out to be an invariant domain and thereon the CCR (2.21) hold rigorously.

As a final remark to conclude, we say that, if $n = 1$, $D(P)$ coincides to the already introduced domain (2.17). In that domain, P is nothing but the weak derivative times the factor $-i$.

(3) The most elementary example of application of Nelson's criterion is in $L^2([0,1], dx)$. Consider $A = -\frac{d^2}{dx^2}$ with dense domain $D(A)$ given by the functions in $C^\infty([0,1];\mathbb{C})$ such that $\psi(0) = \psi(1)$ and $\frac{d\psi}{dx}(0) = \frac{d\psi}{dx}(1)$. A is symmetric thereon as it arises immediately by integration by parts, in particular its domain is dense since it includes the Hilbert basis of exponentials $e^{i2\pi n x}$, $n \in \mathbb{Z}$, which are eigenvectors of A. Thus A is also essentially selfadjoint on the above domain.

A more interesting case is the **Hamiltonian operator of the harmonic oscillator**, H (see the first part) obtained as follows. One starts by

$$H_0 = -\frac{1}{2m}\frac{d^2}{dx^2} + \frac{m\omega^2}{2}x^2$$

with $D(H_0) := \mathcal{S}(\mathbb{R})$. Above, x^2 is the multiplicative operator and $m, \omega > 0$ are constants.

To go on, we start by defining a triple of operators, $a, a^+, \mathcal{N} : \mathcal{S}(\mathbb{R}) \to L^2(\mathbb{R}, dx)$ as

$$a^+ := \sqrt{\frac{m\omega}{2\hbar}} \left(x - \frac{\hbar}{m\omega} \frac{d}{dx} \right), \quad a := \sqrt{\frac{m\omega}{2\hbar}} \left(x + \frac{\hbar}{m\omega} \frac{d}{dx} \right), \quad \mathcal{N} := a^+ a\,.$$

a and a^+ are called creation and annihilation operators. These operators have the same domain which is also invariant: $a(\mathcal{S}(\mathbb{R})) \subset \mathcal{S}(\mathbb{R})$, $a^+(\mathcal{S}(\mathbb{R})) \subset \mathcal{S}(\mathbb{R})$, $\mathcal{N}(\mathcal{S}(\mathbb{R})) \subset \mathcal{S}(\mathbb{R})$. It is also easy to see, using integration by parts that $a^+ \subset a^\dagger$ and that $\mathcal{N}$ (called **number operator**) is Hermitian an also symmetric because $\mathcal{S}(\mathbb{R})$ is dense in $L^2(\mathbb{R}, dx)$. By direct computation, exploiting the given definitions one immediately sees that

$$H_0 = \hbar \left(a^\dagger a + \frac{1}{2} I \right) = \hbar \left(\mathcal{N} + \frac{1}{2} I \right).$$

Finally, we have the commutation relations *on* $\mathcal{S}(\mathbb{R})$

$$[a, a^\dagger]|_{\mathcal{S}(\mathbb{R})} = I|_{\mathcal{S}(\mathbb{R})}\,. \tag{2.38}$$

Supposing that there exists $\psi_0 \in \mathcal{S}(\mathbb{R})$ such that

$$||\psi_0|| = 1\,, \quad a\psi_0 = 0 \tag{2.39}$$

starting form (2.38) and using an inductive procedure on the vectors

$$\psi_n := \frac{1}{\sqrt{n1}} (a^\dagger)^n \psi_0 \tag{2.40}$$

it is quite easy to prove that (e.g., see [5] for details), for $n, m = 0, 1, 2, \ldots$, the relations hold

$$a\psi_n = \sqrt{n}\psi_{n-1}\,, \quad a^\dagger \psi_n = \sqrt{n+1}\psi_{n+1}\,, \quad \langle \psi_n, \psi_m \rangle = \delta_{nm}\,. \tag{2.41}$$

Finally, the ψ_n are eigenvectors of H_0 (and $\mathcal{N}$) since

$$H_0\psi_n = \hbar\omega \left(a^\dagger a \psi_n + \frac{1}{2}\psi_n \right) = \hbar\omega \left(a^\dagger \sqrt{n}\psi_{n-1} + \frac{1}{2}\psi_n \right)$$

$$= \hbar\omega \left(\psi_{n+1} + \frac{1}{2}\psi_n \right) = \hbar\omega \left(n + \frac{1}{2} \right) \psi_n\,. \tag{2.42}$$

As a consequence, $\{\psi_n\}_{n\in\mathbb{N}}$ is an orthonormal set of vectors. This set is actually a Hilbertian basis because (1), a solution in $\mathcal{S}(\mathbb{R})$ of (2.39) exist (an is unique):

$$\psi_0(x) = \frac{1}{\pi^{1/4}\sqrt{s}} e^{-\frac{x^2}{2s^2}}\,, \quad s := \sqrt{\frac{\hbar}{m\omega}}\,,$$

(2) the first identity in (2.41) is a well-known recurrence relation of the Hilbert basis of $L^2(\mathbb{R}, dx)$ made of Schwartz' functions known as *Hermite functions* $\{H_n\}_{n\in\mathbb{N}}$, and $\psi_0(x) = H_0(x)$.

Exploiting Nelson's criterium, we conclude that the symmetric operator H_0 is essentially selfadjoint in $D(H_0) = \mathcal{S}(\mathbb{R})$ and $H := \overline{H_0} = H_0^\dagger$, because H_0 admits a Hilbert basis of eigenvectors with corresponding eigenvalues $\hbar\omega(n + \frac{1}{2})$.

2.2.4 *Spectrum of an operator*

Our goal is to extend (2.4) to a formula valid in the infinite dimensional case. As we shall see shortly, passing to the infinite dimensional case, the sum is replaced by an integral and $\sigma(A)$ must be enlarged with respect to the pure set of eigenvalues of A. This is because, as already noticed in the first section, there are operators which should be decomposed with the prescription (2.4) but they do not have eigenvalues, though they play a crucial role in QM.

Notation 2.2.40. If $A : D(A) \to \mathcal{H}$ is injective, A^{-1} indicates its inverse when the co-domain of A is restricted to $Ran(A)$. In other words, $A^{-1} : Ran(A) \to D(A)$.

The definition of *spectrum* of the operator $A : D(A) \to \mathcal{H}$ extends the notion of set of eigenvalues. The eigenvalues of A are the numbers $\lambda \in \mathbb{C}$ such that $(A - \lambda I)^{-1}$ does not exist. When passing infinite dimensions, topological issues take place. As a matter of fact, even if $(A - \lambda I)^{-1}$ exists, it may be bounded or unbounded and its domain $Ran(A - \lambda I)$ may or may not be dense. These features permit us to define a suitable extension of the notion of a set of eigenvalues.

Definition 2.2.41. Let A be an operator in the complex Hilbert space $\mathcal{H}$. The **resolvent set** of A is the subset of $\mathbb{C}$,

$$\rho(A) := \{\lambda \in \mathbb{C} \,|\, (A - \lambda I) \text{ is injective, } \overline{Ran(A - \lambda I)} = \mathcal{H}\,, (A - \lambda I)^{-1}\text{is bounded}\}$$

The **spectrum** of A is the complement $\sigma(A) := \mathbb{C} \setminus \rho(A)$ and it is given by the union of the following pairwise disjoint three parts:

(i) the **point spectrum**, $\sigma_p(A)$, where $A - \lambda I$ not injective ($\sigma_p(A)$ is the set of *eigenvalues* of A);
(ii) the **continuous spectrum**, $\sigma_c(A)$, where $A - \lambda I$ injective, $\overline{Ran(A - \lambda I)} = \mathcal{H}$ and $(A - \lambda I)^{-1}$ not bounded;
(iii) the **residual spectrum**, $\sigma_r(A)$, where $A - \lambda I$ injective and $\overline{Ran(A - \lambda I)} \neq \mathcal{H}$.

Remark 2.2.42.

(a) *It turns out that* $\rho(A)$ *is always* open, *so that* $\sigma(A)$ *is always* closed *[8; 5; 9].*

(b) *If* A *is closed and normal, in particular, if* A *is either selfadjoint or unitary),* $\sigma_r(A) = \varnothing$ *(e.g., see [5]). Furthermore, if* A *is closed (if* $A \in \mathfrak{B}(\mathcal{H})$ *in particular),* $\lambda \in \rho(A)$ *if and only if* $A - \lambda I$ *admits inverse in* $\mathfrak{B}(\mathcal{H})$ *(see (2) in exercise 2.2.43).*

(c) *If* A *is selfadjoint, one finds* $\sigma(A) \subset \mathbb{R}$ *(see (1) in exercise 2.2.43).*

(d) *If* A *is unitary, one finds* $\sigma(A) \subset \mathbb{T} := \{e^{ia} \mid a \in \mathbb{R}\}$ *(e.g., see [5]).*

(e) *If* $U : \mathcal{H} \to \mathcal{H}$ *is unitary and* A *is any operator in the complex Hilbert space* $\mathcal{H}$*, just by applying the definition, one finds* $\sigma(UAU^\dagger) = \sigma(A)$ *and in particular,*

$$\sigma_p(UAU^\dagger) = \sigma_p(A)\,, \quad \sigma_c(UAU^\dagger) = \sigma_c(A)\,, \quad \sigma_r(UAU^\dagger) = \sigma_r(A)\,. \tag{2.43}$$

The same result holds replacing $U : \mathcal{H} \to \mathcal{H}$ *for* $U : \mathcal{H} \to \mathcal{H}'$ *and* $U^\dagger$ *for* U^{-1}*, where* U *is now a Hilbert space isomorphism (an isometric surjective linear map) and* $\mathcal{H}'$ *another complex Hilbert space.*

To conclude, let us mention two useful technical facts which will turn out to be useful several times in the rest of this part. From the definition of adjoint, one easily has for $A : D(A) \to \mathcal{H}$ densely defined and $\lambda \in \mathbb{C}$,

$$\begin{aligned} Ker(A^\dagger - \lambda^* I) &= [Ran(A - \lambda I)]^\perp\,, \\ Ker(A - \lambda I) &\subset [Ran(A^\dagger - \lambda^* I)]^\perp \end{aligned} \tag{2.44}$$

where the inclusion becomes an identity if $A \in \mathfrak{B}(\mathcal{H})$.

Exercise 2.2.43.

(1) *Prove that if* A *is a selfadjoint operator in the complex Hilbert space* $\mathcal{H}$ *then*

(i) $\sigma(A) \subset \mathbb{R}$;
(ii) $\sigma_r(A) = \varnothing$;
(iii) *eigenvectors with different eigenvalues are orthogonal.*

Solution. Let us begin with (i). Suppose $\lambda = \mu + i\nu$, $\nu \neq 0$ and let us prove $\lambda \in \rho(A)$. If $x \in D(A)$,

$$\langle (A - \lambda I)x, (A - \lambda I)x\rangle = \langle (A - \mu I)x, (A - \mu I)x\rangle + \nu^2\langle x, x\rangle + i\nu[\langle Ax, x\rangle - \langle x, Ax\rangle]\,.$$

The last summand vanishes for A is selfadjoint. Hence

$$||(A - \lambda I)x|| \geq |\nu|\,||x||\,.$$

With a similar argument, we obtain

$$||(A - \lambda^* I)x|| \geq |\nu|\,||x||\,.$$

The operators $A - \lambda I$ and $A - \lambda^* I$ are injective, and $||(A-\lambda I)^{-1}|| \leq |\nu|^{-1}$, where $(A - \lambda I)^{-1} : Ran(A - \lambda I) \to D(A)$. Notice that, from (2.44),

$$\overline{Ran(A - \lambda I)}^{\perp} = [Ran(A - \lambda I)]^{\perp} = Ker(A^{\dagger} - \lambda^* I) = Ker(A - \lambda^* I) = \{0\}\,,$$

where the last equality makes use of the injectivity of $A - \lambda^* I$. Summarising: $A - \lambda I$ in injective, $(A - \lambda I)^{-1}$ bounded and $\overline{Ran(A - \lambda I)}^{\perp} = \{0\}$, i.e. $Ran(A - \lambda I)$ is dense in $\mathcal{H}$; therefore $\lambda \in \rho(A)$, by definition of resolvent set. Let us pass to (ii). Suppose $\lambda \in \sigma(A)$, but $\lambda \notin \sigma_p(A)$. Then $A - \lambda I$ must be one-to-one and $Ker(A - \lambda I) = \{0\}$. Since $A = A^{\dagger}$ and $\lambda \in \mathbb{R}$ by (i), we have $Ker(A^{\dagger} - \lambda^* I) = \{0\}$, so $[Ran(A - \lambda I)]^{\perp} = Ker(A^{\dagger} - \lambda^* I) = \{0\}$ and $\overline{Ran(A - \lambda I)} = \mathcal{H}$. Consequently $\lambda \in \sigma_c(A)$. Proving (iii) is easy: if $\lambda \neq \mu$ and $Au = \lambda u$, $Av = \mu v$, then

$$(\lambda - \mu)\langle u, v\rangle = \langle Au, v\rangle - \langle u, Av\rangle = \langle u, Av\rangle - \langle u, Av\rangle = 0\,;$$

from $\lambda, \mu \in \mathbb{R}$ and $A = A^{\dagger}$. But $\lambda - \mu \neq 0$, so $\langle u, v\rangle = 0$.

(2) *Let $A : D(A) \to \mathcal{H}$ be a closed operator in $\mathcal{H}$ (in particular $A \in \mathfrak{B}(\mathcal{H})$). Prove that $\lambda \in \rho(A)$ if and only if $A - \lambda I$ admits an inverse which belongs to $\mathfrak{B}(\mathcal{H})$.*

Solution. If $(A - \lambda I)^{-1} \in \mathfrak{B}(\mathcal{H})$, it must be $\overline{Ran(A - \lambda I)} = Ran(A - \lambda I) = \mathcal{H}$ and $(A - \lambda I)^{-1}$ is bounded, so that $\lambda \in \rho(A)$ by definition. Let us prove the converse. Suppose that $\lambda \in \rho(A)$, we know that $(A - \lambda I)^{-1}$ is defined on the dense domain $Ran(A - \lambda I)$ and is bounded. To conclude, it is therefore enough proving that $y \in \mathcal{H}$ implies $y \in Ran(A - \lambda I)$. To this end, notice that if $y \in \mathcal{H} = \overline{Ran(A - \lambda I)}$, then $y = \lim_{n\to+\infty}(A - \lambda I)x_n$ for some $x_n \in D(A - \lambda I)$. The sequence of x_n converges. Indeed $\mathcal{H}$ is complete and $\{x_n\}_{n\in\mathbb{N}}$ is Cauchy as (1) $x_n = (A - \lambda I)^{-1}y_n$, (2) $||x_n - x_m|| \leq ||(A - \lambda I)^{-1}||\,||y_n - y_m||$, and (3) $y_n \to y$. To end the proof, we observe that, $A - \lambda I$ is closed since A is such ((b) in remark 2.2.21). It must consequently be ((c) in remark 2.2.21) $x = \lim_{n\to+\infty} x_n \in D(A - \lambda I)$ and $y = (A - \lambda I)x \in Ran(A - \lambda I)$.

Example 2.2.44. The m-axis position operator X_m in $L^2(\mathbb{R}^n, d^n x)$ introduced in (1) of example 2.2.39 satisfies

$$\sigma(X_m) = \sigma_c(X_m) = \mathbb{R}\,. \tag{2.45}$$

The proof can be obtained as follows. First observe that $\sigma(X_m) \subset \mathbb{R}$ since the operator is selfadjoint. However $\sigma_p(X_m) = \varnothing$ as observed in the first section and $\sigma_r(X_m) = \varnothing$ because X_m is selfadjoint ((1) in exercise 2.2.43). Suppose that, for some $r \in \mathbb{R}$, $(X_m - rI)^{-1}$ is bounded. If $\psi \in D(X_m - rI) = D(X_m)$ with $||\psi|| = 1$ we have $||\psi|| = ||(X_m - rI)^{-1}(X_m - rI)\psi||$ and thus $||\psi|| \leq ||(X_m - rI)^{-1}||\,||(X_m -$

$rI)\psi||$. Therefore

$$||(X_m - rI)^{-1}|| \geq \frac{1}{||(X_m - rI)\psi||}$$

For every fixed $\epsilon > 0$, it is simply constructed $\psi \in D(X_m)$ with $||\psi|| = 1$ and $||(X_m - rI)\psi|| < \epsilon$. Therefore $(X_m - rI)^{-1}$ cannot be bounded and thus $r \in \sigma_c(X_m)$. In view of (e) in remark 2.2.42, we also conclude that

$$\sigma(P_m) = \sigma_c(P_m) = \mathbb{R}\,, \tag{2.46}$$

just because the momentum operator P_m is related to the position one by means of a unitary operator given by the Fourier-Plancherel operator $\hat{\mathcal{F}}$ as discussed in (2) of example 2.2.39.

2.2.5 *Spectral measures*

Definition 2.2.45. *Let $\mathcal{H}$ be a complex Hilbert space. $P \in \mathfrak{B}(\mathcal{H})$ is called* **orthogonal projector** *when $PP = P$ and $P^\dagger = P$. $\mathcal{L}(\mathcal{H})$ denotes the set of orthogonal projectors of $\mathcal{H}$.*

We have the well known relation between orthogonal projectors and closed subspaces [8; 5]

Proposition 2.2.46. *If $P \in \mathcal{L}(\mathcal{H})$, then $P(\mathcal{H})$ is a closed subspace. If $\mathcal{H}_0 \subset \mathcal{H}$ is a closed subspace, there exists exactly one $P \in \mathcal{L}(\mathcal{H})$ such that $P(\mathcal{H}) = \mathcal{H}_0$. Finally, $I - P \in \mathcal{L}(\mathcal{H})$ and it projects onto $\mathcal{H}_0^\perp$ (e.g., see [5]).*

We can now state one of the most important definitions in spectral theory.

Definition 2.2.47. *Let $\mathcal{H}$ be a complex Hilbert space and $\Sigma(X)$ a σ-algebra over X. A* **projection-valued measure (PVM)** *(also known as* **resolution of the identity***) on X, P, is a map $\Sigma(X) \ni E \mapsto P_E \in \mathcal{L}(\mathcal{H})$ such that:*

(i) *$P_X = I$;*
(ii) *$P_E P_F = P_{E\cap F}$;*
(iii) *If $N \subset \mathbb{N}$ and $\{E_k\}_{k\in N} \subset \Sigma(X)$ satisfies $E_j \cap E_k = \varnothing$ for $k \neq j$, then*

$$\sum_{j\in N} P_{E_j} x = P_{\cup_{j\in N} E_j} x \quad \textit{for every } x \in \mathcal{H}.$$

(If N is infinite, the sum on the left hand side of (iii) is computed referring to the topology of $\mathcal{H}$.)

Remark 2.2.48.

(a) *(i) and (iii) with* $N = \{1, 2\}$ *imply that* $P_\varnothing = 0$ *using* $E_1 = X$ *and* $E_2 = \varnothing$. *Next (ii) entails that* $P_E P_F = 0$ *if* $E \cap F = \varnothing$. *An important consequence is that for* N *infinite, the vector given by the sum on the left hand side of (iii) is independent from the chosen order because that vector is a sum of pairwise orthogonal vectors* $P_{E_j} x$.

(b) *If* $x, y \in \mathcal{H}$, $\Sigma(X) \ni E \mapsto \langle x, P_E y\rangle =: \mu^{(P)}_{xy}(E)$ *is a* complex measure *whose (finite)* total variation *[8] will be denoted by* $|\mu^{(P)}_{xy}|$. *From the definition of* μ_{xy}, *we immediately have:*

(i) $\mu^{(P)}_{xy}(X) = \langle x, y\rangle$;

(ii) $\mu^{(P)}_{xx}$ *is always positive and finite and* $\mu^{(P)}_{xx}(X) = ||x||^2$;

(iii) *if* $s = \sum_{k=1}^n s_k \chi_{E_k}$ *is a* simple function *[8],* $\int_X s d\mu_{xy} = \langle x, \sum_{k=1}^n s_k P_{E_k} y\rangle$.

Example 2.2.49.

(1) The simplest example of PVM is related to a countable Hilbertian basis N in a separable Hilbert space $\mathcal{H}$. We can define $\Sigma(N)$ as the class of all subsets of N itself. Next, for $E \in \Sigma(N)$ and $z \in \mathcal{H}$ we define

$$P_E z := \sum_{x \in E} \langle x, z\rangle x$$

and $P_\varnothing := 0$. It is easy to prove that the class of all P_E defined this way form a PVM on N. (This definition can be also given if $\mathcal{H}$ is non-separable and N is uncountable, since for every $y \in \mathcal{H}$ only an at most countable subset of elements $x \in E$ satisfy $\langle x, y\rangle \neq 0$). In particular $\mu_{xy}(E) = \langle x, P_E y\rangle = \sum_{z \in E}\langle x|z\rangle\langle z|y\rangle$ and $\mu_{xx}(E) = \sum_{z\in E} |\langle x, z\rangle|^2$.

(2) A more complicated version of (1) consists of a PVM constructed out of a orthogonal Hilbertian decomposition of a separable Hilbert space, $\mathcal{H} = \oplus_{n\in\mathbb{N}} \mathcal{H}_n$, where $\mathcal{H}_n \subset \mathcal{H}$ is a closed subspace and $\mathcal{H}_n \perp \mathcal{H}_m$ if $n \neq m$. Again defining $\Sigma(\mathbb{N})$ as the set of subsets of $\mathbb{N}$, for $E \in \Sigma(N)$ and $z \in \mathcal{H}$ we define

$$P_E z := \sum_{x \in E} Q_n z$$

where Q_n is the orthogonal projector onto $\mathcal{H}_n$ (the reader can easily check that the sum always converges using Bessel's inequality). It is easy to prove that the class of P_Es defined this way form a PVM on $\mathbb{N}$. In particular $\mu_{xy}(E) = \langle x, P_E y\rangle = \sum_{n\in E}\langle x, Q_n y\rangle$ and $\mu_{xx}(E) = \sum_{n\in E} ||Q_n x||^2$.

(3) In $L^2(\mathbb{R}, dx)$ a simple PVM, not related with a Hilbertian basis, is made as follows. To every $E \in \mathcal{B}(\mathbb{R})$, the Borel σ-algebra, associate the orthonormal projector P_E such that, if χ_E is the **characteristic function of** E – $\chi_E(x) = 0$ if $x \notin E$ and $\chi_E(x) = 1$ if $x \in E$ –

$$(P_E\psi)(x) := \chi_E(x)\psi(x) \quad \forall \psi \in L^2(\mathbb{R}, dx) .$$

Moreover $P_\varnothing := 0$. It is easy to prove that the collection of the P_E is a PVM. In particular $\mu_{fg}(E) = \langle f, P_E g\rangle = \int_E f(x)^* g(x) dx$ and $\mu_{ff}(E) = \int_E |f(x)|^2 dx$.

We have the following fundamental result [8; 5; 6; 9].

Proposition 2.2.50. *Let $\mathcal{H}$ be a complex Hilbert space and $P : \Sigma(X) \to \mathcal{L}(\mathcal{H})$ a PVM. If $f : X \to \mathbb{C}$ is measurable, define*

$$\Delta_f := \left\{ x \in \mathcal{H} \;\middle|\; \int_X |f(\lambda)|^2 \mu^{(P)}_{xx}(\lambda) < +\infty \right\}.$$

Δ_f is a dense subspace of $\mathcal{H}$ and there is a unique operator

$$\int_X f(\lambda) dP(\lambda) : \Delta_f \to \mathcal{H} \tag{2.47}$$

such that

$$\left\langle x, \int_X f(\lambda) dP(\lambda) y \right\rangle = \int_X f(\lambda) \mu^{(P)}_{xy}(\lambda) \quad \forall x \in \mathcal{H}\,, \forall y \in \Delta_f \tag{2.48}$$

The operator in (2.47) turns out to be closed and normal. It finally satisfies

$$\left(\int_X f(\lambda)\, dP(\lambda) \right)^\dagger = \int_X \overline{f(\lambda)}\, dP(\lambda) \tag{2.49}$$

and

$$\left|\left| \int_X f(\lambda)\, dP(\lambda) x \right|\right|^2 = \int_X |f(\lambda)|^2 d\mu^{(P)}_{xx}(\lambda) \quad \forall x \in \Delta_f\,. \tag{2.50}$$

Idea of the existence part of the proof. The idea of the proof of existence of the operator in (2.47) relies upon the validity of the inequality ((1) in exercises 2.2.52 below)

$$\int_X |f(\lambda)|\, d|\mu^{(P)}_{xy}|(\lambda) \leq ||x|| \sqrt{\int_X |f(\lambda)|^2 d\mu^{(P)}_{yy}(\lambda)} \quad \forall y \in \Delta_f\,, \forall x \in \mathcal{H}\,. \tag{2.51}$$

This inequality also proves that $f \in L^2(X, d\mu^{(P)}_{yy})$ implies $f \in L^1(X, d|\mu^{(P)}_{xy}|)$ for $x \in \mathcal{H}$, so that (2.48) makes sense. Since from the general measure theory

$$\left| \int_X f(\lambda)\, d\mu^{(P)}_{xy}(\lambda) \right| \leq \int_X |f(\lambda)|\, d|\mu^{(P)}_{xy}|(\lambda)\,,$$

(2.51) implies that $\mathcal{H} \ni x \mapsto \int_X f(\lambda)\, d\mu^{(P)}_{xy}(\lambda)$ is continuous at $x = 0$. This map is also anti-linear as follows from the definition of $\mu_{x,y}$. An elementary use of Riesz' lemma proves that there exists a vector, indicated by $\int_X f(\lambda) dP(\lambda) y$, satisfying (2.48). That is the action of an operator on a vector $y \in \Delta_f$ because $\Delta_f \ni y \mapsto \int_X f(\lambda)\, d\mu^{(P)}_{xy}(\lambda)$ is linear. □

Remark 2.2.51. *Identity (2.50) gives Δ_f a direct meaning in terms of boundedness of $\int_X f(\lambda)\, dP(\lambda)$. Since $\mu_{xx}(X) = ||x||^2 < +\infty$, (2.50) together with the definition of Δ_f immediately implies that: if f is bounded or, more weakly P-*essentially bounded[8] *on X, then*

$$\int_X f(\lambda)\, dP(\lambda) \in \mathfrak{B}(\mathcal{H})$$

and

$$\left|\left|\int_X f(\lambda)\, dP(\lambda)\right|\right| \leq ||f||^{(P)}_\infty \leq ||f||_\infty .$$

The P-essentially boundedness is also a necessary *(not only sufficient) condition for $\int_X f(\lambda)\, dP(\lambda) \in \mathfrak{B}(\mathcal{H})$ [8; 5; 6].*

Exercise 2.2.52.
(1) Prove inequality (2.51).

Solution. Let $x \in \mathcal{H}$ and $y \in \Delta_f$. If $s : X \to \mathbb{C}$ is a *simple function* and $h : X \to \mathbb{C}$ is the *Radon-Nikodym derivative* of μ_{xy} with respect to $|\mu_{xy}|$ so that $|h(x)| = 1$ and $\mu_{xy}(E) = \int_E h d|\mu_{xy}|$ (see, e.g., [5]), we have for an increasing sequence of simple functions $z_n \to h$ pointwise, with $|z_n| \leq |h^{-1}| = 1$, due to the dominate convergence theorem,

$$\int_X |s| d|\mu_{xy}| = \int_X |s| h^{-1} d\mu_{xy} = \lim_{n\to+\infty} \int_X |s| z_n d\mu_{xy} = \lim_{n\to+\infty} \left\langle x, \sum_{k=1}^{N_n} z_{n,k} P_{E_{n,k}} y \right\rangle .$$

In the last step, we have made use of (iii)(b) in remark 2.2.48 for the simple function $|s| z_n = \sum_{k=1}^{N_n} z_{n,k} \chi_{E_{n,k}}$. Cauchy Schwartz inequality immediately yields

$$\int_X |s| d|\mu_{xy}| \leq ||x|| \lim_{n\to+\infty} \left|\left| \sum_{k=1}^{N_n} z_{n,k} P_{E_{n,k}} y \right|\right| = ||x|| \lim_{n\to+\infty} \sqrt{\int_X |s z_n|^2 d\mu_{yy}} ,$$

[8] As usual, $||f||^{(P)}_\infty$ is the infimum of positive reals r such that $P(\{x \in X \,|\, |f(x)| > r\}) = 0$.

where we have used $P^\dagger_{E_{n,k}} P_{E_{n,k'}} = P_{E_{n,k}} P_{E_{n,k'}} = \delta_{kk'} P_{E_{n,k}}$ since $E_{n,k} \cap E_{n,k'} = \varnothing$ for $k \neq k'$. Next observe that, as $|sz_n|^2 \to |sh^{-1}|^2 = |s|^2$, dominate convergence theorem leads to

$$\int_X |s| d|\mu_{xy}| \leq ||x|| \sqrt{\int_X |s|^2 d\mu_{yy}} \, .$$

Finally, replace s above for a sequence of simple functions $|s_n| \to f \in L^2(X, d\mu_{yy})$ pointwise, with $s_n \leq |s_{n+1}| \leq |f|$. Monotone convergence theorem and dominate convergence theorem, respectively applied to the left and right-hand side of the found inequality, produce inequality (2.51).

(2) *Prove that, with the hypotheses of Proposition 2.2.50, it holds*

$$\int_X \chi_E(\lambda) dP(\lambda) = P_E \, , \quad \textit{if } E \in \Sigma(X) \tag{2.52}$$

and in particular

$$\int_X 1 \, dP(\lambda) = I \, . \tag{2.53}$$

Solution. It is sufficient to prove (2.52) since we know that $P_X = I$. To this end, notice that, by direct inspection

$$\langle x, P_E y \rangle = \int_X \chi_E(\lambda) \mu^{(P)}_{xy}(\lambda) \quad \forall x \in \mathcal{H} \, , \; \forall y \in \Delta_{\chi_E} = \mathcal{H} \, .$$

By the uniqueness property stated in Proposition 2.2.50, (2.52) holds.

(3) *Prove that if P, a PVM on $\mathcal{H}$ and T is an operator in $\mathcal{H}$ with $D(T) = \Delta_f$ such that*

$$\langle x, Tx \rangle = \int_X f(\lambda) \mu^{(P)}_{xx}(\lambda) \quad \forall x \in \Delta_f \tag{2.54}$$

then

$$T = \int_X f(\lambda) dP(\lambda) \, .$$

Solution. From the definition of μ_{xy}, we easily have (everywhere omitting $^{(P)}$ for simplicity)

$$4\mu_{xy}(E) = \mu_{x+y,x+y}(E) - \mu_{x-y,x-y}(E) - i\mu_{x+iy,x+iy}(E) + i\mu_{x-iy,x-iy}(E)$$

This identity implies that if $x, y \in \Delta_f$,

$$4 \int_X f d\mu_{xy} = \int_X f d\mu_{x+y,x+y} - \int_X f d\mu_{x-y,x-y}$$
$$- i \int_X f d\mu_{x+iy.x+iy} + i \int_X f d\mu_{x-iy,x-iy}$$

Similarly, from the elementary properties of the scalar product, when $x, y \in D(T)$

$$4\langle x|Ty\rangle = \langle x+y, T(x+y)\rangle - \langle x-y, T(x-y)\rangle - i\langle x+iy, T(x+iy)\rangle + i\langle x-iy, T(x-iy)\rangle.$$

It is then obvious that (2.54) implies

$$\langle x, Ty\rangle = \int_X f(\lambda)\mu_{xy}^{(P)}(\lambda) \quad \forall x, y \in \Delta_f\,,$$

so that

$$\left\langle x, \left(T - \int_X f(\lambda)dP(\lambda)\right) y \right\rangle = 0 \quad \forall x, y \in \Delta_f$$

Since x varies in a dense set Δ_f, $Ty - \int_X f(\lambda)dP(\lambda)y = 0$ for every $y \in \Delta_f$ which is the thesis.

Example 2.2.53.
(1) Referring to the PVM in (2) of example 2.2.49, directly from the definition of $\int_X f(\lambda)dP(\lambda)$ or exploiting (3) in exercises 2.2.52 we have that

$$\int_{\mathbb{N}} f(\lambda)dP(\lambda)z = \sum_{n\in\mathbb{N}} f(n)Q_n z$$

for every $f : \mathbb{N} \to \mathbb{C}$ (which is necessarily measurable with our definition of $\Sigma(\mathbb{N})$). Correspondingly, the domain of $\int_N f(\lambda)dP(\lambda)$ results to be

$$\Delta_f := \left\{ z \in \mathcal{H} \;\middle|\; \sum_{n\in\mathbb{N}} |f(n)|^2 ||Q_n z||^2 < +\infty \right\}$$

We stress that we have found a direct generalization of the expansion (2.4) if the operator A is now hopefully written as

$$Az = \sum_{n\in\mathbb{N}} nQ_n z\,.$$

We shall see below that it is the case.

(2) Referring to the PVM in (3) of example 2.2.49, directly from the definition of $\int_X f(\lambda)dP(\lambda)$ or exploiting (3) in exercises 2.2.52 we have that

$$\left(\int_{\mathbb{R}} f(\lambda)dP(\lambda)\psi\right)(x) = f(x)\psi(x)\,, \quad x \in \mathbb{R}$$

Correspondingly, the domain of $\int_{\mathbb{R}} f(\lambda)dP(\lambda)$ results to be

$$\Delta_f := \left\{ \psi \in L^2(\mathbb{R}, dx) \;\middle|\; \int_{\mathbb{R}} |f(x)|^2|\psi(x)|^2 dx < +\infty \right\}.$$

2.2.6 *Spectral decomposition and representation theorems*

We are in a position to state the fundamental result of the spectral theory of selfadjoint operators, which extend the expansion (2.4) to an integral formula valid also in the infinite dimensional case, and where the set of eigenvalues is replaced by the full spectrum of the selfadjoint operator.

To state the theorem, we preventively notice that (2.49) implies that $\int f(\lambda)dP(\lambda)$ is selfadjoint if f is real: The idea of the theorem is to prove that every selfadjoint operator can be written this way for a specific f and with respect to a PVM on $\mathbb{R}$ associated with the operator itself.

Notation 2.2.54. From now on, $\mathcal{B}(T)$ denotes the Borel σ-algebra on the topological space T.

Theorem 2.2.55 (Spectral decomposition theorem for selfadjoint operators). *Let A be a selfadjoint operator in the complex Hilbert space $\mathcal{H}$.*

(a) *There is a unique PVM, $P^{(A)} : \mathcal{B}(\mathbb{R}) \to \mathcal{L}(\mathcal{H})$, such that*

$$A = \int_{\mathbb{R}} \lambda dP^{(A)}(\lambda) .$$

In particular $D(A) = \Delta_{id}$, where $id : \mathbb{R} \ni \lambda \mapsto \lambda$.

(b) *Defining the* **support** *of $P^{(A)}$, $supp(P^{(A)})$, as the complement in $\mathbb{R}$ of the union of all open sets $O \subset \mathbb{C}$ with $P^{(A)}_O = 0$ it results*

$$supp(P^{(A)}) = \sigma(A)$$

so that

$$P^{(A)}(E) = P^{(A)}(E \cap \sigma(A)) , \quad \forall E \in \mathcal{B}(\mathbb{R}) . \tag{2.55}$$

(c) *$\lambda \in \sigma_p(A)$ if and only if $P^{(A)}(\{\lambda\}) \neq 0$, this happens in particular if λ is an isolated point of $\sigma(A)$.*

(d) *$\lambda \in \sigma_c(A)$ if and only if $P^{(A)}(\{\lambda\}) = 0$ but $P^{(A)}(E) \neq 0$ if $E \ni \lambda$ is an open set of $\mathbb{R}$.*

The proof can be found, e.g., in [8; 5; 9; 6].

Remark 2.2.56. *Theorem 2.2.55 is a particular case, of a more general theorem (see [8; 5; 6] and especially [9]) valid when A is a (densely defined closed) normal operator. The general statement is identical, it is sufficient to replace everywhere $\mathbb{R}$ for $\mathbb{C}$. A particular case is the one of A unitary. In this case, the statement can be rephrased replacing everywhere $\mathbb{R}$ for $\mathbb{T}$ since it includes the spectrum of A in this case ((d) remark 2.2.42).*

Notation 2.2.57. In view of the said theorem, and (b) in particular, if $f : \sigma(A) \to \mathbb{C}$ is measurable (with respect to the σ-algebra obtained by restricting $\mathcal{B}(\mathbb{R})$ to $\sigma(A)$), we use the notation

$$f(A) := \int_{\sigma(A)} f(\lambda) dP^{(A)}(\lambda) := \int_{\mathbb{R}} g(\lambda) dP^{(A)}(\lambda) =: g(A) . \tag{2.56}$$

where $g : \mathbb{R} \to \mathbb{C}$ is the extension of f to the zero function outside $\sigma(A)$ or any other measurable function which coincides with f on $supp(P^{(A)}) = \sigma(A)$. Obviously $g(A) = g'(A)$ if $g, g' : \mathbb{R} \to \mathbb{C}$ coincide in $supp(P^{(A)}) = \sigma(A)$.

Exercise 2.2.58. *Prove that if A is a selfadjoint operator in the complex Hilbert space $\mathcal{H}$, it holds $A \geq 0$ – that is $\langle x|Ax\rangle \geq 0$ for every $x \in D(A)$ – if and only if $\sigma(A) \subset [0, +\infty)$.*

Solution. Suppose that $\sigma(A) \subset [0, +\infty)$. If $x \in D(A)$ we have $\langle x, Ax\rangle = \int_{\sigma(A)} \lambda d\mu_{x,x} \geq 0$ in view of (2.48), the spectral decomposition theorem, since $\mu_{x,x}$ is a positive measure and $\sigma(A) \in [0, +\infty)$. To conclude, we prove that $A \geq 0$ is false if $\sigma(A)$ includes negative elements. To this end, assume that conversely, $\sigma(A) \ni \lambda_0 < 0$. Using (c) and (d) of Theorem 2.2.55, one finds an interval $[a, b] \subset \sigma(A)$ with $[a, b] \subset (-\infty, 0)$ and $P^{(A)}_{[a,b]} \neq 0$ (possibly $a = b = \lambda_0$). If $x \in P^{(A)}_{[a,b]}(\mathcal{H})$ with $x \neq 0$, it holds $\mu_{xx}(E) = \langle x|P_E x\rangle = \langle x, P^\dagger_{[a,b]} P_E x P_{[a,b]}\rangle = \langle x, P_{[a,b]} P_E P_{[a,b]} x\rangle = \langle x, P_{[a,b] \cap E} x\rangle = 0$ if $[a, b] \cap E = \varnothing$. Therefore, $\langle x, Ax\rangle = \int_{\sigma(A)} \lambda d\mu_{x,x} = \int_{[a,b]} \lambda d\mu_{x,x} \leq \int_{[a,b]} b\mu_{x,x} < b||x||^2 < 0$.

Example 2.2.59.

(1) Let us focus on the m-axis position operator X_m in $L^2(\mathbb{R}^n, d^n x)$ introduced in (1) of example 2.2.39. We know that $\sigma(X_m) = \sigma_c(X_m) = \mathbb{R}$ from example 2.2.44.

We are interested in the PVM $P^{(X_m)}$ of X_m defined on $\mathbb{R} = \sigma(X_m)$. Let us fix $m = 1$, where the other cases are analogous. The PVM associated to X_1 is

$$(P^{(X_1)}_E \psi)(x) = \chi_{E \times \mathbb{R}^{n-1}}(x) \psi(x) \quad \psi \in L^2(\mathbb{R}^n, d^n x) , \tag{2.57}$$

where $E \in \mathcal{B}(\mathbb{R})$ is identified with a subset of the first factor of $\mathbb{R} \times \mathbb{R}^{n-1} = \mathbb{R}^n$. Indeed, indicating by P on the right-hand side of (2.57), one easily verifies that $\Delta_{x_1} = D(X_1)$ and[9]

$$\langle \psi|X_1\psi\rangle = \int_{\mathbb{R}} \lambda \mu^{(P)}_{\psi,\psi}(\lambda) \quad \forall \psi \in D(X_1) = \Delta_{x_1}$$

[9]More generally $\int_{\mathbb{R}} \int_{\mathbb{R}^{n-1}} g(x_1)|\psi(x)|^2 dx d^{n-1}x = \int_{\mathbb{R}} g(x_1) d\mu^{(P)}_{\psi,\psi}(x_1)$ is evidently valid for simple functions and then it extends to generic measurable functions when both sides make sense in view of, for instance, Lebesgue's dominate convergence theorem for positive measures.

where $\mu^{(P)}_{\psi,\psi}(E) = \langle\psi, P_E\psi\rangle = \int_{E\times\mathbb{R}^{n-1}} |\psi(x)|^2 d^n x$. (2) in exercise 2.2.49 proves that $X_1 = \int_{\mathbb{R}} \lambda dP(\lambda)$ and thus (2.57) holds true.

(2) Considering that the m-axis momentum operator P_m in $L^2(\mathbb{R}^n, d^n x)$ introduced in (2) of example 2.2.39, taking (2.36) into account where $\hat{\mathcal{F}}$ (and thus $\hat{\mathcal{F}}^\dagger$) is unitary, in view of (i) in Proposition 2.2.66 we immediately have that the PVM of P_m is

$$Q^{(P_m)}_E := \hat{\mathcal{F}}^\dagger P^{(K_m)}_E \hat{\mathcal{F}} .$$

Above, K_m is the operator and X_m represented in $L^2(\mathbb{R}^n, d^n k)$ as in (1) of example 2.2.39.

(3) More complicated cases exist. Considering an operator of the form

$$H := \frac{1}{2m} P^2 + U$$

where P is the momentum operator in $L^2(\mathbb{R}, dx)$, $m > 0$ is a constant and U is a real valued function on $\mathbb{R}$ used as a multiplicative operator. If $U = U_1 + U_2$ with $U_1 \in L^2(\mathbb{R}, dx)$ and $U_2 \in L^\infty(\mathbb{R}, dx)$ real valued, and $D(H) = C^\infty(\mathbb{R}; \mathbb{C})$, H turns out to be (trivially) symmetric but also essentially selfadjoint [5] as a consequence of a well known result (*Kato-Rellich's theorem*). The unique selfadjoint extension $\overline{H} = (H^\dagger)^\dagger$ of H physically represent the Hamiltonian operator of a quantum particle living along $\mathbb{R}$ with a potential energy described by U. In this case, generally speaking, $\sigma(\overline{H})$ has both point and continuous part. $\int_{\sigma_p(\overline{H})} \lambda dP^{(\overline{H})}(\lambda)$ has a form like this

$$\int_{\sigma_p(\overline{H})} \lambda dP^{(\overline{H})}(\lambda) = \sum_{\lambda\in\sigma_p(\overline{H})} \lambda P_\lambda$$

where P_λ is the orthogonal projector onto the eigenspace of $\overline{H}$ with eigenvalue λ. Conversely, $\int_{\sigma_c(\overline{H})} \lambda dP^{(\overline{H})}(\lambda)$ has an expression much more complicated and, under a unitary transform, is similar to the integral decomposition of X.

Remark 2.2.60.

(a) *It is worth stressing that the notion (2.56) of a function of a selfadjoint operator is just an extension of the analogous notion introduced for the finite dimensional case (2.7) and thus may be used in QM applications.*

It is possible to prove that if $f : \sigma(A) \to \mathbb{R}$ is continuous, then

$$\sigma(f(A)) = \overline{f(\sigma(A))} \tag{2.58}$$

where the bar denotes the closure and, if $f : \sigma(A) \to \mathbb{R}$ is measurable,

$$\sigma_p(f(A)) \supset f(\sigma_p(A)) . \tag{2.59}$$

More precise statements based on the notion of essential range *can be found in [5]. It turns out that, for A selfadjoint and $f : \sigma(A) \to \mathbb{C}$ measurable, $z \in \sigma(f(A))$ if and only if $P^{(A)}(E_z) \neq 0$ for some open set $E_z \ni z$. Now $z \in \sigma(f(A))$ is in $\sigma_p(f(A))$ iff $P^{(A)}(f^{-1}(z)) \neq 0$ or it is in $\sigma_c(f(A))$ iff $P^{(A)}(f^{-1}(z)) = 0$.*

(b) *It is fundamental to stress that in QM, (2.58) permits us to adopt the standard operational approach on observables $f(A)$ as the observable whose set of possible values is (the closure of) the set of reals $f(a)$ where a is a possible value of A.*

Proposition 2.2.61. *A selfadjoint operator is bounded (and its domain coincide to the whole $\mathcal{H}$) if and only if $\sigma(A)$ is bounded.*

Proof. It essentially follows from (2.50), restricting the integration space to $X = \sigma(A)$. In fact, if $\sigma(A)$ is bounded and thus compact it being closed, the continuous function $id : \sigma(A) \ni \lambda \to \lambda$ is bounded and (2.50) implies that $A = \int_{\sigma(A)} id dP^{(A)}$ is bounded and the inequality holds

$$||A|| \leq \sup\{|\lambda| \mid \lambda \in \sigma(A)\}\,. \tag{2.60}$$

In this case, it also holds $D(A) = \Delta_{id} = \mathcal{H}$.

If, conversely, $\sigma(A)$ is not bounded, we can find a sequence $\lambda_n \in \sigma(A)$ with $|\lambda_n| \to \infty$ as $n \to +\infty$. With the help of (c) and (d) in Theorem 2.2.55, it is easy to construct vectors x_n with $||x_n|| \neq 0$ and $x_n \in P^{(A)}_{B(\lambda_n)}(\mathcal{H})$ where $B(\lambda_n) := [\lambda_n - 1, \lambda_n + 1]$. (2.50) implies

$$||Ax_n||^2 \geq ||x_n||^2 \inf_{z\in B(\lambda_n)} |id(z)|^2$$

Since $\inf_{z\in B(\lambda_n)} |id(z)|^2 \to +\infty$, we have that $||Ax_n||/||x_n||$ is not bounded and A, in turn, cannot be bounded. In this case, since $A = A^\dagger$, Theorem 2.2.29 entails that $D(A)$ is *strictly* included in $\mathcal{H}$. □

It is possible to prove [5] that (2.60) can be turned into an identity when $A \in \mathfrak{B}(\mathcal{H})$ also if A is not selfadjoint but only normal

$$||A|| = \sup\{|\lambda| \mid \lambda \in \sigma(A)\}\,, \tag{2.61}$$

This is the well known spectral radius formula, *the* **spectral radius** *of $A \in \mathfrak{B}(\mathcal{H})$ being, by definition, the number in the right hand side.*

(d) *The result stated in (c) explains the reason why observables A in QM are very often represented by unbounded selfadjoint operators. $\sigma(A)$ is the set of values of the observable A. When, as it happens very often, that observable is allowed* *to take arbitrarily large values (think of X or P), it cannot be represented by a bounded selfadjoint operator just because its spectrum is not bounded.*

(e) *If P is a PVM on $\mathbb{R}$ and $f : \mathbb{R} \to \mathbb{C}$ is measurable, we can always write*

$$\int_{\mathbb{R}} f(\lambda) dP(\lambda) = f(A)$$

where we have introduced the selfadjoint operator A obtained as

$$A = \int_{\mathbb{R}} id(\lambda) dP(\lambda) , \tag{2.62}$$

due to (2.49) and where $id : \mathbb{R} \ni \lambda \to \lambda$. Evidently $P^{(A)} = P$ due to the uniqueness part of the spectral theorem. This fact leads to the conclusion that, in a complex Hilbert space $\mathcal{H}$, all the PVM over $\mathbb{R}$ are one-to-one associated to all selfadjoint operators in $\mathcal{H}$.

(f) *An element $\lambda \in \sigma_c(A)$ is not an eigenvalue of A. However there is the following known result arising from (d) in Theorem 2.2.55 [5] which proves that we can have approximated eigenvalues with arbitrary precision: With the said hypotheses, for every $\epsilon > 0$ there is $x_\epsilon \in D(A)$ such that*

$$||Ax_\epsilon - \lambda x_\epsilon|| < \epsilon , \quad \textit{but } ||x_\epsilon|| = 1 .$$

(g) *If A is selfadjoint and U unitary, $UAU^\dagger$, with $D(UAU^\dagger) = U(D(A))$, is selfadjoint as well (Proposition 2.2.30). It is very simple to prove that the PVM of $UAU^\dagger$ is nothing but $UP^{(A)}U^\dagger$.*

The next theorem we state here concerns a general explicit form of the integral decomposition $f(A) = \int_{\sigma(A)} f(\lambda) dP^{(A)}(\lambda)$. As a matter of fact, up to multiplicity, one can always reduce to a multiplicative operator in a L^2 space, as it happens for the position operator X. Again, this theorem can be restated for generally normal operators.

Theorem 2.2.62 (Spectral representation theorem for selfadjoint operators). *Let A be a selfadjoint operator in the complex Hilbert space $\mathcal{H}$. The following facts hold.*

(a) *$\mathcal{H}$ may be decomposed as a Hilbert sum*[10] *$\mathcal{H} = \oplus_{a \in S} \mathcal{H}_a$, whose summands $\mathcal{H}_a$ are closed and orthogonal. Moreover:*
(i) *for any $a \in S$,*

$$A(\mathcal{H}_a \cap D(A)) \subset \mathcal{H}_a$$

[10] S is countable, at most, if $\mathcal{H}$ is separable.

and, more generally, for any measurable $f : \sigma(A) \to \mathbb{C}$,

$$f(A)(\mathcal{H}_a \cap D(f(A))) \subset \mathcal{H}_a$$

(ii) *for any* $a \in S$, *there exist a unique finite positive Borel measure* μ_a *on* $\sigma(A) \subset \mathbb{R}$, *and a surjective isometric operator* $U_a : \mathcal{H}_a \to L^2(\sigma(A), \mu_a)$, *such that:*

$$U_a f(A)|_{\mathcal{H}_a} U_a^{-1} = f\cdot$$

for any measurable $f : \sigma(A) \to \mathbb{C}$, *where* $f\cdot$ *is the point-wise multiplication by* f *on* $L^2(\sigma(A), \mu_a)$.

(b) *If* $supp\{\mu_a\}_{a\in S}$ *is the complementary set to the numbers* $\lambda \in \mathbb{R}$ *for which there exists an open set* $O_\lambda \subset \mathbb{R}$ *with* $O_\lambda \ni \lambda$, $\mu_a(O_\lambda) = 0$ *for any* $a \in S$, *then*

$$\sigma(A) = supp\{\mu_a\}_{a\in S} \,.$$

Notice that the theorem encompasses the case of an operator A in $\mathcal{H}$ with $\sigma(A) = \sigma_p(A)$. Suppose in particular that every eigenspace is one-dimensional and the whole Hilbert space is separable. Let $\sigma(A) = \sigma_p(A) = \{\lambda_n \mid n \in \mathbb{N}\}$. In this case

$$A = \sum_{n\in\mathbb{N}} \lambda_n \langle x_n, \ \rangle x_n \,,$$

where x_λ is a unit eigenvector with eigenvalue λ_n. Consider the σ-algebra on $\sigma(A)$ made of all subsets and define $\mu(E) :=$ number of elements of $E \subset \sigma(E)$. In this case, $\mathcal{H}$ is isomorphic to $L^2(\sigma(A), \mu)$ and the isomorphism is $U : \mathcal{H} \ni x \mapsto \psi_x \in L^2(\sigma(A), \mu)$ with $\psi_x(n) := \langle x_n | x\rangle$ if $n \in \mathbb{N}$. With this surjective isometry, trivially

$$Uf(A)U^{-1} = U \int_{\sigma(A)} f(\lambda) dP^{(A)}(\lambda) U^{-1} = U \sum_{n\in\mathbb{N}} f(\lambda_n) \langle x_n, \ \rangle x_n U^{-1} = f\cdot \,.$$

If all eigenspaces have dimension 2, exactly two copies of $L^2(\sigma(A), \mu)$ are sufficient to improve the construction. If the dimension depends on the eigenspace, the construction can be rebuilt exploiting many copies of different $L^2(S_k, \mu_k)$, where the S_k are suitable (not necessarily disjoint) subsets of $\sigma(A)$ and μ_k the measure which counts the elements of S_k.

The last tool we introduce is the notion of *joint spectral measure*. Everything is stated in the following theorem [5; 6].

Theorem 2.2.63 (Joint spectral measure). *Consider selfadjoint operators* $A_1, A_2, \ldots, A_n$ *in the complex Hilbert space* $\mathcal{H}$. *Suppose that the spectral measures of those operators pairwise commute:*

$$P^{(A_k)}_{E_k} P^{(A_h)}_{E_h} = P^{(A_h)}_{E_h} P^{(A_k)}_{E_k} \quad \forall k, h \in \{1, \ldots, n\}\,, \forall E_k, E_h \in \mathcal{B}(\mathbb{R})\,.$$

There is a unique PVM, $P^{(A_1\times\cdots\times A_n)}$, on $\mathbb{R}^n$ such that

$$P^{(A_1\times\cdots\times A_n)}(E_1\times\cdots\times E_n) = P^{(A_1)}_{E_1}\cdots P^{(A_n)}_{E_n}\,, \quad \forall E_1,\ldots,E_n \in \mathcal{B}(\mathbb{R})\,.$$

For every $f:\mathbb{R}\to\mathbb{C}$ measurable, it holds

$$\int_{\mathbb{R}^n} f(x_k)dP^{(A_1\times\cdots\times A_n)}(x) = f(A_k)\,, \quad k=1,\ldots,n \tag{2.63}$$

where $x=(x_1,\ldots,x_k,\ldots,x_n)$.

Definition 2.2.64. *Referring to Theorem 2.2.63, the PVM $P^{(A_1\times\cdots\times A_n)}$ is called the* **joint spectral measure** *of $A_1,A_2,\ldots,A_n$ and its support supp($P^{(A_1\times\cdots\times A_n)}$), i.e. the complement in $\mathbb{R}^n$ to the largest open set A with $P_A=0$, is called the* **joint spectrum** *of $A_1,A_2,\ldots,A_n$.*

Example 2.2.65. The simplest example is provided by considering the n position operators X_m in $L^2(\mathbb{R}^n,d^nx)$. It should be clear that the n spectral measures commute because $P^{(X_k)}_E$, for $E\in\mathcal{B}(\mathbb{R})$, is the multiplicative operator for $\chi_{\mathbb{R}\times\cdots\times\mathbb{R}\times E\times\mathbb{R}\times\cdots\times\mathbb{R}}$ the factor E staying in the k-th position among the n Cartesian factors. In this case, the joint spectrum of the n operators X_m coincides with $\mathbb{R}^n$ itself.

A completely analogous discussion holds for the n momentum operators P_k, since they are related to the position ones by means of the unitary Fourier-Plancherel operator as already seen several times. Again the joint spectrum of the n operators P_m coincides with $\mathbb{R}^n$ itself.

2.2.7 *Measurable functional calculus*

The following proposition states some useful properties of $f(A)$, where A is selfadjoint and $f:\mathbb{R}\to\mathbb{C}$ is Borel measurable. These properties define the so called *measurable functional calculus.* We suppose here that $A=A^\dagger$, but the statements can be reformulated for normal operators [5; 6].

Proposition 2.2.66. *Let A be a selfadjoint operator in the complex Hilbert space $\mathcal{H}$, $f,g:\sigma(A)\to\mathbb{C}$ measurable functions, $f\cdot g$ and $f+g$ respectively denote the point-wise product and the point-wise sum of functions. The following facts hold.*

- **(a)** $f(A)=\sum_{k=0}^n a_kA^k$ *where the right-hand side is defined in its standard domain $D(A^n)$ when $f(\lambda)=\sum_{k=0}^n a_k\lambda^k$ with $a_n\neq 0$;*
- **(b)** $f(A)=P^{(A)}(E)$ *if $f=\chi_E$ the characteristic function of $E\in\mathcal{B}(\sigma(A))$;*
- **(c)** $f(A)^\dagger = f^*(A)$ *where * denotes the complex conjugation;*
- **(d)** $f(A)+g(A)\subset(f+g)(A)$ *and $D(f(A)+g(A))\subset\Delta_f\cap\Delta_g$ (the symbol "$\subset$" can be replaced by "$=$" if and only if $\Delta_{f+g}=\Delta_f\cap\Delta_g$);*

(e) $f(A)f(B) \subset (f \cdot g)(A)$ *and* $D(f(A)f(B)) = \Delta_{f \cdot g} \cap \Delta_g$ *(the symbol "$\subset$" can be replaced by "$=$" if and only if* $\Delta_{f \cdot g} \subset \Delta_g$*);*

(f) $f(A)^\dagger f(A) = |f|^2(A)$ *so that* $D(f(A)^\dagger f(A)) = \Delta_{|f|^2}$*;*

(g) $\langle x, f(A)x\rangle \geq 0$ *for* $x \in \Delta_f$ *if* $f \geq 0$*;*

(h) $||f(A)x||^2 = \int_{\sigma(A)} |f(\lambda)|^2 d\mu_{xx}(\lambda)$*, if* $x \in \Delta_f$*, in particular, if* f *is bounded or* $P^{(A)}$*-essentially bounded*[11] *on* $\sigma(A)$*,* $f(A) \in \mathfrak{B}(\mathcal{H})$ *and*

$$||f(A)|| \leq ||f||_\infty^{P^{(A)}} \leq ||f||_\infty .$$

(i) *If* $U : \mathcal{H} \to \mathcal{H}$ *is unitary,* $Uf(A)U^\dagger = f(UAU^\dagger)$ *and, in particular,* $D(f(UAU^\dagger)) = UD(f(A)) = U(\Delta_f)$*.*

(j) *If* $\phi : \mathbb{R} \to \mathbb{R}$ *is measurable, then* $\mathcal{B}(\mathbb{R}) \ni E \mapsto P'(E) := P^{(A)}(\phi^{-1}(E))$ *is a PVM on* $\mathbb{R}$*. Introducing the selfadjoint operator*

$$A' = \int_{\mathbb{R}} \lambda' dP'(\lambda')$$

such that $P^{(A')} = P'$*, we have*

$$A' = \phi(A) .$$

Moreover, if $f : \mathbb{R} \to \mathbb{C}$ *is measurable,*

$$f(A') = (f \circ \phi)(A) \quad \text{and} \quad \Delta'_f = \Delta_{f \circ \phi} .$$

2.2.8 *Elementary formalism for the infinite dimensional case*

To complete the discussion in the introduction, let us show how practically the physical hypotheses on quantum systems (1)-(3) have to be mathematically interpreted (again reversing the order of (2) and (3) for our convenience) in the general case of infinite dimensional Hilbert spaces. Our general assumptions on the mathematical description of quantum systems are the following ones.

(1) A quantum mechanical system S is always associated to complex Hilbert space $\mathcal{H}$, finite or infinite dimensional;
(2) observables are pictured in terms of (generally unbounded) *selfadjoint* operators A in $\mathcal{H}$;
(3) states are of equivalence classes of *unit* vectors $\psi \in \mathcal{H}$, where $\psi \sim \psi'$ iff $\psi = e^{ia}\psi'$ for some $a \in \mathbb{R}$.

[11] Remark 2.2.51.

Let us show how the mathematical assumptions (1)-(3) permit us to set the physical properties of quantum systems (1)-(3) of Section 2.1.1.2 into mathematically nice form in the general case of an infinite dimesional Hilbert space $\mathcal{H}$.

(1) Randomness. The Borel subset $E \subset \sigma(A)$, represents the outcomes of measurement procedures of the observable associated with the selfadjoint operator A. (In case of continuous spectrum the outcome of a measurement is at least an interval in view of the experimental errors.) Given a state represented by the unit vector $\psi \in \mathcal{H}$, the probability to obtain $E \subset \sigma(A)$ as an outcome when measuring A is

$$\mu_{\psi,\psi}^{(P^{(A)})}(E) := ||P_E^{(A)}\psi||^2 ,$$

where we have used the PVM $P^{(A)}$ of the operator A.

Going along with this interpretation, the **expectation value**, $\langle A \rangle_\psi$, of A when the state is represented by the unit vector $\psi \in \mathcal{H}$, turns out to be

$$\langle A \rangle_\psi := \int_{\sigma(A)} \lambda \, d\mu_{\psi,\psi}^{(P^{(A)})}(\lambda) . \tag{2.64}$$

This identity makes sense provided $id : \sigma(A) \ni \lambda \to \lambda$ belongs to $L^1(\sigma(A), \mu_{\psi,\psi}^{(P^{(A)})})$ (which is equivalent to say that $\psi \in \Delta_{|id|^{1/2}}$ and, in turn, that $\psi \in D(|A|^{1/2})$), otherwise the expectation value is not defined.

Since

$$L^2(\sigma(A), \mu_{\psi,\psi}^{(P^{(A)})}) \subset L^1(\sigma(A), \mu_{\psi,\psi}^{(P^{(A)})})$$

because $\mu_{\psi,\psi}^{(P^{(A)})}$ is finite, we have the popular identity arising from (2.48),

$$\langle A \rangle_\psi = \langle \psi, A\psi \rangle \quad \text{if } \psi \in D(A) . \tag{2.65}$$

The associated **standard deviation**, ΔA_ψ, results to be

$$\Delta A_\psi^2 := \int_{\sigma(A)} (\lambda - \langle A \rangle_\psi)^2 \, d\mu_{\psi,\psi}^{(P^{(A)})}(\lambda) . \tag{2.66}$$

This definition makes sense provided $id \in L^2(\sigma(A), \mu_{\psi,\psi}^{(P^{(A)})})$ (which is equivalent to say that $\psi \in \Delta_{id}$ and, in turn, that $\psi \in D(A)$).

As before, the functional calculus permits us to write the other popular identity

$$\Delta A_\psi^2 = \langle \psi, A^2\psi \rangle - \langle \psi, A\psi \rangle^2 \quad \text{if } \psi \in D(A^2) \subset D(A) . \tag{2.67}$$

We stress that now, Heisenberg inequalities, as established in exercise 2.1.14, are now completely justified as the reader can easily check.

(3) Collapse of the state. If the Borel set $E \subset \sigma(A)$ is the outcome of the (idealized) measurement of A, when the state is represented by the unit vector

$\psi \in \mathcal{H}$, the new state immediately after the measurement is represented by the unit vector

$$\psi' := \frac{P_E^{(A)}\psi}{||P_E^{(A)}\psi||} . \tag{2.68}$$

Remark 2.2.67. *Obviously this formula does not make sense if* $\mu_{\psi,\psi}^{(P^{(A)})}(E) = 0$ *as expected. Moreover the arbitrary phase affecting* ψ *does not give rise to troubles due to the linearity of* $P_E^{(A)}$ *.*

(2) Compatible and Incompatible Observables. Two observables A, B are compatible – i.e. they can be simultaneously measured – if and only if their **spectral measures commute** which means

$$P_E^{(A)}P_F^{(B)} = P_F^{(B)}P_E^{(A)} , \quad E \in \mathcal{B}(\sigma(A)) , \quad F \in \mathcal{B}(\sigma(B)) . \tag{2.69}$$

In this case,

$$||P_E^{(A)}P_F^{(B)}\psi||^2 = ||P_F^{(B)}P_E^{(A)}\psi||^2 = ||P_{E\times F}^{(A,B)}\psi||^2$$

where $P^{(A,B)}$ is the joint spectral measure of A and B, has the natural interpretation of the probability to obtain the outcomes E and F for a simultaneous measurement of A and B. If instead A and B are incompatible it may happen that

$$||P_E^{(A)}P_F^{(B)}\psi||^2 \neq ||P_F^{(B)}P_E^{(A)}\psi||^2 .$$

Sticking to the case of A and B being incompatible, exploiting (2.68),

$$||P_E^{(A)}P_F^{(B)}\psi||^2 = \left|\left|P_E^{(A)}\frac{P_F^{(B)}\psi}{||P_F^{(B)}\psi||}\right|\right|^2 ||P_F^{(B)}\psi||^2 \tag{2.70}$$

has the natural meaning of *the probability of obtaining first* F *and next* E *in a subsequent measurement of* B *and* A*.*

Remark 2.2.68.

(a) *It is worth stressing that the notion of probability we are using here cannot be a classical notion because of the presence of incompatible observables. The theory of conditional probability cannot follows the standard rules. The probability* $\mu_\psi(E_A|F_B)$*, that (in a state defined by a unit vector* ψ*) a certain observable* A *takes the value* E_A *when the observable* B *has the value* F_B*, cannot be computed by the standard procedure*

$$\mu_\psi(E_A|F_B) = \frac{\mu_\psi(E_A \textit{ AND } F_B)}{\mu_\psi(F_B)}$$

if A and B are incompatible, just because, in general, nothing exists which can be interpreted as the event "E_A AND F_B" if $P_E^{(A)}$ and $P_F^{(B)}$ do not commute! The correct formula is

$$\mu_\psi(E_A|F_B) = \frac{\langle\psi, P_F^{(B)} P_E^{(A)} P_F^{(B)} \psi\rangle}{||P_F^{(B)}\psi||^2}$$

which leads to well known different properties with respect to the classical theory, the so called combination of "probability amplitudes" in particular. As a matter of fact, up to now we do not have a clear notion of (quantum) probability. This issue will be clarified in the next section.

(b) *The reason to pass from operators to their spectral measures in defining compatible observables is that, if A and B are selfadjoint and defined on different domains, $AB = BA$ does not make sense in general. Moreover it is possible to find counterexamples (due to Nelson) where commutativity of A and B on common dense invariant subspaces does not imply that their spectral measures commute. However, from general results again due to Nelson, one has the following nice result (see exercise 2.3.76).*

Proposition 2.2.69. *If selfadjoint operators, A and B, in a complex Hilbert space $\mathcal{H}$ commute on a common dense invariant domain D where $A^2 + B^2$ is essentially selfadjoint, then the spectral measures of A and B commute.*

The following result, much easier to prove, is also true [5; 6].

Proposition 2.2.70. *Let A and B be selfadjoint operators in the complex Hilbert space $\mathcal{H}$. If $B \in \mathfrak{B}(\mathcal{H})$ the following facts are equivalent,*

(i) *the spectral measures of A and B commute (i.e. (2.69) holds);*
(ii) $BA \subset AB$;
(iii) $Bf(A) \subset f(A)B$, *if* $f : \sigma(A) \to \mathbb{R}$ *is Borel measurable;*
(iv) $P_E^{(A)} B = BP_E^{(A)}$ *if* $E \in \mathcal{B}(\sigma(A))$.

Another useful result toward the converse direction [5; 6] is the following.

Proposition 2.2.71. *Let A and B be selfadjoint operators in the complex Hilbert space $\mathcal{H}$ such that their spectral measures commute. The following facts hold.*

(a) $ABx = BAx$ *if* $x \in D(AB) \cap D(BA)$;
(b) $\langle Ax, By\rangle = \langle Bx, Ay\rangle$ *if* $x, y \in D(A) \cap D(B)$.

2.2.9 *Technical interemezzo: three operator topologies*

In QM, there are at least 7 relevant topologies [13] which enter the game discussing sequences of operators. Here we limit ourselves to quickly illustrate the three most important ones. We assume that $\mathcal{H}$ is a complex Hilbert space though the illustrated examples may be extended to more general context with some re-adaptation.

(a) The strongest topology is the **uniform operator topology** in $\mathfrak{B}(\mathcal{H})$: It is the topology induced by the operator norm $|| \ ||$ defined in (2.29).

As a consequence of the definition of this topology, a sequence of elements $A_n \in \mathfrak{B}(\mathcal{H})$ is said to **uniformly** converge to $A \in \mathfrak{B}(\mathcal{H})$ when $||A_n - A|| \to 0$ for $n \to +\infty$.

We already know that $\mathfrak{B}(\mathcal{H})$ is a Banach algebra with respect to that norm and also a C^* algebra.

(b) If $\mathfrak{L}(D;\mathcal{H})$ with $D \subset \mathcal{H}$ a subspace, denotes the complex vector space of the operators $A : D \to \mathcal{H}$, the **strong operator topology** on $\mathfrak{L}(D;\mathcal{H})$ is the topology induced by the seminorms p_x with $x \in D$ and $p_x(A) := ||Ax||$ if $A \in \mathfrak{L}(D;\mathcal{H})$.

As a consequence of the definition of this topology, a sequence of elements $A_n \in \mathfrak{L}(D;\mathcal{H})$ is said to **strongly** converge to $A \in \mathfrak{L}(D;\mathcal{H})$ when $||(A_n - A)x|| \to 0$ for $n \to +\infty$ for every $x \in D$.

It should be evident that, if we restrict ourselves to work in $\mathfrak{B}(\mathcal{H})$, the uniform operator topology is stronger than the strong operator topology.

(c) The **weak operator topology** on $\mathfrak{L}(D;\mathcal{H})$ is the topology induced by the seminorms $p_{x,y}$ with $x \in \mathcal{H}$, $y \in D$ and $p_{x,y}(A) := |\langle x, Ay\rangle|$ if $A \in \mathfrak{L}(D;\mathcal{H})$.

As a consequence of the definition of this topology, a sequence of elements $A_n \in \mathfrak{L}(D;\mathcal{H})$ is said to **weakly** converge to $A \in \mathfrak{L}(D;\mathcal{H})$ when $|\langle x, (A_n - A)y\rangle|| \to 0$ for $n \to +\infty$ for every $x \in \mathcal{H}$ and $y \in D$.

It should be evident that, the strong operator topology is stronger than the weak operator topology.

Example 2.2.72.
(1) If $f : \mathbb{R} \to \mathbb{C}$ is Borel measurable, and A a selfadjoint operator in $\mathcal{H}$, consider the sets

$$R_n := \{r \in \mathbb{R} \,|\, |f(r)| < n\} \quad \text{for } n \in \mathbb{N}\,.$$

It is clear that $\chi_{R_n} f \to f$ pointwise as $n \to +\infty$ and that $|\chi_{R_n} f|^2 \leq |f|^2$. As a consequence of restricting the operators on the left-hand side to Δ_f,

$$\left.\int_{\sigma(A)} \chi_{R_n} f dP^{(A)}\right|_{\Delta_f} \to f(A) \quad \text{strongly, for } n \to +\infty\,,$$

as an immediate consequence of Lebesgue's dominate convergence theorem and the first part of (h) in Proposition 2.2.66.

(2) If in the previous example f is bounded on $\sigma(A)$, and $f_n \to f$ uniformly on $\sigma(A)$, (or P-essentially uniformly $||f - f_n||_\infty^{(P^{(A)})} \to 0$ for $n \to +\infty$) then

$$f_n(A) \to f(A) \quad \text{uniformly, as } n \to +\infty\,,$$

again for the (second part of (h) in Proposition 2.2.66.

Exercise 2.2.73. *Prove that a selfadjoint operator A in the complex Hilbert $\mathcal{H}$ admits a dense set of analytic vectors in its domain.*

Solution. Consider the class of functions $f_n = \chi_{[-n,n]}$ where $n \in \mathbb{N}$. As in (1) of example 2.2.72, we have $\psi_n := f_n(A)\psi = \int_{[-n,n]} 1 dP^{(A)}\psi \to \int_{\mathbb{R}} 1 dP^{(A)}\psi = P_{\mathbb{R}}^{(A)}\psi = \psi$ for $n \to +\infty$. Therefore the set $D := \{\psi_n | \psi \in \mathcal{H}, n \in \mathbb{N}\}$ is dense in $\mathcal{H}$. The elements of D are analytic vectors for A as we go to prove. Clearly $\psi_n \in D(A^k)$ since $\mu_{\psi_n,\psi_n}^{(P^{(A)})}(E) = \mu_{\psi,\psi}^{(P^{(A)})}(E \cap [-n,n])$ as immediate consequence of the definition of the measure $\mu_{x,y}^{(P)}$, therefore $\int_{\mathbb{R}} |\lambda^k|^2 d\mu_{\psi_n,\psi_n}^{(P^{(A)})}(\lambda) = \int_{[-n,n]} |\lambda|^{2k} d\mu_{\psi,\psi}^{(P^{(A)})}(\lambda) \leq \int_{[-n,n]} |n|^{2k} d\mu_{\psi,\psi}^{(P^{(A)})}(\lambda) \leq |n|^{2k} \int_{\mathbb{R}} d\mu_{\psi,\psi}^{(P^{(A)})}(\lambda) = |n|^{2k}||\psi||^2 < +\infty$. Similarly $||A^k\psi_n||^2 = \langle A^k\psi_n, A^k\psi_n\rangle = \langle \psi_n, A^{2k}\psi_n\rangle = \int_{\mathbb{R}} \lambda^{2k} d\mu_{\psi_n,\psi_n}^{(P^{(A)})}(\lambda) \leq |n|^{2k}||\psi||^2$. We conclude that $\sum_{k=0}^{+\infty} \frac{(it)^k}{k!}||A^k\psi_n||$ conveges for every $t \in \mathbb{C}$ as it is dominated by the series $\sum_{k=0}^{+\infty} \frac{|t|^k}{k!}|n|^{2k}||\psi||^2 = e^{|t|\,|n|^2}||\psi||^2$.

2.3 More Fundamental Quantum Structures

The question we want to answer now is the following:

Is there anything more fundamental behind the phenomenological facts (1), (2), and (3) discussed in the first section and their formalization presented in Sect. 2.2.8?

An appealing attempt to answer that question and justify the formalism based on the spectral theory is due to von Neumann [7] (and subsequently extended by Birkhoff and von Neumann). This section is devoted to quickly review an elementary part of those ideas, adding however several more modern results (see also [11] for a similar approach).

2.3.1 *The Boolean logic of CM*

Consider a classical Hamiltonian system described in symplectic manifold (Γ, ω), where $\omega = \sum_{k=1}^n dq^k \wedge dp_k$ in any system of local symplectic coordinates

$q^1, \ldots, q^n, p_1, \ldots, p_n$. The state of the system at time t is a point $s \in \Gamma$, in local coordinates $s = (q^1, \ldots, q^n, p_1, \ldots, p_n)$, whose evolution $\mathbb{R} \ni t \mapsto s(t)$ is a solution of the *Hamiltonian equation* of motion. Always in local symplectic coordinates, they read

$$\frac{dq^k}{dt} = \frac{\partial H(t,q,p)}{\partial p_k}, \quad k = 1, \ldots, n, \tag{2.71}$$

$$\frac{dp_k}{dt} = -\frac{\partial H(t,q,p)}{\partial q^k}, \quad k = 1, \ldots, n, \tag{2.72}$$

H being the Hamiltonian function of the system, depending on the (inertial) reference frame. Every physical *elementary property*, E, that the system may possess at a certain time t, i.e., which can be true or false at that time, can be identified with a subset $E \subset \Gamma$. The property is true if $s \in E$ and it is not if $s \notin E$. From this point of view, the standard set theory operations $\cap$, $\cup$, $\subset$, $\neg$ (where $\neg E := \Gamma \setminus E$ from now on is the **complement operation**) have a logical interpretation:

(i) $E \cap F$ corresponds to the property "E AND F";
(ii) $E \cup F$ corresponds to the property "E OR F";
(iii) $\neg E$ corresponds to the property "NOT F";
(iv) $E \subset F$ means "E IMPLIES F".

In this context:

(v) Γ is the property which is always true;
(vi) $\varnothing$ is the property which is always false.

This identification is possible because, as is well known, the logical operations have the same algebraic structure of the set theory operations.

As soon as we admit the possibility to construct statements including *countably infinite number of disjunctions or conjunctions*, we can enlarge our interpretation towards the abstract measure theory, interpreting the states as *probability Dirac measures* supported on a single point. To this end, we first restrict the class of possible elementary properties to the Borel σ-algebra of Γ, $\mathcal{B}(\Gamma)$. For various reasons, this class of sets seems to be sufficiently large to describe physics (in particular $\mathcal{B}(\Gamma)$ includes the preimages of measurable sets under continuous functions). A state at time t, $s \in \Gamma$, can be viewed as a Dirac measure, δ_s, supported on s itself. If $E \in \mathcal{B}(\Gamma)$, $\delta_s(E) = 0$ if $s \notin E$ or $\delta_s(E) = 1$ if $s \in E$.

If we do not have a perfect knowledge of the system, as for instance it happens in *statistical mechanics*, the state at time t, μ, is a proper probability measure on $\mathcal{B}(\Gamma)$ which now, is allowed to attain all values of $[0, 1]$. If $E \in \mathcal{B}(\Gamma)$ is an elementary property of the physical system, $\mu(E)$ denotes the probability that the property E is true for the system at time t.

Remark 2.3.1. *The evolution equation of μ, in statistical mechanics is given by the well-known* Liouville's equation *associate with the Hamiltonian flow. In that case, μ is proportional to the natural symplectic volume measure of Γ, $\Omega = \omega \wedge \cdots \wedge \omega$ (n-times, where $2n = dim(\Gamma)$). In fact, we have $\mu = \rho\Omega$, where the non-negative function ρ is the so-called* **Liouville density** *satisfying the famous* Liouville's equation. *In symplectic local coordinates that equation reads*

$$\frac{\partial \rho(t,q,p)}{\partial t} + \sum_{k=1}^{n} \left(\frac{\partial \rho}{\partial q^k}\frac{\partial H}{\partial p_k} - \frac{\partial \rho}{\partial p_k}\frac{\partial H}{\partial q^k} \right) = 0 \, .$$

We shall not deal any further with this equation in this section.

More complicated classical quantities of the system can be described by *Borel measurable* functions $f : \Gamma \to \mathbb{R}$. Measurability is a good requirement as it permits one to perform physical operations like computing, for instance, the *expectation value* (at a given time) when the state is μ:

$$\langle f \rangle_\mu = \int_\Gamma f\mu \, .$$

Also elementary properties can be pictured by measurable functions, in fact they are one-to-one identified with all the Borel measurable functions $g : \Gamma \to \{0,1\}$. The Borel set E_g associated to g is $g^{-1}(\{1\})$ and in fact $g = \chi_{E_g}$.

A generic physical quantity, a measurable function $f : \Gamma \to \mathbb{R}$, is completely determined by the class of Borel sets (elementary properties) $E_B^{(f)} := f^{-1}(B)$ where $B \in \mathcal{B}(\mathbb{R})$. The meaning of $E_B^{(f)}$ is

$$E_B^{(f)} = \text{“the value of } f \text{ belongs to } B\text{''} \, . \tag{2.73}$$

It is possible to prove [5] that the map $\mathcal{B}(\mathbb{R}) \ni B \mapsto E_B^{(f)}$ permits one to reconstruct the function f. The sets $E_B^{(f)} := f^{-1}(B)$ form a σ-algebra as well and the class of sets $E_B^{(f)}$ satisfies the following elementary properties when B ranges in $\mathcal{B}(\mathbb{R})$.

(Fi) $E_{\mathbb{R}}^{(f)} = \Gamma$;
(Fii) $E_B^{(f)} \cap E_C^{(f)} = E_{B\cap C}^{(f)}$;
(Fiii) If $N \subset \mathbb{N}$ and $\{B_k\}_{k\in N} \subset \mathcal{B}(\mathbb{R})$ satisfies $B_j \cap B_k = \varnothing$ if $k \neq j$, then

$$\cup_{j\in N} E_{B_j}^{(f)} = E_{\cup_{j\in N} B_j}^{(f)} \, .$$

These conditions just say that $\mathcal{B}(\mathbb{R}) \ni B \mapsto E_B^{(f)}$ is a **homomorpism of σ-algebras**.

For future convenience, we observe that our model of *classical* elementary properties can be also viewed as another mathematical structure, when referring to the notion of *lattice*.

Definition 2.3.2. *A partially ordered set* $(X, \geq)$ *is a* **lattice** *when, for* $a, b \in X$,

(a) $\sup\{a, b\}$ *exists, denoted* $a \vee b$ *(sometimes called "join");*
(b) $\inf\{a, b\}$ *exists, written* $a \wedge b$ *(sometimes "meet").*
(The partially ordered set is not required to be totally ordered.)

Remark 2.3.3.
(a) *In our considered concrete case,* $X = \mathcal{B}(\mathbb{R})$ *and* $\geq$ *is nothing but* $\supset$ *and thus* $\vee$ *means* $\cup$ *and* $\wedge$ *has the meaning of* $\cap$.
(b) *In the general case,* $\vee$ *and* $\wedge$ *turn out to be separately* associative, *therefore it make sense to write* $a_1 \vee \cdots \vee a_n$ *and* $a_1 \wedge \cdots \wedge a_n$ *in a lattice. Moreover they are also separately* commutative *so*

$$a_1 \vee \cdots \vee a_n = a_{\pi(1)} \vee \cdots \vee a_{\pi(n)} \quad \text{and} \quad a_1 \wedge \cdots \wedge a_n = a_{\pi(1)} \wedge \cdots \wedge a_{\pi(n)}$$

for every permutation $\pi : \{1, \ldots, n\} \to \{1, \ldots, n\}$.

Let us pass to some relevant definitions.

Definition 2.3.4. *A lattice* $(X, \geq)$ *is said to be:*

(a) **distributive** *if* $\vee$ *and* $\wedge$ *distribute over one another: for any* $a, b, c \in X$,

$$a \vee (b \wedge c) = (a \vee b) \wedge (a \vee c)\,, \quad a \wedge (b \vee c) = (a \wedge b) \vee (a \wedge c)\,;$$

(b) **bounded** *if it admits a minimum* $\mathbf{0}$ *and a maximum* $\mathbf{1}$ *(sometimes called "bottom" and "top");*
(c) **orthocomplemented** *if bounded and equipped with a mapping* $X \ni a \mapsto \neg a$, *where* $\neg a$ *is the* **orthogonal complement** *of* a, *such that:*

(i) $a \vee \neg a = \mathbf{1}$ *for any* $a \in X$;
(ii) $a \wedge \neg a = \mathbf{0}$ *for any* $a \in X$;
(iii) $\neg(\neg a) = a$ *for any* $a \in X$;
(iv) $a \geq b$ *implies* $\neg b \geq \neg a$ *for any* $a, b \in X$;

(d) σ**-complete**, *if every countable set* $\{a_n\}_{n \in \mathbb{N}} \subset X$ *admits least upper bound* $\vee_{n \in \mathbb{N}} a_n$.

A lattice with properties (a), (b) and (c) is called a **Boolean algebra**. *A Boolean algebra satisfying (d) is a* **Boolean** σ**-algebra**.

Definition 2.3.5. *If X, Y are lattices, a map $h : X \to Y$ is a* **(lattice) homomorphism** *when*

$$h(a \vee_X b) = h(a) \vee_Y h(b)\,, \quad h(a \wedge_X b) = h(a) \wedge_Y h(b)\,, \quad a, b \in X$$

(with the obvious notations.) If X and Y are bounded, a homomorphism h is further required to satisfy

$$h(\mathbf{0}_X) = \mathbf{0}_Y\,, \quad h(\mathbf{1}_X) = \mathbf{1}_Y\,.$$

If X and Y are orthocomplemented, a homomorphism h also satisfies

$$h(\neg_X a) = \neg_Y h(x)\,.$$

If X, Y are σ-complete, h further fulfills

$$h(\vee_{n\in\mathbb{N}} a_n) = \vee_{n\in\mathbb{N}} h(a_n)\,, \textit{if } \{a_n\}_{n\in\mathbb{N}} \subset X\,.$$

In all cases (bounded, orthocomplemented, σ-complete lattices, Boolean (σ-)algebras) if h is bijective it is called **isomorphism** *of the relative structures.*

It is clear that, just because it is a concrete σ-algebra, the lattice of the elementary properties of a classical system is a lattice which is *distributive*, *bounded* (here $0 = \varnothing$ and $1 = \Gamma$), *orthocomplemented* (the orthocomplement being the complement with respect to Γ) and *σ-complete.* Moreover, as the reader can easily prove, the above map, $\mathcal{B}(\mathbb{R}) \ni B \mapsto E_B^{(f)}$, is also a homomorphism of Boolean σ-algebras.

Remark 2.3.6. *Given an abstract Boolean σ-algebra X, does there exist a concrete σ-algebra of sets that is isomorphic to the previous one? In this respect the following general result holds, known as* Loomis-Sikorski theorem.[12] *This guarantees that every Boolean σ-algebra is isomorphic to a quotient Boolean σ-algebra $\Sigma/\mathcal{N}$, where Σ is a concrete σ-algebra of sets over a measurable space and $\mathcal{N} \subset \Sigma$ is closed under countable unions; moreover, $\varnothing \in \mathcal{N}$ and for any $A \in \Sigma$ with $A \subset N \in \mathcal{N}$, then $A \in \mathcal{N}$. The equivalence relation is $A \sim B$ iff $A \cup B \setminus (A \cap B) \in \mathcal{N}$, for any $A, B \in \Sigma$. It is easy to see the coset space $\Sigma/\mathcal{N}$ inherits the structure of Boolean σ-algebra from Σ with respect to the (well-defined) partial order relation $[A] \geq [B]$ if $A \supset B$, $A, B \in \Sigma$.*

[12] Sikorski S.: *On the representation of Boolean algebras as field of sets.* Fund. Math. **35**, 247-256 (1948).

2.3.2 *The non-Boolean logic of QM, the reason why observables are selfadjoint operators*

It is evident that the classical like picture illustrated in Sect. 2.3.1 is untenable if referring to quantum systems. The deep reason is that there is pair of elementary properties E, F of quantum systems which are incompatible. Here an elementary property is an observable which, if measured by means of a corresponding experimental apparatus, can only attain two values: 0 if it is false or 1 if it is true. For instance, $E =$ "the component S_x of the electron is $\hbar/2$" and $F =$ "the component S_y of the electron is $\hbar/2$". There is no physical instrument capable to establish if E AND F is true or false. We conclude that some of elementary observables of quantum systems cannot be logically combined by the standard operation of the logic. The model of Borel σ-algebra seems not to be appropriate for quantum systems. However one could try to use some form of lattice structure different form the classical one. The fundamental ideas by von Neumann were the following pair.

(vN1) Given a quantum system, there is a complex separable Hilbert space $\mathcal{H}$ such that the **elementary observables** – the ones which only assume values in $\{0, 1\}$ – are one-to-one represented by all the elements of $\mathcal{L}(\mathcal{H})$, the orthogonal projectors in $\mathfrak{B}(\mathcal{H})$.

(vN2) Two elementary observables P, Q are compatible if and only if they commute as projectors.

Remark 2.3.7.

(a) *As we shall see later (vN1) has to be changed for those quantum systems which admit* superselection rules. *For the moment we stick to the above version of (vN1).*

(b) *The technical requirement of separability will play a crucial role in several places.*

Let us analyze the reasons for von Neumann's postulates. First of all we observe that $\mathcal{L}(\mathcal{H})$ is in fact a lattice if one remembers the relation between orthogonal projectors and closed subspaces stated in Proposition 2.2.46.

Notation 2.3.8. Referring to Proposition 2.2.46, if $P, Q \in \mathcal{L}(\mathcal{H})$, we write $P \geq Q$ if and only if $P(\mathcal{H}) \supset Q(\mathcal{H})$.

$P(\mathcal{H}) \supset Q(\mathcal{H})$ is equivalent to $PQ = Q$. Indeed, if $P(\mathcal{H}) \supset Q(\mathcal{H})$ then there is a Hilbert basis of $P(\mathcal{H})$ $N_P = N_Q \cup N'_Q$ where N_Q ia a Hilbert basis of $Q(\mathcal{H})$ and N'_Q of $Q(\mathcal{H})^{\perp_P}$, the notion of orthogonal being referred to the Hilbert space $P(\mathcal{H})$. From $Q = \sum_{z \in N_Q} \langle z, \cdot \rangle z$ and $P = Q + \sum_{z \in N'_Q} \langle z, \cdot \rangle z$ we have $PQ = Q$.

The converse implication is obvious.

As preannounced, it turns out that $(\mathcal{L}(\mathcal{H}), \geq)$ is a lattice and, in particular, it enjoys the following properties (e.g., see [5]) whose proof is direct.

Proposition 2.3.9. *Let $\mathcal{H}$ be a complex separable Hilbert space and, if $P \in \mathcal{L}(\mathcal{H})$, define $\neg P := I - P$ (the orthogonal projector onto $P(\mathcal{H})^\perp$). With this definition, $(\mathcal{L}(\mathcal{H}), \geq, \neg)$ turns out to be bounded, orthocomplemented, σ-complete lattice which is not distributive if $\dim(\mathcal{H}) \geq 2$.*
More precisely,

(i) *$P \vee Q$ is the orthogonal projector onto $\overline{P(\mathcal{H}) + Q(\mathcal{H})}$.*
The analogue holds for a countable set $\{P_n\}_{n\in\mathbb{N}} \subset \mathcal{P}(\mathcal{H})$, $\vee_{n\in\mathbb{N}} P_n$ is the orthogonal projector onto $\overline{+_{n\in\mathbb{N}} P_n(\mathcal{H})}$.
(ii) *$P \wedge Q$ is the orthogonal projector on $P(\mathcal{H}) \cap Q(\mathcal{H})$.*
The analogue holds for a countable set $\{P_n\}_{n\in\mathbb{N}} \subset \mathcal{P}(\mathcal{H})$, $\wedge_{n\in\mathbb{N}} P_n$ is the orthogonal projector onto $\cap_{n\in\mathbb{N}} P_n(\mathcal{H})$.
(iii) *The bottom and the top are respectively 0 and I.*
Referring to (i) and (ii), it turns out that

$$\vee_{n\in\mathbb{N}} P_n = \lim_{k\to+\infty} \vee_{n\leq k} P_n \quad \text{and} \quad \wedge_{n\in\mathbb{N}} P_n = \lim_{k\to+\infty} \wedge_{n\leq k} P_n$$

with respect to the strong operator topology.

Remark 2.3.10. *The fact that the distributive property does not hold is evident from the following elementary counterexample in $\mathbb{C}^2$ (so that it is valid for every dimension > 1). Let $\{e_1, e_2\}$ be the standard basis of $\mathbb{C}^2$ and define the subspaces $\mathcal{H}_1 := span(e_1)$, $\mathcal{H}_2 := span(e_2)$, $\mathcal{H}_3 := span(e_1 + e_2)$. Finally P_1, P_2, P_3 respectively denote the orthogonal projectors onto these spaces. By direct inspection one sees that $P_1 \wedge (P_2 \vee P_3) = P_1 \wedge I = P_1$ and $(P_1 \wedge P_2) \vee (P_1 \wedge P_3) = 0 \vee 0 = 0$, so that $P_1 \wedge (P_2 \vee P_3) \neq (P_1 \wedge P_2) \vee (P_1 \wedge P_3)$.*

The crucial observation is that, nevertheless $(\mathcal{L}(\mathcal{H}), \geq, \neg)$ includes lots of Boolean σ algebras, and precisely the maximal sets of pairwise compatible projectors [5].

Proposition 2.3.11. *Let $\mathcal{H}$ be a complex separable Hilbert space and consider the lattice $(\mathcal{L}(\mathcal{H}), \geq, \neg)$. If $\mathcal{L}_0 \subset \mathcal{L}(\mathcal{H})$ is a maximal subset of pairwise commuting elements, then $\mathcal{L}_0$ contains 0, I is $\neg$-closed and, if equipped with the restriction of the lattice structure of $(\mathcal{L}(\mathcal{H}), \geq, \neg)$, turns out to be a Boolean σ-algebra.*

In particular, if $P, Q \in \mathcal{L}_0$,

(i) *$P \vee Q = P + Q - PQ$;*
(ii) *$P \wedge Q = PQ$.*

Proof. $\mathcal{L}_0$ includes both 0 and I because $\mathcal{L}_0$ is maximally commutative. Having (i) and (ii), due to (iii) in proposition 2.3.9, the sup and the inf of a sequence of projectors of $\mathcal{L}_0$ commute with the elements of $\mathcal{L}_0$, maximality implies that they belong to $\mathcal{L}_0$. Finally (i) and (ii) prove by direct inspection that $\vee$ and $\wedge$ are mutually distributive. Let us prove (ii) and (i) to conclude. If $PQ = QP$, PQ is an orthogonal projector and $PQ(\mathcal{H}) = QP(\mathcal{H}) \subset P(\mathcal{H}) \cap Q(\mathcal{H})$. On the other hand, if $x \in P(\mathcal{H}) \cap Q(\mathcal{H})$ then $Px = x$ and $x = Qx$ so that $PQx = x$ and thus $P(\mathcal{H}) \cap Q(\mathcal{H}) \subset PQ(\mathcal{H})$ and (ii) holds. To prove (i) observe that $\overline{< P(\mathcal{H}), Q(\mathcal{H}) >}^{\perp} = P(\mathcal{H})^{\perp} \cap Q(\mathcal{H})^{\perp}$. Using (ii), this can be rephrased as $I - P \vee Q = (I - P)(I - Q)$ which entails (i) immediately. □

Remark 2.3.12.

(a) *Every set of pairwise commuting orthogonal projectors can be completed to a maximal set as an elementary application of Zorn's lemma. However, since the commutativity property is* not *transitive, there are many possible maximal subsets of pairwise commuting elements in $\mathcal{L}(\mathcal{H})$ with non-empty intersection.*

(b) *As a consequence of the stated proposition, the symbols $\vee$, $\wedge$ and $\neg$ have the same properties in $\mathcal{L}_0$ as the corresponding symbols of classical logic OR, AND and NOT. Moreover $P \geq Q$ can be interpreted as "Q IMPLIES P".*

(c) *There has been many attempts to interpret $\vee$ and $\wedge$ as connectives of a new non-distributive logic when dealing with the whole $\mathcal{L}(\mathcal{H})$:* a quantum logic. *The first noticeable proposal was due to Birkhoff and von Neumann [14]. Nowadays there are lots of quantum logics [15; 16] all regarded with suspicion by physicists. Indeed, the most difficult issue is the physical operational interpretation of these connectives taking into account the fact that they put together incompatible propositions, which cannot be measured simultaneously. An interesting interpretative attempt, due to Jauch, relies upon a result by von Neumann (e.g., [5])*

$$(P \wedge Q)x = \lim_{n \to +\infty} (PQ)^n x \quad \textit{for every } P, Q \in \mathcal{L}(\mathcal{H}) \textit{ and } x \in \mathcal{H}\,.$$

Notice that the result holds in particular if P and Q do not commute, so they are incompatible elementary observables. The right hand side of the identity above can be interpreted as the consecutive and alternated measurement of an infinite sequence of elementary observables P and Q. As

$$||(P \wedge Q)x||^2 = \lim_{n \to +\infty} ||(PQ)^n x||^2 \quad \textit{for every } P, Q \in \mathcal{L}(\mathcal{H}) \textit{ and } x \in \mathcal{H}\,,$$

the probabilty that $P \wedge Q$ is true for a state represented by the unit vector $x \in \mathcal{H}$ is the probabilty that the infinite sequence of consecutive alternated measurements of P and Q produce is true at each step.

We are in a position to clarify why, in this context, observables are PVMs. Exactly as in CM, an observable A is a collection of elementary observables $\{P_E\}_{E\in\mathcal{B}(\mathbb{R})}$ labelled on the Borel sets E of $\mathbb{R}$. Exactly as for classical quantities, (2.73) we can say that the meaning of P_E is

$$P_E = \text{"the value of the observable belongs to } E\text{"}\,. \tag{2.74}$$

We expect that all those elementary observables are pairwise compatible and that they satisfy the same properties (Fi)-(Fiii) as for classical quantities. We can complete $\{P_E\}_{E\in\mathcal{B}(\mathbb{R})}$ to a maximal set of compatible elementary observables. Taking Proposition 2.3.11 into account (Fi)-(Fiii) translate into

(i) $P_{\mathbb{R}} = I$;
(ii) $P_E P_F = P_{E\cap F}$;
(iii) If $N \subset \mathbb{N}$ and $\{E_k\}_{k\in N} \subset \mathcal{B}(\mathbb{R})$ satisfies $E_j \cap E_k = \varnothing$ for $k \neq j$, then

$$\sum_{j\in N} P_{E_j}x = P_{\cup_{j\in N} E_j}x \quad \text{for every } x \in \mathcal{H}\,.$$

(The presence of x is due to the fact that the convergence of the series if N is infinite is in the strong operator topology as declared in the last statement of Proposition 2.3.9.) *In other words we have just found Definition 2.2.47, specialized to PVM on $\mathbb{R}$: Observables in QM are PVM over $\mathbb{R}$!*

We know that all PVM over $\mathbb{R}$ are one-to-one associated to all selfadjoint operators in view of the results presented in the previous section (see (e) in remark 2.2.60). We conclude that, adopting von Neumann's framework, in QM observables are naturally described by selfadjoint operators, whose spectra coincide with the set of values attained by the observables.

2.3.3 *Recovering the Hilbert space structure*

A reasonable question to ask is whether there are better reasons for choosing to describe quantum systems via a lattice of orthogonal projectors, other than the kill-off argument "it works". To tackle the problem we start by listing special properties of the lattice of orthogonal projectors, whose proof is elementary.

Theorem 2.3.13. *The bounded, orthocomplemented, σ-complete lattice $\mathcal{L}(\mathcal{H})$ of Propositions 2.3.9 and 2.3.11 satisfies these additional properties:*

(i) **separability** *(for $\mathcal{H}$ separable): if $\{P_a\}_{a\in A} \subset \mathcal{L}(\mathcal{H})$ satisfies $P_iP_j = 0$, $i \neq j$, then A is at most countable;*
(ii) **atomicity and atomisticity**: *there exist elements in $A \in \mathcal{L}(\mathcal{H}) \setminus \{0\}$, called* **atoms**, *for which $0 \leq P \leq A$ implies $P = 0$ or $P = A$; for any*

$P \in \mathcal{L}(\mathcal{H}) \setminus \{0\}$ there exists an atom A with $A \leq P$ ($\mathcal{L}(\mathcal{H})$ is then called **atomic***); For every $P \in \mathcal{L}(\mathcal{H}) \setminus \{0\}$, P is the* sup *of the set of atoms $A \leq P$ ($\mathcal{L}(\mathcal{H})$ is then called* **atomistic***);*

(iii) **orthomodularity***: $P \leq Q$ implies $Q = P \vee ((\neg P) \wedge Q)$;*

(iv) **covering property***: if $A, P \in \mathcal{L}(\mathcal{H})$, with A an atom, satisfy $A \wedge P = 0$, then (1) $P \leq A \vee P$ with $P \neq A \vee P$, and (2) $P \leq Q \leq A \vee P$ implies $Q = P$ or $Q = A \vee P$;*

(v) **irreducibility***: only 0 and I commute with every element of $\mathcal{L}(\mathcal{H})$.*

The orthogonal projectors onto one-dimensional spaces are the only atoms of $\mathcal{L}(\mathcal{H})$.

Irreducibility can easily be proven observing that if $P \in \mathcal{L}(\mathcal{H})$ commutes with all projectors along one-dimensional subspaces, $Px = \lambda_x x$ for every $x \in \mathcal{H}$. Thus $P(x+y) = \lambda_{x+y}(x+y)$ but also $Px + Py = \lambda_x x + \lambda_y y$ and thus $(\lambda_x - \lambda_{x+y})x = (\lambda_{x+y} - \lambda_y)y$, which entails $\lambda_x = \lambda_y$ if $x \perp y$. If $N \subset \mathcal{H}$ is a Hilbert basis, $Pz = \sum_{x \in N} \langle x, z \rangle \lambda x = \lambda z$ for some fixed $\lambda \in \mathbb{C}$. Since $P = P^\dagger = PP$, we conclude that either $\lambda = 0$ or $\lambda = 1$, i.e., either $P = 0$ or $P = I$, as wanted. Orthomodularity is a weaker version of distributivity of $\vee$ with respect to $\wedge$ that we know to be untenable in $\mathcal{L}(\mathcal{H})$.

Actually each of the listed properties admits a physical operational interpretation (e.g. see [15].) So, based on the experimental evidence of quantum systems, we could try to prove, in the absence of any Hilbert space, that elementary propositions with experimental outcome in $\{0, 1\}$ form a poset. More precisely, we could attempt to find a bounded, orthocomplemented σ-complete lattice that verifies conditions (i)–(v) above, and then prove this lattice is described by the orthogonal projectors of a Hilbert space.

The partial order relation of elementary propositions can be defined in various ways. But it will always correspond to the logical implication, in some way or another. Starting from [17] a number of approaches (either of essentially physical nature, or of formal character) have been developed to this end: in particular, those making use of the notion of (quantum) *state*, which we will see in a short while for the concrete case of propositions represented by orthogonal projectors. The object of the theory is now [17] the pair $(\mathcal{O}, \mathcal{S})$, where $\mathcal{O}$ is the class of observables and $\mathcal{S}$ the one of states. The elementary propositions form a subclass $\mathcal{L}$ of $\mathcal{O}$ equipped with a natural poset structure $(\mathcal{L}, \leq)$ (also satisfying a weaker version of some of the conditions (i)–(v)). A state $s \in \mathcal{S}$, in particular, defines the probability $m_s(P)$ that P is true for every $P \in \mathcal{L}$ [17]. As a matter of fact, if $P, Q \in \mathcal{L}$, $P \leq Q$ means by definition that the probability $m_s(P) \leq m_s(Q)$ for every state $s \in \mathcal{S}$. More difficult is to justify that the poset thus obtained is a lattice, i.e. that it admits a greatest lower bound $P \vee Q$ and a least upper bound

$P \wedge Q$ for every P, Q. There are several proposals, very different in nature, to introduce this lattice structure (see [15] and [16] for a general treatise) and make the physical meaning explicit in terms of measurement outcome. See Aerts in [16] for an abstract but operational viewpoint and Sect. 21.1 of [15] for a summary on several possible ways to introduce the lattice structure on the partially ordered set of abstract elementary propositions $\mathcal{L}$.

If we accept the lattice structure on elementary propositions of a quantum system, then we may define the operation of orthocomplementation by the familiar logical/physical negation. Compatible propositions can then be defined in terms of commuting propositions, i.e. commuting elements of a orthocomplemented lattice as follows.

Definition 2.3.14. *Let* $(\mathcal{L}, \geq, \neg)$ *an orthocomplemented lattice. Two elements* $a, b \in \mathcal{L}$ *are said to be:*

(i) **orthogonal** *written* $a \perp b$, *if* $\neg a \geq b$ *(or equivalently* $\neg b \geq a$*);*
(ii) **commuting**, *if* $a = c_1 \vee c_3$ *and* $b = c_2 \vee c_3$ *with* $c_i \perp c_j$ *if* $i \neq j$.

These notions of orthogonality and compatibility make sense beacuse, *a posteriori*, they turn out to be the usual ones when propositions are interpreted via projectors. As the reader may easily prove, two elements $P, Q \in \mathcal{L}(\mathcal{H})$ are orthogonal in accordance with Definition 2.3.14 if and only if $PQ = QP = 0$ (in other words they project onto mutually orthogonal subspaces), and commute in accordance with Definition 2.3.14 if and only if $PQ = QP$. (If $P = P_1 + P_3$ and $Q = P_2 + P_3$ where the orthogonal projectors satisfy $P_i \perp P_j = 0$ for $i \neq j$, we trivially have $PQ = QP$. If conversely, $PQ = QP$, the said decomposition arises for $P_3 := PQ$, $P_1 := P(I - Q)$, $P_2 := Q(I - P)$.)

Now fully-fledged with an orthocomplemented lattice and the notion of compatible propositions, we can attach a physical meaning (an interpretation backed by experimental evidence) to the requests that the lattice be orthocomplemented, complete, atomistic, irreducible and that it have the covering property [15]. Under these hypotheses and assuming there exist at least 4 pairwise-orthogonal atoms, Piron ([18; 19], Sect. 21 of [15], Aerts in [16]) used projective geometry techniques to show that the lattice of quantum propositions can be canonically identified with the closed (in a generalized sense) subsets of a generalized Hilbert space of sorts. In the latter: (a) the field is replaced by a division ring (usually not commutative) equipped with an involution, and (b) there exists a certain non-singular Hermitian form associated with the involution. It has been conjectured by many people (see [15]) that if the lattice is also orthomodular and separable, the division ring can only be picked among $\mathbb{R}, \mathbb{C}$ or $\mathbb{H}$ (quaternion algebra). More recently

Solèr[13], Holland[14] and Aerts–van Steirteghem[15] have found sufficient hypotheses, in terms of the existence of infinite orthogonal systems, for this to happen. Under these hypotheses, if the ring is $\mathbb{R}$ or $\mathbb{C}$, we obtain precisely the lattice of orthogonal projectors of the separable Hilbert space. In the case of $\mathbb{H}$, one gets a similar generalized structure. In all these arguments, the assumption of irreducibility is not really crucial: if property (v) fails, the lattice can be split into irreducible sublattices [20; 15]. Physically-speaking this situation is natural in the presence of *superselection rules*, of which more will be explained soon.

It is worth stressing that the covering property in Theorem 2.3.13 is a crucial property. Indeed there are other lattices relevant in physics verifying all the remaining properties in the aforementioned theorem. Remarkably the family of the so-called *causally closed sets* in a general spacetime satisfies all the said properties but the covering one[16]. This obstruction prevents one from endowing a spacetime with a natural (generalized) Hilbert space structure, while it suggests some ideas towards a formulation of quantum gravity.

2.3.4 *States as measures on $\mathcal{L}(\mathcal{H})$: Gleason's theorem*

Let us introduce an important family of operators. This family will play a decisive rôle in the issue concerning a possible justification of the fact that quantum states are elements of the projective space $P\mathcal{H}$.

2.3.4.1 *Trace class operators*

Definition 2.3.15. *If $\mathcal{H}$ is a complex Hilbert space, $\mathfrak{B}_1(\mathcal{H}) \subset \mathfrak{B}(\mathcal{H})$ denotes the set of* **trace class** *or* **nuclear** *operators, i.e. the operators $T \in \mathfrak{B}(\mathcal{H})$ satisfying*

$$\sum_{z\in N} \langle z, |T|z\rangle < +\infty \tag{2.75}$$

for some Hilbertian basis $N \in \mathcal{H}$ and where $|T| := \sqrt{T^\dagger T}$ defined via functional calculus.

[13] Solèr, M. P.: Characterization of Hilbert spaces by orthomodular spaces. *Communications in Algebra*, **23**, 219-243 (1995).
[14] Holland, S.S.: Orthomodularity in infinite dimensions; a theorem of M. Solèr. *Bulletin of the American Mathematical Society*, **32**, 205-234, (1995).
[15] Aerts, D., van Steirteghem B.: Quantum Axiomatics and a theorem of M.P. Solér. *International Journal of Theoretical Physics*. **39**, 497-502, (2000).
[16] See H. Casini, *The logic of causally closed spacetime subsets*, Class. Quant. Grav. **19**, 2002, 6389-6404.

Remark 2.3.16. *Notice that, above, $T^\dagger T$ is selfadjoint and $\sigma(T^\dagger T) \in [0, +\infty)$, so that $\sqrt{T^\dagger T}$ is well defined as a function of $T^\dagger T$.*

Trace class operators enjoy several remarkable properties [5; 6]. Here we only mention the ones relevant for these lecture notes.

Proposition 2.3.17. *Let $\mathcal{H}$ be a complex Hilbert space, $\mathfrak{B}_1(\mathcal{H})$ satisfying the following properties.*

(a) *If $T \in \mathfrak{B}_1(\mathcal{H})$ and $N \subset \mathcal{H}$ is any Hilbertian basis, then (2.75) holds and thus*

$$||T||_1 := \sum_{z \in N} \langle z, |T| z \rangle$$

is well defined.

(b) *$\mathfrak{B}_1(\mathcal{H})$ is a subspace of $\mathfrak{B}(\mathcal{H})$ which is moreover a two-sided $*$-ideal, namely*

(i) *$AT, TA \in \mathfrak{B}_1(\mathcal{H})$ if $T \in \mathfrak{B}_1(\mathcal{H})$ and $A \in \mathfrak{B}(\mathcal{H})$;*
(ii) *$T^\dagger \in \mathfrak{B}_1(\mathcal{H})$ if $T \in \mathfrak{B}_1(\mathcal{H})$.*

(c) *$||\ ||_1$ is a norm on $\mathfrak{B}_1(\mathcal{H})$ making it a Banach space and satisfying*

(i) *$||TA||_1 \leq ||A||\, ||T||_1$ and $||AT||_1 \leq ||A||\, ||T||_1$ if $T \in \mathfrak{B}_1(\mathcal{H})$ and $A \in \mathfrak{B}(\mathcal{H})$;*
(ii) *$||T||_1 = ||T^\dagger||_1$ if $T \in \mathfrak{B}_1(\mathcal{H})$.*

(d) *If $T \in \mathfrak{B}_1(\mathcal{H})$, the* **trace** *of T,*

$$tr\, T := \sum_{z \in N} \langle z, Tz \rangle \in \mathbb{C}$$

is well defined, does not depend on the choice of the Hilbertian basis N and the sum converges absolutely (so can be arbitrarily re-ordered).

Remark 2.3.18.

(1) *Obviously we have $tr\, |T| = ||T||_1$ if $T \in \mathfrak{B}_1(\mathcal{H})$.*

(2) *The trace just possesses the properties one expects from the finite dimensional case. In particular, [5; 6],*

(i) *it is linear on $\mathfrak{B}_1(\mathcal{H})$;*
(ii) *$tr\, T^\dagger = \overline{tr\, T}$ if $T \in \mathfrak{B}_1(\mathcal{H})$;*
(iii) *the trace satisfies the* **cyclic property**,

$$tr(T_1 \cdots T_n) = tr(T_{\pi(1)} \cdots T_{\pi(n)}) \tag{2.76}$$

if at least one of the T_k belongs to $\mathfrak{B}_1(\mathcal{H})$, the remaining ones are in $\mathfrak{B}(\mathcal{H})$, and $\pi : \{1, \ldots, n\} \to \{1, \ldots, n\}$ is a cyclic permutation.

The trace of $T \in \mathfrak{B}_1(\mathcal{H})$ can be computed on a basis of eigenvectors in view of the following further result [5; 6]. Actually (d) and (e) easily follow from (a), (b), (c), (d), and the spectral theory previously developed.

Proposition 2.3.19. *Let $\mathcal{H}$ be a complex Hilbert space and $T^\dagger = T \in \mathfrak{B}_1(\mathcal{H})$. The following facts hold.*

(a) $\sigma(T) \setminus \{0\} = \sigma_p(T) \setminus \{0\}$. *If $0 \in \sigma(T)$ it may be either the unique element of $\sigma_c(T)$ or an element of $\sigma_p(T)$.*
(b) *Every eigenspace $\mathcal{H}_\lambda$ has finite dimension d_λ provided $\lambda \neq 0$.*
(c) *$\sigma_p(T)$ is made of at most countable number of reals such that*

(i) 0 *is unique possible accumulation point;*
(ii) $||T|| = \max_{\lambda \in \sigma_p(T)} |\lambda|$.

(d) *There is a Hilbert basis of eigenvectors $\{x_{\lambda,a}\}_{\lambda \in \sigma_p(T), a=1,2,\ldots,d_\lambda}$ (d_0 may be infinite) and*

$$tr(T) = \sum_{\lambda \in \sigma_p(T)} d_\lambda \lambda \,,$$

where the sum converges absolutely (and thus can be arbitrarily re-ordered).
(e) *Referring to the basis presented in (d), the spectral decomposition of T reads*

$$T = \sum_{\lambda \in \sigma_p(T)} \lambda P_\lambda$$

where $P_\lambda = \sum_{a=1,2,\ldots,d_\lambda} \langle x_{\lambda,s},\ \rangle x_{\lambda,a}$ and the sum is computed in the strong operator topology and can be re-ordered arbitarily. The convergence holds in the uniform topology too if the set of eigenspaces are suitably ordered in the count.

Corollary 2.3.20. *$tr : \mathfrak{B}_1(\mathcal{H}) \to \mathbb{C}$ is continuous with respect to the norm $||\ ||_1$ because $|trT| \leq tr|T| = ||T||_1$ if $T \in \mathfrak{B}_1(\mathcal{H})$.*

Proof. If $T \in \mathfrak{B}(\mathcal{H})$, we have the *polar decomposition* $T = U|T|$ (see, e.g., [5]) where $U \in \mathfrak{B}(\mathcal{H})$ is isometric on $Ker(T)^\perp$ and $Ker(U) = Ker(T) = Ker(|T|)$ so that, in particular $||U|| \leq 1$. Let N be a Hilbertian basis of $\mathcal{H}$ made of eigenvectors of $|T|$ (it exists for the previous theorem since $|T|$ is trace class). We have

$$|tr\, T| = \left|\sum_{u\in N} \langle u, U\,|T|u\rangle\right| = \left|\sum_{u\in N} \langle u, Uu\rangle \lambda_u\right| \leq \sum_{u\in N} |\lambda_u|\, |\langle u, Uu\rangle| \,.$$

Next observe that $|\lambda_u| = \lambda_u$ because $|T| \geq 0$ and $|\langle u, Uu\rangle| \leq ||u||||Uu|| \leq 1||Uu|| \leq ||u|| = 1$ and thus, $|tr\, T| \leq \sum_{u\in N} \lambda_u = \sum_{u\in N} \langle u, |T|u\rangle = tr|T| = ||T||_1$. □

2.3.4.2 *The notion of quantum state and the crucial theorem by Gleason*

As commented in (a) in remark 2.2.68, the probabilistic interpretation of quantum states is not well defined because there is no true probability measure in view of the fact that there are incompatible observables. The idea is to redefine the notion of probability in the bounded, orthocomplemented, σ-complete lattice like $\mathcal{L}(\mathcal{H})$ instead of on a σ-algebra. Exactly as in CM, where the generic states are probability measures on Boolean lattice $\mathcal{B}(\Gamma)$ of the elementary properties of the system (Sect. 2.3.1), we can think of states of a quantum system as σ-additive probability measures over the non-Boolean lattice of the elementary observables $\mathcal{L}(\mathcal{H})$.

Definition 2.3.21. *Let $\mathcal{H}$ be a complex Hilbert space. A* **quantum state** *in $\mathcal{H}$ is a map $\rho : \mathcal{L}(\mathcal{H}) \to [0, 1]$ such that the following requirements are satisfied.*

(1) $\rho(I) = 1$.
(2) *If $\{P_n\}_{n\in N} \subset \mathcal{L}(\mathcal{H})$, for N at most countable satisfies $P_k(\mathcal{H}) \perp P_h(\mathcal{H})$ when $h \neq k$ for $h, k \in N$, then*

$$\rho(\vee_{k\in N} P_k) = \sum_{k\in N} \rho(P_k)\,. \tag{2.77}$$

The set of the states in $\mathcal{H}$ will be denoted by $\mathfrak{S}(\mathcal{H})$.

Remark 2.3.22.

(a) *The condition $P_k(\mathcal{H}) \perp P_h(\mathcal{H})$ is obviously equivalent to $P_k P_h = 0$. Since (taking the adjoint) we also obtain $P_h P_k = 0 = P_k P_h$, we conclude that we are dealing with pairwise compatible elementary observables. Therefore Proposition 2.3.11 permits us to equivalently re-write the σ-additivity (2) as follows.*

(2) If $\{P_n\}_{n\in N} \subset \mathcal{L}(\mathcal{H})$, for N at most countable satisfies $P_k P_h = 0$ when $h \neq k$ for $h, k \in N$, then

$$\rho\left(\sum_{k\in N} P_k\right) = \sum_{k\in N} \rho(P_k)\,, \tag{2.78}$$

the sum on the left-hand side being computed with respect to the strong operator topology if N is infinite.

(b) *Requirement (2), taking (1) into account implies $\rho(0) = 0$.*

(c) *Quantum states do exist. It is immediately proven that, in fact, $\psi \in \mathcal{H}$ with $||\psi|| = 1$ defines a quantum state ρ_ψ as*

$$\rho_\psi(P) = \langle \psi, P\psi \rangle \quad P \in \mathcal{L}(\mathcal{H}) . \tag{2.79}$$

This is in nice agreement with what we already know and proves that these types of quantum states are one-to-one with the elements of $P\mathcal{H}$ as well known.

However these states do not exhaust $\mathfrak{S}(\mathcal{H})$. In fact, it immediately arises from Definition 2.3.21 that the set of the states is convex: If $\rho_1, \ldots, \rho_n \in \mathfrak{S}(\mathcal{H})$ then $\sum_{j=1}^{n} p_k \rho_k \in \mathfrak{S}(\mathcal{H})$ if $p_k \geq 0$ and $\sum_{k=1}^{n} p_k = 1$. These convex combinations of states generally do not have the form ρ_ψ.

(d) *Restricting ourselves to a maximal set $\mathcal{L}_0$ of pairwise commuting projectors, which in view of Proposition 2.3.11 has the abstract structure of a σ-algebra, a quantum state ρ reduces thereon to a standard probability measure. In this sense the "quantum probability" we are considering extends the classical notion. Differences show up just when one deals with conditional probability involving incompatible elementary observables.*

An interesting case of (c) in the remark above is a convex combination of states induced by unit vectors as in (2.79), where $\langle \psi_k, \psi_h \rangle = \delta_{hk}$,

$$\rho = \sum_{k=1}^{n} p_k \rho_{\psi_k} .$$

By direct inspection, completing the finite orthonormal system $\{\psi_k\}_{k=1,\ldots,n}$ to a full Hilbertian basis of $\mathcal{H}$, one quickly proves that, defining

$$T = \sum_{k=1}^{n} p_k \langle \psi_k , \ \rangle \psi_k \tag{2.80}$$

$\rho(P)$ can be computed as

$$\rho(P) = tr(TP) \quad P \in \mathcal{L}(\mathcal{H})$$

In particular it turns out that T is in $\mathfrak{B}_1(\mathcal{H})$, satisfies $T \geq 0$ (so it is selfadjoint for (3) in exercise 2.2.31) and $tr\, T = 1$. As a matter of fact, (2.80) is just the spectral decomposition of T, whose spectrum is $\{p_k\}_{k=1,\ldots,n}$. This result is general [5; 6]

Proposition 2.3.23. *Let $\mathcal{H}$ be a complex Hilbert space and let $T \in \mathfrak{B}_1(\mathcal{H})$ satisfy $T \geq 0$ and $Tr\, T = 1$, then the map*

$$\rho_T : \mathcal{L}(\mathcal{H}) \ni P \mapsto tr(TP)$$

is well defined and $\rho_T \in \mathfrak{S}(\mathcal{H})$.

The very remarkable fact is that these operators exhaust $\mathfrak{S}(\mathcal{H})$ if $\mathcal{H}$ is separable with dimension $\neq 2$. As established by Gleason in a celebrated theorem, we restate re-adapting it to these lecture notes (see [5] for a the original statement and [21] for a general treatise on the subject).

Theorem 2.3.24 (Gleason's theorem)**.** *Let $\mathcal{H}$ be a complex Hilbert space of finite dimension $\neq 2$, or infinite dimensional and separable. If $\rho \in \mathfrak{S}(\mathcal{H})$ there exists a unique operator $T \in \mathfrak{B}_1(\mathcal{H})$ with $T \geq 0$ and $tr\, T = 1$ such that $tr(TP) = \rho(P)$ for every $P \in \mathcal{L}(\mathcal{H})$.*

Concerning the existence of T, Gleason's proof works for real Hilbert spaces too. If the Hilbert space is complex, the operator T associated to ρ is unique for the following reason. Any other T' of trace class such that $\rho(P) = tr(T'P)$ for any $P \in \mathcal{L}(\mathcal{H})$ must also satisfy $\langle x, (T - T')x \rangle = 0$ for any $x \in \mathcal{H}$. If $x = 0$ this is clear, while if $x \neq 0$ we may complete the vector $x/||x||$ to a basis, in which $tr((T - T')P_x) = 0$ reads $||x||^{-2}\langle x, (T - T')x \rangle = 0$, where P_x is the projector onto $span(x)$. By (3) in exercise 2.2.31, we obtain $T - T' = 0$.[17]

Remark 2.3.25.

(a) *Imposing* $\dim \mathcal{H} \neq 2$ *is mandatory, a well-known counterexample can be found, e.g. in [5].*

(b) *Particles with spin* $1/2$*, like electrons, admit a Hilbert space – in which the observable spin is defined – of dimension* 2*. The same occurs to the Hilbert space in which the polarization of light is described (cf. helicity of photons). When these systems are described in full, however, for instance including degrees of freedom relative to position or momentum, they are representable on a separable Hilbert space of infinite dimension.*

Gleason's characterization of states has an important consequence known as the *Bell-Kochen-Specker theorem.* It proves in particular (see [6] for an extended discussion) that in QM there are no states assigning probability 1 to some elementary observables and 0 to the remaining ones, differently to what happens in CM.

---

[17] In a real Hilbert space $\langle x, Ax \rangle = 0$ for all x does not imply $A = 0$. Think of real antisymmetric matrices in $\mathbb{R}^n$ equipped with the standard scalar product.

Theorem 2.3.26 (Bell-Kochen-Specker theorem). *Let $\mathcal{H}$ be a complex Hilbert space of finite dimension $\neq 2$, or infinite dimensional and separable. There is no quantum state $\rho : \mathcal{L}(\mathcal{H}) \to [0,1]$, in the sense of Def. 2.3.21, such that $\rho(\mathcal{L}(\mathcal{H})) = \{0,1\}$*

Proof. Define $\overline{S} := \{x \in \mathcal{H} \mid ||x|| = 1\}$ endowed with the topology induced by $\mathcal{H}$, and let $T \in \mathfrak{B}_1(\mathcal{H})$ be the representative of ρ using Gleason's theorem. The map $f_\rho : \overline{S} \ni x \mapsto \langle x, Tx\rangle = \rho(\langle x, \ \rangle x) \in \mathbb{C}$ is continuous because T is bounded. We have $f_\rho(\overline{S}) \subset \{0,1\}$, where $\{0,1\}$ is equipped with the topology induced by $\mathbb{C}$. Since $\overline{S}$ is connected its image must be connected also. So either $f_\rho(\overline{S}) = \{0\}$ or $f_\rho(\overline{S}) = \{1\}$. In the first case $T = 0$ which is impossible because $trT = 1$, in the second case $tr\,T \neq 2$ which is similarly impossible. □

This negative result produces no-go theorems in some attempts to explain QM in terms of CM introducing *hidden variables* [6].

Remark 2.3.27. *In view of Proposition 2.3.23 and Theorem 2.3.24, assuming that $\mathcal{H}$ has finite dimension or is separable, we henceforth identify $\mathfrak{S}(\mathcal{H})$ with the subset of $\mathfrak{B}_1(\mathcal{H})$ of positive operators with unit trace. We simply disregard the states in $\mathcal{H}$ with dimension 2 which are not of this form especially taking (b) in remark 2.3.25 into account.*

We are in a position to state some definitions of interest for physicists, especially the distinction between pure and mixed states, so we proceed to analyse the structure of the space of the states. To this end, we remind the reader that, if C is a convex set in a vector space, $e \in C$ is called **extreme** if it cannot be written as $e = \lambda x + (1-\lambda)y$, with $\lambda \in (0,1)$, $x, y \in C \setminus \{e\}$.

We have the following simple result whose proof can be found in [5].

Proposition 2.3.28. *Let $\mathcal{H}$ be a complex separable Hilbert space.*

(a) *$\mathfrak{S}(\mathcal{H})$ is a convex closed subset in $\mathfrak{B}_1(\mathcal{H})$ whose extreme points are those of the form: $\rho_\psi := \langle \psi, \ \rangle \psi$ for every vector $\psi \in \mathcal{H}$ with $||\psi|| = 1$. (This sets up a bijection between extreme states and elements of $P\mathcal{H}$.)*

(b) *A state $\rho \in \mathfrak{S}(\mathcal{H})$ is extreme if and only if $\rho\rho = \rho$. (All the elements of $\mathfrak{S}(\mathcal{H})$ however satisfy $\langle x, \rho\rho x\rangle \leq \langle x, \rho x\rangle$ for all $x \in \mathcal{H}$.)*

(c) *Any state $\rho \in \mathfrak{S}(\mathcal{H})$ is a linear combination of extreme states, including infinite combinations in the strong operator topology. In particular there is always a decomposition*

$$\rho = \sum_{\phi \in N} p_\phi \langle \phi, \ \rangle \phi,$$

where N is an eigenvector basis for ρ, $p_\phi \in [0,1]$ for any $\phi \in N$, and

$$\sum_{\phi \in N} p_\phi = 1 .$$

The stated proposition allows us to introduce some notions and terminology relevant in physics. First of all, extreme elements in $\mathfrak{S}(\mathcal{H})$ are usually called **pure states** by physicists. We shall denote their set is denoted $\mathfrak{S}_p(\mathcal{H})$. Non-extreme states are instead called **mixed states, mixtures** or **non-pure states**. If

$$\psi = \sum_{i \in I} a_i \phi_i ,$$

with I finite or countable (and the series converges in the topology of $\mathcal{H}$ in the second case), where the vectors $\phi_i \in \mathcal{H}$ are all non-null and $0 \neq a_i \in \mathbb{C}$, physicists say that the state $\langle \psi | \ \rangle \psi$ is called an **coherent superposition** of the states $\langle \phi_i, \ \rangle \phi_i / ||\phi_i||^2$.

The possibility of creating pure states by non-trivial combinations of vectors associated to other pure states is called, in the jargon of QM, **superposition principle of (pure) states**.

There is however another type of superposition of states. If $\rho \in \mathfrak{S}(\mathcal{H})$ satisfies:

$$\rho = \sum_{i \in I} p_i \rho_i$$

with I finite, $\rho_i \in \mathfrak{S}(\mathcal{H})$, $0 \neq p_i \in [0,1]$ for any $i \in I$, and $\sum_i p_i = 1$, the state ρ is called **incoherent superposition** of states ρ_i (possibly pure).

If $\psi, \phi \in \mathcal{H}$ satisfy $||\psi|| = ||\phi|| = 1$ the following terminology is very popular: The complex number $\langle \psi, \phi \rangle$ is the **transition amplitude** or **probability amplitude** of the state $\langle \phi, \ \rangle \phi$ on the state $\langle \psi, \ \rangle \psi$, moreover the non-negative real number $|\langle \psi, \phi \rangle|^2$ is the **transition probability** of the state $\langle \phi, \ \rangle \phi$ on the state $\langle \psi, \ \rangle \psi$. We make some comments about these notions. Consider the pure state $\rho_\psi \in \mathfrak{S}_p(\mathcal{H})$, written $\rho_\psi = \langle \psi, \ \rangle \psi$ for some $\psi \in \mathcal{H}$ with $||\psi|| = 1$. What we want to emphasise is that this pure state is also an orthogonal projector $P_\psi := \langle \psi, \ \rangle \psi$, so it must correspond to an elementary observable of the system (an *atom* using the terminology of Theorem 2.3.13). The naïve and natural interpretation of that observable is this: *"the system's state is the pure state given by the vector ψ"*. We can therefore interpret the square modulus of the transition amplitude $\langle \phi, \psi \rangle$ as follows. If $||\phi|| = ||\psi|| = 1$, as the definition of transition amplitude imposes, $tr(\rho_\psi P_\phi) = |\langle \phi, \psi \rangle|^2$, where $\rho_\psi := \langle \psi, \ \rangle \psi$ and $P_\phi = \langle \phi, \ \rangle \phi$. Using (4) we conclude: *$|\langle \phi, \psi \rangle|^2$ is the probability that the state, given (at time t) by the vector ψ, following a measurement (at time t) on the system becomes determined by ϕ.*

Notice $|\langle\phi,\psi\rangle|^2 = |\langle\psi,\phi\rangle|^2$, so the probability transition of the state determined by ψ on the state determined by ϕ coincides with the analogous probability where the vectors are swapped. This fact is, *a priori*, highly non-evident in physics.

Since we have introduced a new notion of state, the axiom concerning the collapse of the state (Sect. 2.2.8) must be improved in order to encompass all states of $\mathfrak{S}(\mathcal{H})$. The standard formulation of QM assumes the following axiom (introduced by von Neumann and generalized by Lüders) about what occurs to the physical system, in state $\rho \in \mathfrak{S}(\mathcal{H})$ at time t, when subjected to the measurement of an elementary observable $P \in \mathcal{L}(\mathcal{H})$, if the latter is true (so in particular $tr(\rho P) > 0$, prior to the measurement). We are referring to *non-destructive* testing, also known as *indirect measurement* or *first-kind measurement*, where the physical system examined (typically a particle) is not absorbed/annihilated by the instrument. They are idealised versions of the actual processes used in labs, and only in part they can be modelled in such a way.

Collapse of the state revisited. If the quantum system is in state $\rho \in \mathfrak{S}(\mathcal{H})$ at time t and proposition $P \in \mathcal{L}(\mathcal{H})$ is true after a measurement at time t, the system's state immediately afterwards is:

$$\rho_P := \frac{P\rho P}{tr(\rho P)}.$$

In particular, if ρ is pure and determined by the unit vector ψ, the state immediately after measurement is still pure, and determined by:

$$\psi_P = \frac{P\psi}{||P\psi||}.$$

Obviously, in either case ρ_P and ψ_P define states. In the former, in fact, ρ_P is positive of trace class, with unit trace, while in the latter $||\psi_P|| = 1$.

Remark 2.3.29.

(a) *Measuring a property of a physical quantity goes through the interaction between the system and an instrument (supposed to be macroscopic and obeying the laws of classical physics). Quantum Mechanics, in its standard formulation, does not establish what a measuring instrument is, it only says they exist; nor is it capable of describing the interaction of instrument and quantum system set out in the von Neumann Lüders' postulate quoted above. Several viewpoints and conjectures exist on how to complete the physical description of the measuring process; these are called, in the slang of QM,* **collapse**, *or* **reduction**, **of the state** *or* **of the wavefunction** *(see [5] for references).*

(b) *Measuring instruments are commonly employed to* prepare a system in a certain pure state. *Theoretically-speaking the preparation of a* pure *state is carried out like this. A finite collection of* compatible *propositions* $P_1, \dots, P_n$ *is chosen so that the projection subspace of* $P_1 \wedge \cdots \wedge P_n = P_1 \cdots P_n$ *is* one-dimensional. *In other words* $P_1 \cdots P_n = \langle \psi, \ \rangle \psi$ *for some vector with* $||\psi|| = 1$. *The existence of such propositions is seen in practically all quantum systems used in experiments. (From a theoretical point of view these are* atomic *propositions) Then propositions* P_i *are simultaneously measured on several identical copies of the physical system of concern (e.g., electrons), whose initial states, though, are unknown. If for one system the measurements of all propositions are successful, the post-measurement state is determined by the vector* ψ*, and the system was* **prepared** *in that particular pure state.*
Normally each projector P_i *belongs to the PVM* $P^{(A)}$ *of an observable* A_i *whose spectrum is made of isolated points (thus a pure point spectrum) and* $P_i = P^{(A)}_{\{\lambda_i\}}$ *with* $\lambda_i \in \sigma_p(A_i)$.

(c) *Let us finally explain how to practically obtain non-pure states from pure ones. Consider* q_1 *identical copies of system* S *prepared in the pure state associated to* ψ_1, q_2 *copies of* S *prepared in the pure state associated to* ψ_2 *and so on, up to* ψ_n. *If we mix these states each one will be in the non-pure state:* $\rho = \sum_{i=1}^{n} p_i \langle \psi_i, \ \rangle \psi_i$, *where* $p_i := q_i / \sum_{i=1}^{n} q_i$. *In general,* $\langle \psi_i, \psi_j \rangle$ *is not zero if* $i \neq j$, *so the above expression for* ρ *is not the decomposition with respect to an eigenvector basis for* ρ. *This procedure hints at the existence of two different types of probability, one intrinsic and due to the quantum nature of state* ψ_i, *the other epistemic, and encoded in the probability* p_i. *But this is not true: once a non-pure state has been created, as above, there is no way, within QM, to distinguish the states forming the mixture. For example, the same* ρ *could have been obtained mixing other pure states than those determined by the* ψ_i. *In particular, one could have used those in the decomposition of* ρ *into a basis of its eigenvectors. For physics, no kind of measurement would distinguish the two mixtures.*

Another delicate point is that, dealing with mixed states, definitions (2.64) and (2.66) for, respectively the expectation value $\langle A \rangle_\psi$ and the standard deviation ΔA_ψ of an observable A referred to the pure state $\langle \psi, \ \rangle \psi$ with $||\psi|| = 1$ are no longer valid. We just say that extended natural definitions can be stated referring to the probability measure associated to both the mixed state $\rho \in \mathfrak{B}_1(\mathcal{H})$ (with $\rho \geq 0$ and tr $\rho = 1$) and the observable,

$$\mu^{(A)}_\rho : \mathcal{B}(\mathbb{R}) \ni E \mapsto tr(\rho P^{(A)}_E) \, .$$

We refer the reader to [5; 6] for a technical discussion on these topics.

2.3.5 *von Neumann algebra of observables, superselection rules*

The aim of this section is to focus on the class of observables of a quantum system, described in the complex Hilbert space $\mathcal{H}$, exploiting some elementary results of the theory of *von Neuman algebras.* Up to now, we have tacitly supposed that *all* selfadjoint operators in $\mathcal{H}$ represent observables, *all* orthogonal projectors represent elementary observables, *all* normalized vectors represent pure states. This is not the case in physics due to the presence of the so-called *superselection rules.* Within the Hilbert space approach the modern tool to deal with this notion is the mathematical structure of a *von Neumann algebra.* For this reason we spend the initial part of this section to introduce this mathematical tool.

2.3.5.1 *von Neumann algebras*

Before we introduce it, let us first define the *commutant* of an operator algebra and state an important preliminary theorem. If $\mathfrak{M} \subset \mathfrak{B}(\mathcal{H})$ is a subset in the algebra of bounded operators on the complex Hilbert space $\mathfrak{B}(\mathcal{H})$, the **commutant** of $\mathfrak{M}$ is:

$$\mathfrak{M}' := \{T \in \mathfrak{B}(\mathcal{H}) \mid TA - AT = 0 \quad \text{for any } A \in \mathfrak{M}\} . \tag{2.81}$$

If $\mathfrak{M}$ is closed under the adjoint operation (i.e. $A^\dagger \in \mathfrak{M}$ if $A \in \mathfrak{M}$) the commutant $\mathfrak{M}'$ is certainly a *-algebra with unit. In general: $\mathfrak{M}_1' \subset \mathfrak{M}_2'$ if $\mathfrak{M}_2 \subset \mathfrak{M}_1$ and $\mathfrak{M} \subset (\mathfrak{M}')'$, which imply $\mathfrak{M}' = ((\mathfrak{M}')')'$. Hence we cannot reach beyond the second commutant by iteration.

The continuity of the product of operators in the uniform topology says that the commutant $\mathfrak{M}'$ is closed in the uniform topology, so if $\mathfrak{M}$ is closed under the adjoint operation, its commutant $\mathfrak{M}'$ is a C^*-algebra (C^*-subalgebra) in $\mathfrak{B}(\mathcal{H})$.

$\mathfrak{M}'$ has other pivotal topological properties in this general setup. It is easy to prove that $\mathfrak{M}'$ is both strongly and weakly closed. This holds, despite the product of operators is not continuous with respect to the strong operator topology, because separate continuity in each variable is sufficient.

In the sequel, we shall adopt the standard convention used for von Neumann algebras and write $\mathfrak{M}''$ in place of $(\mathfrak{M}')'$ *etc.* The next crucial result is due to von Neumann (see e.g. [5; 6]).

Theorem 2.3.30 (von Neumann's double commutant theorem). *If $\mathcal{H}$ is a complex Hilbert space and $\mathfrak{A}$ an unital *-subalgebra in $\mathfrak{B}(\mathcal{H})$, the following statements are equivalent.*

(a) $\mathfrak{A} = \mathfrak{A}''$;
(b) $\mathfrak{A}$ *is weakly closed;*
(c) $\mathfrak{A}$ *is strongly closed.*

At this juncture we are ready to define von Neumann algebras.

Definition 2.3.31. *Let $\mathcal{H}$ be a complex Hilbert space. A* **von Neumann algebra** *in $\mathfrak{B}(\mathcal{H})$ is a *-subalgebra of $\mathfrak{B}(\mathcal{H})$, with unit, that satisfies any of the equivalent properties appearing in von Neumann's theorem 2.3.30.*

In particular $\mathfrak{M}'$ is a von Neumann algebra provided $\mathfrak{M}$ is a *-closed subset of $\mathfrak{B}(\mathcal{H})$, because $(\mathfrak{M}')'' = \mathfrak{M}'$ as we saw above. Note how, by construction, a von Neumann algebra in $\mathfrak{B}(\mathcal{H})$ is a C^*-algebra with unit, or better, a C^*-subalgebra with unit of $\mathfrak{B}(\mathcal{H})$.

It is not hard to see that the intersection of von Neumann algebras is a von Neumann algebra. If $\mathfrak{M} \subset \mathfrak{B}(\mathcal{H})$ is closed under the adjoint operation, $\mathfrak{M}''$ turns out to be the smallest (set-theoretically) von Neumann algebra containing $\mathfrak{M}$ as a subset [13]. Thus $\mathfrak{M}''$ is called the **von Neumann algebra generated** by $\mathfrak{M}$.

Since in QM it is natural to deal with unbounded selfadjoint operators, the definition of commutant is extended to the case of a set of generally unbounded selfadjoint operators, exploiting the fact that these operators admit spectral measures made of bounded operators.

Definition 2.3.32. *If $\mathfrak{N}$ is a set of (generally unbounded) selfadjoint operators in the complex Hilbert space $\mathcal{H}$, the* **commutant** *$\mathfrak{N}'$ of $\mathfrak{N}$, is defined as the commutant in the sense of (2.81) of the set of all the spectral measures $P^{(A)}$ of every $A \in \mathfrak{N}$. The von Neuman algebra $\mathfrak{N}''$ generated by $\mathfrak{N}$ is defined as $(\mathfrak{N}')'$, where the external prime is the one of definition (2.81).*

Remark 2.3.33. *Notice that, if the selfadjoint operators are all bounded, $\mathfrak{N}'$ obtained this way coincides with the one already defined in (2.81) as a consequence of of (ii) and (iv) of Proposition 2.2.70 (for a bounded selfadjoint operator A). Thus $\mathfrak{N}'$ is well-defined and gives rises to a von Neumann algebra because the set of spectral measures is *-closed. $\mathfrak{N}''$ is a von Neumann algebra too for the same reason.*

We are in a position to state a technically important result which concerns both the spectral theory and the notion of von Neumann algebra [5; 6].

Proposition 2.3.34. *Let $\mathbf{A} = \{A_1, \ldots, A_n\}$ be a finite collection of selfadjoint operators in the separable Hilbert space $\mathcal{H}$ whose spectral measures commute. The von Neumann algebra $\mathbf{A}''$ coincides with the collection of operators*

$$f(A_1, \ldots, A_n) := \int_{supp(P^{(\mathbf{A})})} f(x_1, \ldots, x_n) dP^{(\mathbf{A})} ,$$

with $f : supp(P^{(\mathbf{A})}) \to \mathbb{C}$ measurable and bounded.

2.3.5.2 *Lattices of von Neumann algebras*

To conclude this elementary mathematical survey, we will say some words about von Neumann algebras and their associated lattices of orthogonal projectors.

Consider a von Neumann algebra $\mathfrak{R}$ on the complex Hilbert space $\mathcal{H}$. It is easy to prove that the set $\mathcal{L}_{\mathfrak{R}}(\mathcal{H}) \subset \mathfrak{R}$ of the orthogonal projectors included in $\mathfrak{R}$ form a lattice, which is bounded by 0 and I, orthocomplemented with respect to the orthocomplementation operation of $\mathcal{L}(\mathcal{H})$ and σ-complete (because this notion involves only the strong topology ((iii) in Proposition 2.3.9) and $\mathfrak{R}$ is closed with respect to that topology in view of Theorem 2.3.30. Moreover $\mathcal{L}_{\mathfrak{R}}(\mathcal{H})$ is orthomodular, and separable like the whole $\mathcal{L}(\mathcal{H})$, assuming that $\mathcal{H}$ is separable. It is interesting to note that, as expected, $\mathcal{L}_{\mathfrak{R}}(\mathcal{H})$ contains all information about $\mathfrak{R}$ itself since the following result holds.

Proposition 2.3.35. *Let $\mathfrak{R}$ be a von Neumann algebra on the complex Hilbert space $\mathcal{H}$ and consider the lattice $\mathcal{L}_{\mathfrak{R}}(\mathcal{H}) \subset \mathfrak{R}$ of the orthogonal projectors in $\mathfrak{R}$. then the equality $\mathcal{L}_{\mathfrak{R}}(\mathcal{H})'' = \mathfrak{R}$ holds.*

Proof. Since $\mathcal{L}_{\mathfrak{R}}(\mathcal{H}) \subset \mathfrak{R}$, we have $\mathcal{L}_{\mathfrak{R}}(\mathcal{H})' \supset \mathfrak{R}'$ and $\mathcal{L}_{\mathfrak{R}}(\mathcal{H})'' \subset \mathfrak{R}'' = \mathfrak{R}$. Let us prove the other inclusion. $A \in \mathfrak{R}$ can always be decomposed as a linear combination of two selfadjoint operators of $\mathfrak{R}$, $A + A^\dagger$ and $i(A - A^\dagger)$. So we can restrict ourselves to the case of $A^\dagger = A \in \mathfrak{R}$, proving that $A \in \mathcal{L}_{\mathfrak{R}}(\mathcal{H})''$ if $A \in \mathfrak{R}$. The PVM of A belongs to $\mathfrak{R}$ because of (ii) and (iv) of Proposition 2.2.70: $P^{(A)}$ commutes with every bounded operator B which commutes with A. So $P^{(A)}$ commutes, in particular, with the elements of $\mathfrak{R}'$ because $\mathfrak{R} \ni A$. We conclude that every $P^{(A)}_E \in \mathfrak{R}'' = \mathfrak{R}$. Finally, there is a sequence of simple functions s_n uniformly converging to id in a compact $[-a, a] \supset \sigma(A)$ (e.g, see [5; 6]). By construction $\int_{\sigma(A)} s_n dP^{(A)} \in \mathcal{L}_{\mathfrak{R}}(\mathcal{H})''$ because it is a linear combination of elements of $P^{(A)}$ and $\mathcal{L}_{\mathfrak{R}}(\mathcal{H})''$ is a linear space. Finally $\int_{\sigma(A)} s_n dP^{(A)} \to A$ for $n \to +\infty$ uniformly, and thus strongly, as seen in (2) of example 2.2.72. Since $\mathcal{L}_{\mathfrak{R}}(\mathcal{H})''$ is closed with respect to the strong topology, we must have $A \in \mathcal{L}_{\mathfrak{R}}(\mathcal{H})''$, proving that $\mathcal{L}_{\mathfrak{R}}(\mathcal{H}) \supset \mathfrak{R}$ as wanted. □

2.3.5.3 *General algebra of observables and its center*

Let us pass to physics and we apply these notions and results. Relaxing the hypothesis that all selfadjoint operators in the separable Hilbert space $\mathcal{H}$ associated to a quantum system represent observables, there are many reasons to assume that the observables of a quantum system are represented (in the sense we are going to illustrate) by the selfadjoint elements of an algebra of von Neumann, we hereafter indicated by $\mathfrak{R}$, called the **von Neumann algebra of observables** (though only the selfadjoint elements are observables). Including non-selfadjoint elements $B \in \mathfrak{R}$ is armless, as they can always be one-to-one decomposed into a pair of selfadjoint elements

$$B = B_1 + iB_2 = \frac{1}{2}(B + B^\dagger) + i\frac{1}{2i}(B - B^\dagger) .$$

The fact that the elements of $\mathfrak{R}$ are bounded does not seem a physical problem. If $A = A^\dagger$ is unbounded and represents an observable it does not belong to $\mathfrak{R}$. Nevertheless the associated *class* of bounded selfadjoint operators $\{A_n\}_{n\in\mathbb{N}}$ where

$$A_n := \int_{[-n,n]\cap\sigma(A)} \lambda dP^{(A)}(\lambda) ,$$

embodies the same information as A itself. A_n is bounded due to Proposition 2.2.61 because the support of its spectral measures is included in $[-n, n]$. Physically speaking, we can say that A_n is nothing but the observable A when it is measured with an instrument unable to produce outcomes larger than $[-n, n]$. All real measurement instruments are similarly limited. We can safely assume that every A_n belongs to $\mathfrak{R}$. Mathematically speaking, the whole (unbounded) observable A is recovered as the limit in the *strong operator topology* $A = \lim_{n\to+\infty} A_n$ ((1) in examples 2.2.72). Moreover the union of the spectral measures of all the A_n is that of A. Finally the spectral measure of A belongs to $\mathfrak{R}$ since the spectral measure of every $A_n \in \mathfrak{R}$ does, as has been established in the proof of Proposition 2.3.35 above.

Within this framework the orthogonal projectors $P \in \mathfrak{R}$ represent all elementary observables of the system. The lattice of these projectors, $\mathcal{L}_{\mathfrak{R}}(\mathcal{H})$, encompass the amount of information about observables as established in Proposition 2.3.35. As said above $\mathcal{L}_{\mathfrak{R}}(\mathcal{H}) \subset \mathfrak{R}$ is bounded, orthocomplemented, σ-complete, orthomodular and separable like the whole $\mathcal{L}(\mathcal{H})$ (assuming that $\mathcal{H}$ is separable) but there is no guarantee for the validity of the other properties listed in Theorem 2.3.13. The natural question is whether $\mathfrak{R}$ is *-isomorphic to $\mathfrak{B}(\mathcal{H}_1)$ for a suitable complex Hilbert space $\mathcal{H}_1$, which would automatically imply that also the remaining properties were true. In particular there would exist atomic elements in $\mathcal{L}_{\mathfrak{R}}(\mathcal{H})$ and the covering property would be satisfied. A necessary condition is that, exactly as it happens for $\mathfrak{B}(\mathcal{H}_1)$, there are no non-trivial elements in $\mathfrak{R}\cap\mathfrak{R}'$, since $\mathfrak{B}(\mathcal{H}_1) \cap \mathfrak{B}(\mathcal{H}_1)' = \mathfrak{B}(\mathcal{H}_1)' = \{cI\}_{c\in\mathbb{C}}$.

Definition 2.3.36. *A von Neumann algebra $\mathfrak{R}$ is a* **factor** *when its* **center**, *the subset $\mathfrak{R}\cap\mathfrak{R}'$ of elements commuting with the whole algebra, is trivial: $\mathfrak{R}\cap\mathfrak{R}' = \{cI\}_{c\in\mathbb{C}}$.*

Remark 2.3.37. *It is possible to prove that a von Neumann algebra is always a direct sum or a direct integral of factors. Therefore factors play a crucial role. The classification of factors, started by von Neumann and Murray, is one of the key chapters in the theory of operator algebras, and has enormous consequences in the algebraic theory of quantum fields. The factors isomorphic to $\mathfrak{B}(\mathcal{H}_1)$ for some complex Hilbert space $\mathcal{H}_1$, are called of* type I. *These factors admit atoms, fulfil the covering property (orthomodularity and irreducibility are always true). Regarding separability, it depends on separability of $\mathcal{H}_1$ and requires a finer classification in factors of* type I_n *where n is a cardinal number. There are however factors of type II and III which do not admit atoms and are not important in elementary QM.*

The center of the von Neumann algebra of observables enters the physical theory in a nice way. A common situation dealing with quantum systems is the existence of a **maximal set of compatible observables**, i.e. a finite maximal class $\mathfrak{A} = \{A_1, \ldots, A_n\}$ of pairwise compatible observables. The notion of maximality here means that, if a (bounded) selfadjoint operator commutes with all the observables in $\mathfrak{A}$, then it is a *function* of them. In perticular it is an observable as well. In view of proposition 2.3.34 the existence of a maximal set of compatible observables is equivalent to say that there is a finite set of observables $\mathfrak{A}$ such that $\mathfrak{A}' = \mathfrak{A}''$. We have the following important consequence

Proposition 2.3.38. *If a quantum physical system admits a maximal set of compatible observables, then the commutant $\mathfrak{R}'$ of the von Neumann algebra of observables $\mathfrak{R}$ is Abelian and coincides with the center of $\mathfrak{R}$.*

Proof. As the spectral measures of each $A \in \mathfrak{A}$ belong to $\mathfrak{R}$, it must be (i) $\mathfrak{A}'' \subset \mathfrak{R}$. Since $\mathfrak{A}' = \mathfrak{A}''$, (i) yields $\mathfrak{A}' \subset \mathfrak{R}$ and thus, taking the commutant, (ii) $\mathfrak{A}'' \supset \mathfrak{R}'$. Comparing (i) and (ii) we have $\mathfrak{R}' \subset \mathfrak{R}$. In other words $\mathfrak{R}' = \mathfrak{R}' \cap \mathfrak{R}$. In particular, $\mathfrak{R}'$ must be Abelian. □

Example 2.3.39.
(1) Considering a quantum particle without spin and referring to the rest space $\mathbb{R}^3$ of an inertial reference frame, $\mathcal{H} = L^2(\mathbb{R}^3, d^3x)$. A maximal set of compatible observables is the set of the three position operators $\mathfrak{A}_1 = \{X_1, X_2, X_3\}$ or the the set of the three momenta operators $\mathfrak{A}_2 = \{P_1, P_2, P_3\}$. $\mathfrak{R}$ is the von Neumann algebra generated by $\mathfrak{A}_1 \cup \mathfrak{A}_2$. It is possible to prove that the commutant (which coincides with the center) of this von Neumann algebra is trivial (as it includes a unitary irreducible representation of the Weyl-Heisenberg group) so that $\mathfrak{R} = \mathfrak{B}(\mathcal{H})$ (see also Theorem 2.3.78).

(2) If adding the spin space (for instance dealing with an electron "without charge"), we have $\mathcal{H} = L^2(\mathbb{R}^3, d^3x) \otimes \mathbb{C}^2$. Referring to (2.11) a maximal set of compatible observables is, for instance, $\mathfrak{A}_1 = \{X_1 \otimes I, X_2 \otimes I, X_3 \otimes I, I \otimes S_z\}$, another is $\mathfrak{A}_2 = \{P_1 \otimes I, P_2 \otimes I, P_3 \otimes I, I \otimes S_x\}$. As before $(\mathfrak{A}_1 \cup \mathfrak{A}_2)''$ is the von Neumann algebra of observables of the system (changing the component of the spin passing from $\mathfrak{A}_1$ to $\mathfrak{A}_2$ is crucial for this result). Also in this case, it turns out that the commutant of the von Neumann algebra of observables is trivial yielding $\mathfrak{R} = \mathfrak{B}(\mathcal{H})$.

2.3.5.4 *Superselection charges and coherent sectors*

We must have accumulated enough formalism to successfully investigate the structure of the Hilbert space (always supposed to be separable) and the algebra of the observables when not all selfadjoint operators represent observables and not all orthogonal projectors are intepreted as elementary observables. Re-adapting the approach by Wightman [22] to our framework, we make two assumptions generally describing the so called *superselection rules* for QM formulated in a (separable) Hilbert space where $\mathfrak{R}$ denotes the von Neumann algebra of observables.

(SS1) There is a maximal set of compatible observables in $\mathfrak{R}$, so that $\mathfrak{R}' = \mathfrak{R}' \cap \mathfrak{R}$.

(SS2) $\mathfrak{R}' \cap \mathfrak{R}$ contains a finite class of observables $\mathfrak{Q} = \{Q_1, \ldots, Q_n\}$, with $\sigma(Q_k) = \sigma_p(Q_k)$, $k = 1, 2, \ldots, n$, generating the centre: $\mathfrak{Q}'' \supset \mathfrak{R}' \cap \mathfrak{R}$.
(If the Q_k are unbounded, $\mathfrak{Q} \subset \mathfrak{R}' \cap \mathfrak{R}$ means that the PVM of the Q_j are included in $\mathfrak{R}' \cap \mathfrak{R}$.)

The Q_k are called **superselection charges.**

As the reader can easily prove, the joint spectral measure $P^{(\mathfrak{Q})}$ in $\mathbb{R}^n$ has support given exactly by $\times_{k=1}^n \sigma_p(Q_k)$ and, if $E \subset \mathbb{R}^n$,

$$P_E^{(\mathfrak{Q})} = \sum_{(q_1,\ldots,q_n) \in \times_{k=1}^n \sigma_p(Q_k) \cap E} P_{\{q_1\}}^{(Q_1)} \cdots P_{\{q_n\}}^{(Q_n)} \tag{2.82}$$

We have the following remarkable result where we occasionally adopt the notation $\mathbf{q} := (q_1, \ldots, q_n)$ and $\sigma(\mathfrak{Q}) := \times_{k=1}^n \sigma_p(Q_k)$.

Proposition 2.3.40. *Let $\mathcal{H}$ be a complex separable Hilbert and suppose that the von Neumann algebra $\mathfrak{R}$ in $\mathcal{H}$ satisfies* **(SS1)** *and* **(SS2)**. *The following facts hold.*

(a) *$\mathcal{H}$ admits the following direct decomposition into closed pairwise orthogonal subspaces, called* **superselection sectors** *or* **coherent sectors,**

$$\mathcal{H} = \bigoplus_{\mathbf{q} \in \sigma(\mathfrak{Q})} \mathcal{H}_{\mathbf{q}} \tag{2.83}$$

where

$$\mathcal{H}_{\mathbf{q}} := P^{(\mathfrak{Q})}_{\mathbf{q}} \mathcal{H} .$$

and each $\mathcal{H}_{\mathbf{q}}$ is invariant and irreducible under $\mathfrak{R}$.

(b) *An analogous direct decomposition occurs for $\mathfrak{R}$.*

$$\mathfrak{R} = \bigoplus_{\mathbf{q} \in \sigma(\mathfrak{Q})} \mathfrak{R}_{\mathbf{q}} \tag{2.84}$$

where

$$\mathfrak{R}_{\mathbf{q}} := \left\{ A|_{\mathcal{H}_{\mathbf{q}}} \;\middle|\; A \in \mathfrak{R} \right\}$$

is a von Neumann algebra on $\mathcal{H}_{\mathbf{q}}$ considered as Hilbert space in its own right. Finally,

$$\mathfrak{R}_{\mathbf{q}} = \mathfrak{B}(\mathcal{H}_{\mathbf{q}})$$

(c) *Each map*

$$\mathfrak{R} \ni A \mapsto A|_{\mathcal{H}_{\mathbf{q}}} \in \mathfrak{R}_{\mathbf{q}}$$

is an *-algebra representation of $\mathfrak{R}$ (Def.2.2.18). Representations associated with different values of $\mathbf{q}$ are (unfaithful and) unitarily inequivalent: In other words there is no isometric surjective map $U : \mathcal{H}_{\mathbf{q}} \to \mathcal{H}_{\mathbf{q}'}$ such that

$$UA|_{\mathcal{H}_{\mathbf{q}}} U^{-1} = A|_{\mathcal{H}_{\mathbf{q}'}}$$

when $\mathbf{q} \neq \mathbf{q}'$.

Proof. (a) Since $P^{(\mathfrak{Q})}_{\mathbf{q}} P^{(\mathfrak{Q})}_{\mathbf{s}} = 0$ if $\mathbf{q} \neq \mathbf{s}$ and $\sum_{\mathbf{q} \in \sigma_p(\mathfrak{Q})} P^{(\mathfrak{Q})}_{\mathbf{q}} = I$, $\mathcal{H}$ decomposes as in (2.83). Since $P^{(\mathfrak{Q})}_{\mathbf{q}}$ belongs to the centre of $\mathfrak{R}$, the subspaces of the decomposition are invariant under the action of each element of $\mathfrak{R}$. Let us pass to the irreducibility. If $P \in \mathfrak{R}' \cap \mathfrak{R}$ is an orthogonal projector it must be a function of the Q_k by hypotheses: $P = \int_{\mathbb{R}^n} f(x) dP^{(\mathfrak{Q})}(x)$ since $P = PP \geq 0$ and $P = P^\dagger$, exploiting the measurable functional calculus, we easily find that $f(x) = \chi_E(x)$ for some $E \subset supp(P^{(\mathfrak{Q})})$. In other words P is an element of the joint PVM of $\mathfrak{Q}$: that PVM exhausts all orthogonal projectors in $\mathfrak{R}' \cap \mathfrak{R}$. Now, if $\{0\} \neq \mathcal{K} \subset \mathcal{H}_{\mathbf{s}}$ is an invariant closed subspace for $\mathfrak{R}$, its orthogonal projector $P_{\mathcal{K}}$ must commute with $\mathfrak{R}$, so it must belong to the centre for (SS2) and thus it belongs to $P^{(\mathfrak{Q})}$ for (SS1) and, more precisely it must be of the form $P_{\mathcal{K}} = P^{(\mathfrak{Q})}_{\mathbf{s}}$ because $P_{\mathcal{K}} \leq P^{(\mathfrak{Q})}_{\mathbf{s}}$ by hypothesis, but there are no projectors smaller than $P^{(\mathfrak{Q})}_{\mathbf{s}}$ in the PVM of $\mathfrak{Q}$. So $\mathcal{K} = \mathcal{H}_{\mathbf{s}}$.

(b) $\mathfrak{R}_{\mathbf{q}} := \left\{ A|_{\mathcal{H}_{\mathbf{q}}} \;\middle|\; A \in \mathfrak{R} \right\}$ is a von Neumann algebra on $\mathcal{H}_{\mathbf{s}}$ considered as a Hilbert space in its own right as it arises by direct inspection. (2.84) holds by definition. Since $\mathcal{H}_{\mathbf{q}}$ is irreducible for $\mathfrak{R}_{\mathbf{q}}$, we have $\mathfrak{R}_{\mathbf{s}} = \mathfrak{R}''_{\mathbf{s}} = \mathfrak{B}(\mathcal{H}_{\mathbf{s}})$. Each map $\mathfrak{R} \ni A \mapsto A|_{\mathcal{H}_{\mathbf{q}}} \in \mathfrak{R}_{\mathbf{q}}$ is a representation of *-algebras as follows by direct check. If $\mathbf{q} \neq \mathbf{q}'$ –for instance $q_1 \neq q'_1$– there is no isometric surjective map $U : \mathcal{H}_{\mathbf{q}} \to \mathcal{H}_{\mathbf{q}'}$ such that

$$UA|_{\mathcal{H}_{\mathbf{q}}}U^{-1} = A|_{\mathcal{H}_{\mathbf{q}'}}$$

If such an operator existed one would have, contrarily to our hypothesis $q_1 \neq q'_1$, $q_1 I_{\mathcal{H}_{\mathbf{q}'}} = UQ_1|_{\mathcal{H}_{\mathbf{q}}}U^{-1} = Q_1|_{\mathcal{H}_{\mathbf{q}'}} = q'_1 I_{\mathcal{H}_{\mathbf{q}'}}$ so that $q_1 = q'_1$. □

We have found that, in the presence of superselection charges, the Hilbert space decomposes into pairwise orthogonal subspaces which are invariant and irreducible with respect to the algebra of the observables, giving rise to inequivalent representations of the algebra itself. Restricting ourselves to each such subspace, QM takes its standard form as all orthogonal projectors are representatives of elementary observables, differently from what happens in the whole Hilbert space where there are orthogonal projectors which cannot represent observables: These are the projectors which do not commute with $P^{(\mathfrak{Q})}$.

There are several superselection structures as the one pointed out in physics. The three most known are of very different nature: The superselection structure of the *electric charge*, the superselection structure of *integer/semi integers values of the angular momentum*, and the one related to the mass in non-relativistic physics, i.e., *Bargmann's superselection rule*.

Example 2.3.41. The electric charge is the typical example of superselction charge. For instance, referring to an electron, its Hilbert space is $L^2(\mathbb{R}^3, d^3x) \otimes \mathcal{H}_s \otimes \mathcal{H}_e$. The space of the electric charge is $\mathcal{H}_e = \mathbb{C}^2$ and therein $Q = e\sigma_z$ (see (2.12)). Many other observables could exist in $\mathcal{H}_e$ in principle, but the elecrtic charge superselection rule imposes that the only possible observables are functions of σ_z. The centre of the algebra of observables is $I \otimes I \otimes f(\sigma_3)$ for every function $f : \sigma(\sigma_z) = \{1, 1\} \to \mathbb{C}$. We have the decomposition in coherent sectors

$$\mathcal{H} = (L^2(\mathbb{R}^3, d^3x) \otimes \mathcal{H}_s \otimes \mathcal{H}_+) \bigoplus (L^2(\mathbb{R}^3, d^3x) \otimes \mathcal{H}_s \otimes \mathcal{H}_-)\,,$$

where $\mathcal{H}_\pm$ are respectively the eigenspaces of Q with eigenvalue $\pm e$.

Remark 2.3.42.

(a) *A fundamental requirement is that the superselection charges have punctual spectrum. If instead $\mathfrak{R} \cap \mathfrak{R}'$ includes an operator A with a continuous part in its spectrum (A may also be the strong limit on $D(A)$ of a sequence of elements in $\mathfrak{R} \cap \mathfrak{R}'$), the established proposition does not hold. $\mathcal{H}$ cannot be decomposed into a direct sum of closed subspaces. In this case, it decomposes into a direct integral and we find a much more complicated structure whose physical meaning seems dubious.*

(b) *The representations $\mathfrak{R} \ni A \mapsto A|_{\mathcal{H}_{\mathbf{q}}} \in \mathfrak{R}_{\mathbf{q}}$ are not faithful (injective), because both I and $P_{\mathbf{s}}^{(\mathfrak{Q})}$ have the same image under the representation.*

(c) *The discussed picture is not the most general one though we only deal with it in these notes. There are quantum physical systems such that their $\mathfrak{R}'$ is not Abelian (think of chromodynamics where $\mathfrak{R}'$ includes a faithful representation of $SU(3)$) so that the center of $\mathfrak{R}$ does not contain the full information about $\mathfrak{R}'$. In this case, the non-Abelian group of the unitary operators in $\mathfrak{R}'$ is called the* **gauge group** *of the theory. The existence of a gauge group is compatible with the presence of superselection rules which are completely described by the center $\mathfrak{R}' \cap \mathfrak{R}$. The only difference is that now $\mathfrak{R}_{\mathbf{q}} = \mathfrak{B}(\mathcal{H}_{\mathbf{q}})$ cannot be possible for every coherent subspace otherwise we would have $\mathfrak{R}' = \mathfrak{R} \cap \mathfrak{R}'$.*

2.3.5.5 *States in the presence of superselection rules*

Let us come to the problem to characterize the states when a superselection structure is assumed on a complex separable Hilbert space $\mathcal{H}$ in accordance with **(SS1)** and **(SS2)**. In principle we can extend Definition 2.3.21 already given for the case of $\mathfrak{R}$ with trivial center. As usual $\mathcal{L}_{\mathfrak{R}}(\mathcal{H})$ indicates the lattice of orthogonal projectors in $\mathfrak{R}$, which we know to be bounded by 0 and I, orthocomplemented, σ-complete, orthomodular and separable, but not atomic and it does not satisfy the covering property in general. The atoms are one dimensional projectors exactly as pure sates, so we may expect some difference at that level when $\mathfrak{R} \neq \mathfrak{B}(\mathcal{H})$.

Definition 2.3.43. *Let $\mathcal{H}$ be a complex separable Hilbert space. A* **quantum state** *in $\mathcal{H}$, for a quantum sistem with von Neumann algebra of observables $\mathfrak{R}$, is a map $\rho : \mathcal{L}_{\mathfrak{R}}(\mathcal{H}) \to [0, 1]$ such that the following requirement are satisfied.*

(1) $\rho(I) = 1$.

(2) *If $\{Q_n\}_{n\in N} \subset \mathcal{L}_{\mathfrak{R}}(\mathcal{H})$, for N at most countable satisfies $Q_k \wedge Q_h = 0$ when $h, k \in N$, then*

$$\rho(\vee_{k\in N} Q_k) = \sum_{k\in N} \rho(Q_k)\,. \tag{2.85}$$

The set of the states will be denoted by $\mathfrak{S}_{\mathfrak{R}}(\mathcal{H})$.

If there is a superselection structure, we have the decompositions so we re-write down into a simpler version,

$$\mathcal{H} = \bigoplus_{k\in K} \mathcal{H}_k\,, \quad \mathfrak{R} = \bigoplus_{k\in K} \mathfrak{R}_k\,, \quad \mathfrak{R}_k = \mathfrak{B}(\mathcal{H}_k)\,,\; k \in K \tag{2.86}$$

where K is some finite or countable set. The lattice $\mathcal{L}_{\mathfrak{R}}(\mathcal{H})$, as a consequence of (2.85), decomposes as (the notation should be obvious)

$$\mathcal{L}_{\mathfrak{R}}(\mathcal{H}) = \bigvee_{k\in K} \mathcal{L}_{\mathfrak{R}_k}(\mathcal{H}_k) = \bigvee_{k\in K} \mathcal{L}(\mathcal{H}_k) \tag{2.87}$$

where

$$\mathcal{L}_{\mathfrak{R}_k}(\mathcal{H}_k) \bigwedge \mathcal{L}_{\mathfrak{R}_h}(\mathcal{H}_h) = \{0\} \quad \text{if } k \neq h\,.$$

In other words $Q \in \mathcal{L}_{\mathfrak{R}}(\mathcal{H})$ can uniquely be written as $Q = +_{k\in K} Q_k$ where $Q_k \in \mathcal{L}(\mathfrak{B}(\mathcal{H}_k))$. In fact $Q_k = P_k Q_k$, where P_k is the orthogonal projector onto $\mathcal{H}_k$.

In this framework, it is possible to readapt Gleason's result simply observing that a state ρ on $\mathcal{L}_{\mathfrak{R}}(\mathcal{H})$ as above defines a state ρ_k on $\mathcal{L}_{\mathfrak{R}_k}(\mathcal{H}_k) = \mathcal{L}(\mathcal{H}_k)$ by

$$\rho_k(P) := \frac{1}{\rho(P_k)}\rho(P)\,, \quad P \in \mathcal{L}(\mathcal{H}_k)\,.$$

If $dim(\mathcal{H}_k) \neq 2$ we can exploit Gleason's theorem.

Theorem 2.3.44. *Let $\mathcal{H}$ be a complex separable Hilbert space and assume that the von Neumann algebra $\mathfrak{R}$ in $\mathcal{H}$ satisfies* **(SS1)** *and* **(SS2)**, *so that the decomposition (2.86) in coherent sectors is valid where we suppose $dim\mathcal{H}_k \neq 2$ for every $k \in K$. The following facts hold.*

(a) *If $T \in \mathfrak{B}_1(\mathcal{H})$ satisfies $T \geq 0$ and $tr\, T = 1$ then*

$$\rho_T : \mathcal{L}_{\mathfrak{R}}(\mathcal{H}) \ni P \mapsto tr(TP)$$

is an elemeont of $\mathfrak{S}_{\mathfrak{R}}(\mathcal{H})$ that is a state on $\mathcal{L}_{\mathfrak{R}}(\mathcal{H})$.

(b) *For $\rho \in \mathfrak{S}_{\mathfrak{R}}(\mathcal{H})$ there is a $T \in \mathfrak{B}_1(\mathcal{H})$ satisfies $T \geq 0$ and $tr\, T = 1$ such that $\rho = \rho_T$.*

(c) *If $T_1, T_2 \in \mathfrak{B}_1(\mathcal{H})$ satisfy same hypotheses as T in (a), then $\rho_{T_1} = \rho_{T_2}$ is valid if and only if $P_k T_1 P_k = P_k T_2 P_k$ for all $k \in K$, P_k being the orthogonal projector onto $\mathcal{H}_k$.*

(d) *A unit vector $\psi \in \mathcal{H}$ defines a pure state only if it belongs to a coherent sector. More precisely, a state $\rho \in \mathfrak{S}_{\mathfrak{R}}(\mathcal{H})$ is pure, that is extremal, if and only if there is $k_0 \in K$, $\psi \in \mathcal{H}_{k_0}$ with $||\psi|| = 1$ such that*

$$\rho(P) = 0 \quad \textit{if } P \in \mathcal{L}(\mathcal{H}_k),\ k \neq k_0 \quad \textit{and} \quad \rho(P) = \langle\psi, P\psi\rangle \textit{ if } P \in \mathcal{L}(\mathcal{H}_{k_0})$$

Proof. **(a)** is obvious from Proposition 2.3.23, as restricting a state ρ on $\mathcal{L}(\mathcal{H})$ to $\mathcal{L}_{\mathfrak{R}}(\mathcal{H})$ we still obtain a state as one can immediately verify. Let us prove (b). Evidently, every $\rho|_{\mathcal{L}(\mathcal{H}_k)}$ is a positive measure with $0 \leq \rho(P_k) \leq 1$. We can apply Gleason's theorem finding $T_k \in \mathfrak{B}(\mathcal{H}_k)$ with $T_k \geq 0$ and $Tr\, T_k = \rho(P_k)$ such that $\rho(Q) = tr(T_k Q)$ if $Q \in \mathcal{L}(\mathcal{H}_k)$. Notice also that $||T_k|| \leq \rho(P_k)$ because

$$||T_k|| = \sup_{\lambda \in \sigma_p(T_k)} |\lambda| = \sup_{\lambda \in \sigma_p(T_k)} \lambda \leq \sum_{\lambda \in \sigma_p(T_k)} d_\lambda \lambda = Tr\, T_k = \rho(P_k)\,.$$

If $Q \in \mathcal{L}_{\mathfrak{R}}(\mathcal{H})$, $Q = \sum_k Q_k$, where $Q_k := P_k Q \in \mathcal{L}(\mathcal{H}_k)$, $Q_k Q_h = 0$ if $k \neq h$ and thus, by σ-additivity,

$$\rho(Q) = \sum_k \rho(Q_k) = \sum_k tr(T_k Q_k)$$

since $\mathcal{H}_k \perp \mathcal{H}_h$, this identity can be rewritten as

$$\rho(Q) = tr(TQ)$$

provided $T := \oplus_k T_k \in \mathfrak{B}_1(\mathcal{H})$. It is clear that $T \in \mathfrak{B}(\mathcal{H})$ because, if $x \in \mathcal{H}$ and $||x|| = 1$ then, as $x = \sum_k x_k$ with $x_k \in \mathcal{H}_k$, $||Tx|| \leq \sum_k ||T_k||\,||x_k|| \leq \sum_k ||T_k|| 1 \leq \sum_k \rho(P_k) = 1$. In particular $||T|| \leq 1$. $T \geq 0$ because each $T_k \geq 0$. Hence $|T| = \sqrt{T^\dagger T} = \sqrt{TT} = T$ via functional calculus, and also $|T_k| = T_k$. Moreover, using the spectral decomposition of T, whose PVM commutes with each P_k, one easily has $|T| = \oplus_k |T_k| = \oplus_k T_k$. The condition

$$1 = \rho(I) = \sum_k \rho(P_k) = \sum_k tr(T_k P_k) = \sum_k tr(|T_k| P_k)$$

is equivalent to say that $tr\, |T| = 1$ using a Hilbertian basis of $\mathcal{H}$ made of the union of bases in each $\mathcal{H}_k$. We have obtained, as wanted, that $T \in \mathfrak{B}_1(\mathcal{H})$, $T \geq 0$, $tr\, T = 1$ and $\rho(Q) = tr(TQ)$ for all $Q \in \mathcal{L}_{\mathfrak{R}}(\mathcal{H})$.

(c) For, the proof straightforwardly follows the form $\mathcal{L}_{\mathfrak{R}_k}(\mathcal{H}_k) = \mathcal{L}(\mathfrak{B}(\mathcal{H}_k))$ because $\mathfrak{R}_k = \mathfrak{B}(\mathcal{H}_k)$ and, evidently, $\rho_{T_1} = \rho_{T_2}$ if and only if $\rho_{T_1}|_{\mathcal{L}(\mathfrak{B}(\mathcal{H}_k))} = \rho_{T_2}|_{\mathcal{L}(\mathfrak{B}(\mathcal{H}_k))}$ for all $k \in K$. Regarding (d) it is clear that if ρ encompasses more than one component $\rho|_{\mathcal{L}(\mathcal{H}_k)} \neq 0$ cannot be extremal because it is, by construction, a convex combination of other states which vanishes in some of the given coherent subspace. Therefore only states such that only one restriction $\rho|_{\mathcal{L}(\mathcal{H}_{k_0})}$ does not vanish may be extremal. Now (a) of Proposition 2.3.28 implies that, among these states, the extremal ones are precisely those of the form said in (d) of the thesis. □

Remark 2.3.45.

(a) *Take* $\psi = \sum_{k \in K} c_k \psi_k$ *where the* $\psi_k \in \mathcal{H}_k$ *are unit vectors and also suppose that* $||\psi||^2 = \sum_k |c_k|^2 = 1$. *This vector induces a state* ρ_ψ *on* $\mathfrak{R}$ *by means of* *the standard procedure (which is nothing but the trace procedure with respect to $T_\psi := \langle \psi, \ \rangle \psi$!)*

$$\rho_\psi(P) = \langle \psi, P\psi \rangle \quad P \in \mathcal{L}_{\mathfrak{R}}(\mathcal{H}) \,.$$

In this case however, since $PP_k = P_kP$ and $\psi_k = P_k\psi_k$ we have

$$\rho_\psi(P) = \sum_k \sum_h \overline{c_k} c_h \langle \psi_k, P_k P P_h \psi_k \rangle = \sum_k \sum_h \overline{c_k} c_h \langle \psi_k, P P_k P_h \psi_k \rangle$$

$$= \sum_k \sum_h \overline{c_k} c_h \langle \psi, P P_k \psi \rangle \delta_{kh} = \sum_k |c_k|^2 \langle \psi_k, P\psi_k \rangle = tr(T'_\psi P)$$

where

$$T'_\psi = \sum_{k \in K} |c_k|^2 \langle \psi_k, \ \rangle \psi_k \,.$$

We conclude that the apparent pure state ψ and the mixed state T'_ψ cannot be distinguished, just because the algebra $\mathfrak{R}$ is too small to make a difference. Actually they define the same state at all and this is an elementary case of (c) in the above theorem with $T_1 = \langle \psi, \ \rangle \psi$ and $T_2 = T'_\psi$.

This discussion, in the language of physicist is often stated as follows:

No coherent superpositions $\psi = \sum_{k \in K} c_k \psi_k$ of pure states $\psi_k \in \mathcal{H}_k$ of different coherent sectors are possible, only incoherent superpositions $\sum_{k \in K} |c_k|^2 \langle \psi_k, \ \rangle \psi_k$ are allowed.

(b) *It should be clear that the one-to-one correspondence between pure states and atomic elementary observables (one dimensional projectors) here does not work. Consequently, notions like* probability amplitude *must be handled with great care. In general, however, everything goes right if staying in a fixed superselection sector $\mathcal{H}_k$ where the said correspondence exists.*

2.3.6 *Quantum symmetries: unitary projective representations*

The notion of symmetry in QM is quite abstract. Actually there are three distinct ideas, respectively by Wigner, Kadison and Segal [23]. Here we focus on the first pair only. Physically speaking, a *symmetry* is an active transformation on the quantum system changing its state. It is supposed that this transformation preserves some properties of the physical system and here we have to distinguish between the two aforementioned cases. However in both cases the transformation is required to be reversible (injective) and to cover (surjective) the space of the states. Symmetries are supposed to mathematically describe some concrete transformation acting on the physical system. Sometimes their action, in practice, can be cancelled by simply changing the reference frame. This is not the general case however, even if this class of symmetries play a relevant role in physics.

2.3.6.1 *Wigner and Kadison theorems, groups of symmetries*

Consider a quantum system described in the complex Hilbert space $\mathcal{H}$. We assume that either $\mathcal{H}$ is the whole Hilbert space in the absence of superselection charges or it denotes a single coherent sector. Let $\mathfrak{S}(\mathcal{H})$ and $\mathfrak{S}_p(\mathcal{H})$ respectively indicate the convex body of the quantum states and the set of pure states, referred to the sector $\mathcal{H}$ if it is the case.

Definition 2.3.46. *If $\mathcal{H}$ is a complex Hilbert space, we have the following definitions.*

(a) *A* **Wigner symmetry** *is a bijective map*

$$s_W : \mathfrak{S}_p(\mathcal{H}) \ni \langle \psi, \ \rangle\psi \to \langle \psi', \ \rangle\psi' \in \mathfrak{S}_p(\mathcal{H})$$

which preserves the probabilties of transition. In other words

$$|\langle \psi_1, \psi_2 \rangle|^2 = |\langle \psi_1', \psi_2' \rangle|^2 \quad \text{if} \quad \psi_1 \,, \psi_2 \in \mathcal{H} \ \text{ with } \ ||\psi_1|| = ||\psi_2|| = 1 \,.$$

(b) *A* **Kadison symmetry** *is a bijective map*

$$s_K : \mathfrak{S}(\mathcal{H}) \ni \rho \to \rho' \in \mathfrak{S}(\mathcal{H})$$

which preserves the convex structure of the space of the states. In other words

$$(p\rho_1 + q\rho_2)' = p\rho_1' + q\rho_2' \quad \text{if} \quad \rho_1, \rho_2 \in \mathfrak{S}(\mathcal{H}) \quad \text{and} \quad p, q \geq 0 \text{ with } p + q = 1 \,.$$

We observe that the first definition is well-posed even if unit vectors define pure states just up to a phase, as the reader can immediately prove, because transition probabilities are not affected by that ambiguity.

Though the definitions are evidently of different nature, they lead to the same mathematical object, as established in a pair of famous characterization theorems we quote into a unique statement. We need a preliminary definition.

Definition 2.3.47. *Let $\mathcal{H}$ be a complex Hilbert space. A map $U : \mathcal{H} \to \mathcal{H}$ is said to be an* **antiunitary operator** *if it is surjective, isometric and $U(ax + by) = \overline{a}Ux + \overline{b}Uy$ when $x, y \in \mathcal{H}$ and $a, b \in \mathbb{C}$.*

We come to the celebrated theorem. The last statement is obvious, the difficult parts are (a) and (b) (see, e.g., [5]).

Theorem 2.3.48 (Wigner and Kadison theorems). *Let $\mathcal{H} \neq \{0\}$ be a complex Hilbert space. The following facts hold.*

(a) *For every Wigner symmetry s_W there is an operator $U : \mathcal{H} \to \mathcal{H}$, which can be either unitary or antiunitary, and this choice is fixed by s_w if $\dim \mathcal{H} \neq \infty$, such that*

$$s_w : \langle\psi, \ \rangle\psi \to \langle U\psi, \ \rangle U\psi \,, \quad \forall \langle\psi, \ \rangle\psi \in \mathfrak{S}_p(\mathcal{H}) \,. \tag{2.88}$$

U and U' are associated to the same s_W if and only if $U' = e^{ia}U$ for $a \in \mathbb{R}$.

(b) *For every Kadison symmetry s_K there is an operator $U : \mathcal{H} \to \mathcal{H}$, which can be either unitary or anti unitary, and this choice is fixed by s_w if $\dim \mathcal{H} \neq \infty$, such that*

$$s_w : \rho \to U\rho U^{-1} \,, \quad \forall \rho \in \mathfrak{S}(\mathcal{H}) \,. \tag{2.89}$$

U and U' are associated to the same s_K if and only if $U' = e^{ia}U$ for $a \in \mathbb{R}$.

(c) *$U : \mathcal{H} \to \mathcal{H}$, either unitary or antiunitary, simultaneously defines a Wigner and a Kadison symmetry by means of (2.88) and (2.89) respectively.*

Remark 2.3.49.

(a) *It is worth stressing that the Kadison notion of symmetry is an extension of the Wigner one, after the result above. In fact, a Kadison symmetry $\rho \mapsto U\rho U^{-1}$ restricted to one dimensional projector preserves the probability transitions, as immediately follows from the identity $|\langle\psi, \phi\rangle|^2 = tr(\rho_\psi \rho_\phi)$ and the cyclic property of the trace, where we use the notation $\rho_\chi = \langle\chi, \ \rangle\chi$. In particular we can use the same operator U to represent also the found Wigner symmetry.*

(b) *If superselection rules are present, in general, quantum symmetries are described in a similar way with unitary or antiunitary operators acting in a single coherent sector or also swapping different sectors [5].*

If a unitary or antiunitary operator V represents a symmetry s, it has an action on observables, too. If A is an observable (a selfadjoint operator on $\mathcal{H}$), we define the **transformed observable** along the action of s as

$$s^*(A) := VAV^{-1} \,. \tag{2.90}$$

Obviously $D(s^*(A)) = V(D(A))$. It is evident that this definition is not affected by the ambiguity of the arbitrary phase in the choice of V when s is given.

According with (i) in Proposition 2.2.66 the spectral measure of $s^*(A)$ is

$$P_E^{(s^*(A))} = VP_E^{(A)}V^{-1} = s^*(P_E^{(A)})$$

as expected.

The meaning of $s^*(A)$ should be evident: the probability that the observable $s^*(A)$ produces the outcome E when the state is $s(\rho)$ (namely $tr(P_E^{(s^*(A))}s(\rho))$) is the same as the probability that the observable A produces the outcome E when the state is ρ (that is $tr(P_E^{(A)}\rho)$). Changing simultaneously and coherently observables and states nothing changes. Indeed

$$tr(P_E^{(s^*(A))}s(\rho)) = tr(VP_E^{(A)}V^{-1}V\rho V^{-1}) = tr(VP_E^{(A)}\rho V^{-1})$$
$$= tr(P_E^{(A)}\rho V^{-1}V) = tr(P_E^{(A)}\rho)\,.$$

Example 2.3.50.

(1) Fixing an inertial reference frame, the pure state of a quantum particle is defined, up to phases, as a unit norm element ψ of $L^2(\mathbb{R}^3, d^3x)$, where $\mathbb{R}^3$ stands for the rest three space of the reference frame. The group of isometries $IO(3)$ of $\mathbb{R}^3$ equipped with the standard Euclidean structure acts on states by means of symmetries the sense of Wigner and Kadison. If $(R,t) : x \mapsto Rx+t$ is the action of the generic element of $IO(3)$, where $R \in O(3)$ and $t \in \mathbb{R}^3$, the associated quantum (Wigner) symmetry $s_{(R,t)}(\langle\psi,\ \rangle\psi) = \langle U_{(R,t)}\psi,\ \rangle U_{(R,t)}\psi$ is completely fixed by the unitary operators $U_{(R,t)}$. They are defined as

$$(U_{(R,t)}\psi)(x) := \psi((R,t)^{-1}x)\,, \quad x \in \mathbb{R}^3\,, \psi \in L^2(\mathbb{R}^3, d^3x)\,, \quad ||\psi|| = 1\,.$$

The fact that the Lebesgue measure is invariant under $IO(3)$ immediately proves that $U_{(R,t)}$ is unitary. It is furthermore easy to prove that, with the given definition

$$U_{(I,0)} = I\,, \quad U_{(R,t)}U_{(R',t')} = U_{(R,t)\circ(R',t')}\,, \quad \forall (R,t),(R',t') \in IO(3)\,. \qquad (2.91)$$

(2) The so-called *time reversal* transformation classically corresponds to invert the sign of all the velocities of the physical system. It is possible to prove [5] (see also (3) in exercise 2.3.69 below) that, in QM and for systems whose energy is bounded below but not above, the time reversal symmetry cannot be represented by unitary transformations, but only antiunitary. In the most elementary situation as in (1), the time reversal is defined by means of the antiunitary operator

$$(T\psi)(x) := \overline{\psi(x)}\,, \quad x \in \mathbb{R}^3\,, \quad \psi \in L^2(\mathbb{R}^3, d^3x)\,, \quad ||\psi|| = 1\,.$$

(3) According to the example in (1), let us focus on the subgroup of $IO(3)$ of displacements along x_1 parametrized by $u \in \mathbb{R}$,

$$\mathbb{R}^3 \ni x \mapsto x + u\mathbf{e}_1\,,$$

where $\mathbf{e}_1$ denotes the unit vector in $\mathbb{R}^3$ along x_1. For every value of the parameter u, we indicate by s_u the corresponding (Wigner) quantum symmetry, $s_u(\langle\psi,\ \rangle\psi) = \langle U_u\psi,\ \rangle U_u\psi$ with

$$(U_u\psi)(x) = \psi(x - u\mathbf{e}_1)\,, \quad u \in \mathbb{R}\,,$$

The action of this symmetry on the observable X_k turns out to be

$$s_u^*(X_k) = U_u X_k U_u^{-1} = X_k + u\delta_{k1} I \,, \quad u \in \mathbb{R} \,.$$

2.3.6.2 *Groups of quantum symmetries*

As in (1) in the example above, very often in physics one deals with groups of symmetries. In other words, there is a certain group G, with unit element e and group product $\cdot$, and one associates each element $g \in G$ to a symmetry s_g (if Kadison or Wigner is immaterial here, in view of the above discussion). In turn, s_g is associated to an operator U_g, unitary or antiunitary.

Remark 2.3.51. *In the rest of this section, we assume that all the U_g are unitary.*

It would be nice to fix these operators U_g in order that the map $G \ni g \mapsto U_g$ be a **unitary representation** of G on $\mathcal{H}$, that is

$$U_e = I \,, \quad U_g U_{g'} = U_{g\cdot g'} \quad g, g' \in G \tag{2.92}$$

The identities (2.91) found in (1) in example 2.3.50 shows that it is possible at least in certain cases. In general the requirement (2.92) does not hold. What we know is that $U_{g\cdot g'}$ equals $U_g U_{g'}$ just *up to phases*:

$$U_g U_{g'} U_{g\cdot g'}^{-1} = \omega(g, g') I \quad \text{with } \omega(g, g') \in U(1) \text{ for all } g, g' \in G \,. \tag{2.93}$$

For $g = e$ this identity gives in particular

$$U_e = \omega(e, e) I \,. \tag{2.94}$$

The numbers $\omega(g, g')$ are called **multipliers**. They cannot be completely arbitrary, indeed associativity of composition of operators $(U_{g_1} U_{g_2}) U_{g_3} = U_{g_1} (U_{g_2} U_{g_3})$ yields the identity

$$\omega(g_1, g_2)\omega(g_1 \cdot g_2, g_3) = \omega(g_1, g_2 \cdot g_3)\omega(g_2, g_3) \,, \quad g_1, g_2, g_3 \in G \tag{2.95}$$

which also implies

$$\omega(g, e) = \omega(e, g) = \omega(g', e) \,, \quad \omega(g, g^{-1}) = \omega(g^{-1}, g) \,, \quad g, g' \in G \,. \tag{2.96}$$

Definition 2.3.52. *If G is a group, a map $G \ni g \mapsto U_g$ – where the U_g are unitary operators in the complex Hilbert space $\mathcal{H}$ – is named a* **unitary projective representation** *of G on $\mathcal{H}$ if (2.93) holds (so that also (2.94) and (2.95) are valid). Moreover,*

(i) *two unitary projective representation $G \ni g \mapsto U_g$ and $G \ni g \mapsto U'_g$ are said to be* **equivalent** *if $U'_g = \chi_g U_g$, where $\chi_g \in U(1)$ for every $g \in G$.*

That is the same as requiring that there are numbers $\chi_h \in U(1)$, if $h \in G$, such that

$$\omega'(g,g') = \frac{\chi_{g\cdot g'}}{\chi_g \chi_{g'}}\, \omega(g,g') \quad \forall g, g' \in G \tag{2.97}$$

with obvious notation;

(ii) *a unitary projective representation with $\omega(e,e) = \omega(g,e) = \omega(e,g) = 1$ for every $g \in G$ is said to be* **normalized***.*

Remark 2.3.53.

(a) *It is easily proven that every unitary projective representation is always equivalent to a normalized representation.*

(b) *It is clear that two projective unitary representations are equivalent if and only if they are made of the same Wigner (or Kadison) symmetries.*

(c) *In case of superselection rules, continuous symmetries representing a connected topological group do not swap different coherent sectors when acting on pure states [5].*

(d) *One may wonder if it is possible to construct a group representation $G \ni g \mapsto V_g$ where the operators V_g may be both unitary or antiunitary. If every $g \in G$ can be written as $g = h \cdot h$ for some h depending on g – and this is the case if G is a connected Lie group – all the operators U_g must be unitary because $U_g = U_h U_h$ is necessarily linear no matter if U_h is linear or anti linear. The presence of arbitrary phases does not change the result.*

Given a unitary projective representation, a technical problem is to check if it is equivalent to a unitary representation, because unitary representations are much simpler to handle. This is a difficult problem [11; 5] which is tackled especially when G is a *topological group* (or *Lie group*) and the representation satisfies the following natural *continuity property*

Definition 2.3.54. *A unitary projective representation of the topological group G, $G \ni g \mapsto U_g$ on the Hilbert space $\mathcal{H}$ is said to be* **continuous** *if the map*

$$G \ni g \mapsto |\langle \psi, U_g \phi \rangle|$$

is continuous for every $\psi, \phi \in \mathcal{H}$.

The notion of continuity defined above is natural as it regards continuity of probability transitions. A well-known co-homological condition assuring that a unitary projective representation of Lie groups is equivalent to a unitary one is due to Bargmann [24; 5].

Theorem 2.3.55 (Bargmann's criterion)**.** *Let G be a connected and simply connected (real finite dimensional) Lie group with Lie algebra $\mathfrak{g}$. Every continuous* *unitary projective representation of G in a complex Hilbert space is equivalent to a strongly continuous unitary representation of G if, for every bilinear antisymmetric map $\Theta : \mathfrak{g} \times \mathfrak{g} \to \mathbb{R}$ such that*

$$\Theta([u,v],w) + \Theta([v,w],u) + \Theta([w,u],v) = 0\,, \quad \forall u,v,w \in \mathfrak{g}$$

there is a linear map $\alpha : \mathfrak{g} \to \mathbb{R}$ such that $\Theta(u,v) = \alpha([u,v])$, for all $u, v \in \mathfrak{g}$.

Remark 2.3.56. *The condition is equivalent to require that the second cohomology group $H^2(G,\mathbb{R})$ is trivial. $SU(2)$ for instance satisfies the requirement.*

However, non-unitarizable unitary projective representations do exist and one has to deal with them. There is nevertheless a way to circumvent the technical problem. Given a unitary projective representation $G \ni g \mapsto U_g$ with multiplicators ω, let us put on $U(1) \times G$ the group structure arising by the product $\circ$

$$(\chi, g) \circ (\chi', g') = (\chi\chi'\omega(g,g'), g \cdot g')$$

and indicate by $\hat{G}_\omega$ the obtained group. The map

$$\hat{G}_\omega \ni (\chi, g) \mapsto \chi U_g =: V_{(\chi,g)}$$

is a *unitary representation* of $\hat{G}_\omega$. If the initial representation is normalized, $\hat{G}_\omega$ is said to be a **central extension of** G by means of $U(1)$ [11; 5]. Indeed, the elements (χ, e), $\chi \in U(1)$, commute with all the elements of $\hat{G}_\omega$ and thus they belong to the center of the group.

Remark 2.3.57. *These types of unitary representations of central extensions play a remarkable role in physics. Sometimes $\hat{G}_\omega$ with a particular choice for ω is seen as the* true *group of symmetries at quantum level, when G is the* classical group *of symmetries. There is a very important case. If G is the* Galilean group *– the group of transformations between inertial reference frames in classical physics, viewed as* active *transformations – as clarified by Bargmann [5] the only physically relevant unitary projective representations in QM are just the ones which are* not *equivalent to unitary representations! The multiplicators embody the information about the* mass *of the system. This phenomenon gives also rise to a famous superselection structure in the Hilbert space of quantum systems admitting the Galilean group as a symmetry group, known as* Bargmann's superselection rule *[5].*

To conclude, we just state a technically important result [5] which introduces the one-parameter strongly continuous unitary groups as crucial tool in QM.

Theorem 2.3.58. *Let $\gamma : \mathbb{R} \ni r \mapsto U_r$ be a continuous unitary projective representation of the additive topological group $\mathbb{R}$ on the complex Hilbert space $\mathcal{H}$. The following facts hold.*

(a) *γ is equivalent to a strongly continuous unitary representation $\mathbb{R} \ni r \mapsto V_r$ of the same topological additive group on $\mathcal{H}$.*

(b) *A strongly continuous unitary representation $\mathbb{R} \ni r \mapsto V'_r$ is equivalent to γ if and only if*

$$V'_r = e^{icr} V_r$$

for some constant $c \in \mathbb{R}$ and all $r \in \mathbb{R}$.

The above unitary representation can also be defined as a *strongly continuous one-parameter unitary group.*

Definition 2.3.59. *If $\mathcal{H}$ is a Hilbert space, $V : \mathbb{R} \ni r \mapsto V_r \in \mathfrak{B}(\mathcal{H})$, such that:*

(i) *V_r is unitary for every $r \in \mathbb{R}$;*

(ii) *$V_r V_s = V_{r+s}$ for all $r, s \in \mathbb{R}$*
is called **one-parameter unitary group**. *It is called* **strongly continuous one-parameter unitary group** *if in addition to (i) and (ii) we also have*

(iii) *V is continuous referring to the strong operator topology. In other words $V_r\psi \to V_{r_0}\psi$ for $r \to r_0$ and every $r_0 \in \mathbb{R}$ and $\psi \in \mathcal{H}$.*

Remark 2.3.60.

(a) *It is evident that, in view of the group structure, a one-parameter unitary group $\mathbb{R} \ni r \mapsto V_r \in \mathfrak{B}(\mathcal{H})$ is strongly continuous if and only if it is strongly continuous for $r = 0$.*

(b) *It is a bit less evident but true that a one-parameter unitary group $\mathbb{R} \ni r \mapsto V_r \in \mathfrak{B}(\mathcal{H})$ is strongly continuous if and only if it is* weakly continuous *at $r = 0$. Indeed, if V is weakly continuous at $r = 0$, for every $\psi \in \mathcal{H}$, we have*

$$\begin{aligned}||U_r\psi - \psi||^2 &= ||U_r\psi||^2 + ||\psi||^2 - \langle\psi, U_r\psi\rangle - \langle U_r\psi, \psi\rangle \\ &= 2||\psi||^2 - \langle\psi, U_r\psi\rangle - \langle U_r\psi, \psi\rangle \to 0\end{aligned}$$

for $r \to 0$.

2.3.6.3 *One-parameter strongly continuous unitary groups: von Neumann and Stone theorems*

Theorem 2.3.58 establishes that, dealing with continuous unitary projective representation of the additive topological group $\mathbb{R}$, one can always reduce to work with proper strongly continuous one-parameter unitary groups. So, for instance, the action on a quantum system of rotations around an axis can always described by means of strongly continuous one-parameter unitary groups. There is a couple of technical results of very different nature which are very useful in QM. The former is due to von Neumann (see, e.g., [5]) and proves that the one-parameter unitary group which are not strongly continuous are not so many in separable Hilbert spaces.

Theorem 2.3.61. *If $\mathcal{H}$ is a separable complex Hilbert space and $V : \mathbb{R} \ni r \mapsto V_r \in \mathfrak{B}(\mathcal{H})$ is a one parameter unitary group, it is strongly continuous if and only if the maps $\mathbb{R} \ni r \mapsto \langle \psi, U_r \phi \rangle$ are Borel measurable for all $\psi, \phi \in \mathcal{H}$.*

The second proposition we quote [5; 6] is a celebrated result due to Stone (and later extend to the famous *Hille-Yoshida theorem* in Banach spaces). We start by noticing that, if A is a selfadjoint operator in a Hilbert space, $U_t := e^{itA}$, for $t \in \mathbb{R}$, defines a strongly continuous one-parameter unitary group as one easily proves using the functional calculus. The result is remarkably reversible.

Theorem 2.3.62 (Stone theorem)**.** *Let $\mathbb{R} \ni t \mapsto U_t \in \mathfrak{B}(\mathcal{H})$ be a strongly continuous one-parameter unitary group in the complex Hilbert space $\mathcal{H}$. The following facts hold.*

(a) *There exists a unique selfadjoint operator, called the* **(selfadjoint) generator** *of the group, $A : D(A) \to \mathcal{H}$ in $\mathcal{H}$, such that*

$$U_t = e^{-itA} , \quad t \in \mathbb{R} . \tag{2.98}$$

(b) *The generator is determined as*

$$A\psi = i \lim_{t \to 0} \frac{1}{t}(U_t - I)\psi \tag{2.99}$$

and $D(A)$ is made of the vectors $\psi \in \mathcal{H}$ such that the right hand side of (2.99) exists in $\mathcal{H}$.

(c) *$U_t(D(A)) \subset D(A)$ for all $t \in \mathbb{R}$ and*

$$AU_t\psi = U_t A\psi \quad \text{if } \psi \in D(A) \text{ and } t \in \mathbb{R}.$$

Remark 2.3.63. *For a selfadjoint operator A, the expansion*

$$e^{-itA}\psi = \sum_{n=0}^{+\infty} \frac{(-it)^n}{n!} A^n \psi$$

generally does not *work for $\psi \in D(A)$. It works in two cases however: (i) if ψ is an* analytic *vector of A (Def. 2.2.36 and this result is due to Nelson), (ii) if $A \in \mathfrak{B}(\mathcal{H})$ which is equivalent to say that $D(A) = \mathcal{H}$. In the latter case, one more strongly finds $e^{-itA} = \sum_{n=0}^{+\infty} \frac{(-it)^n}{n!} A^n$, referring to the uniform operator topology. [5].*

One parameter unitary group generated by selfadjoint operators can be used to check if the associated observables are compatible in view of the following nice result [5; 6].

Proposition 2.3.64. *If A and B are selfadjoint operators in the complex Hilbert space $\mathcal{H}$, the identity holds*

$$e^{-itA}e^{-isB} = e^{-isB}e^{-itA} \quad \forall t, s \in \mathbb{R}$$

if and only if the spectral measures of A and B commute.

2.3.6.4 *Time evolution, Heisenberg picture and quantum Noether theorem*

Consider a quantum system described in the Hilbert space $\mathcal{H}$ when an inertial reference frame is fixed. Suppose that, physically speaking, the system is either isolated or interacts with some external stationary environment. With these hypotheses, the time evolution of states is axiomatically described by a continuous symmetry, more precisely, by a continuous one-parameter group of unitary projective operators $\mathbb{R} \ni t \mapsto V_t$. In view of Theorems 2.3.58 and 2.3.62, this group is equivalent to a strongly continuous one-parameter group of unitary operators $\mathbb{R} \ni t \mapsto U_t$ and, up to additive constant, there is a unique selfadjoint operator H, called the **Hamiltonian operator** such that (notice the sign in front of the exponent)

$$U_t = e^{-itH}, \quad t \in \mathbb{R}. \tag{2.100}$$

U_t is called **evolution operator** The observable represented by H is usually identified with *the energy of the system* in the considered reference frame.

Within this picture, if $\rho \in \mathfrak{S}(\mathcal{H})$ is the state of the system at $t = 0$, as usual described by a positive trace-class operator with unit trace, the state at time t is $\rho_t = U_t \rho U_t^{-1}$. If the initial state is pure and represented by the unit vector $\psi \in \mathcal{H}$, the state at time t is $\psi_t := U_t\psi$. In this case, if $\psi \in D(H)$ we have that $\psi_t \in D(H)$ for every $t \in \mathbb{R}$ in view of (c) in Theorem 2.3.62 and furthermore, for (b) of the same theorem

$$-iH\psi_t = \frac{d\psi_t}{dt}. \tag{2.101}$$

where the derivative is computed wit respect to the topology of $\mathcal{H}$. One recognises in Eq. (2.101) the general form of **Schrödinger equation.**

Remark 2.3.65. *It is possible to study quantum systems interacting with some external system which is not stationary. In this case, the Hamiltonian observable depends parametrically on time as already introduced in remark 2.1.8. In these cases, a Schrödinger equation is assumed to describe the time evolution of the* *system giving rise to a groupoid of unitary operators [5]. We shall not enter into the details of this technical issue here.*

Adopting the above discussed framework, observables do not evolve and states do. This framework is called **Schrödinger picture**. There is however another approach to describe time evolution called **Heisenberg picture**. In this representation states do not evolve in time but observables do. If A is an observable at $t = 0$, its evolution at time t is the observable

$$A_t := U_t^{-1} A U_t \,.$$

Obviously $D(A_t) = U_t^{-1}(D(A)) = U_{-t}(D(A)) = U_t^\dagger(D(A))$. According with (i) in Proposition 2.2.66 the spectral measure of A_t is

$$P_E^{(A_t)} = U_t^{-1} P_E^{(A)} U_t$$

as expected. The probability that, at time t, the observable A produces the outcome E when the state is ρ at $t = 0$, can equivalently be computed both using the standard picture, where states evolve as $tr(P_E^{(A)} \rho_t)$, or Heisenberg picture where observables do obtaining $tr(P_E^{(A_t)} \rho)$. Indeed

$$tr(P_E^{(A)} \rho_t) = tr(P_E^{(A)} U_t^{-1} \rho U_t) = tr(U_t P_E^{(A)} U_t^{-1} \rho) = tr(P_E^{(A_t)} \rho)\,.$$

The two pictures are completely equivalent to describe physics. Heisenberg picture permits to give the following important definition

Definition 2.3.66. *In the complex Hilbert space $\mathcal{H}$ equipped with a strongly continuous unitary one-parameter group representing the time evolution $\mathbb{R} \ni t \mapsto U_t$, an observable represented by the selfadjoint operator A is said to be a* **constant of motion** *with respect to U, if $A_t = A_0$.*

The meaning of the definition should be clear: Even if the state evolve, the probability to obtain an outcome E, measuring a constant of motion A, remains stationary. Also expectation values and standard deviations do not change in time. We are now in a position to state the equivalent of the *Noether theorem* in QM.

Theorem 2.3.67 (Noether quantum theorem). *Consider a quantum system described in the complex Hilbert space $\mathcal{H}$ equipped with a strongly continuous unitary one-parameter group representing the time evolution $\mathbb{R} \ni t \mapsto U_t$. If A is an observable represented by a (generally unbounded) selfadjoint operator A in $\mathcal{H}$, the following facts are equivalent.*

(a) *A is a constant of motion: $A_t = A_0$ for all $t \in \mathbb{R}$.*

(b) *The one-parameter group of symmetries generated by A, $\mathbb{R} \ni s \mapsto e^{-isA}$ is a* **group of dynamical symmetries***: It commutes with time evolution*

$$e^{-isA}U_t = U_t e^{-isA} \quad \text{for all } s, t \in \mathbb{R}\,. \tag{2.102}$$

In particular transforms evolutions of pure states into evolutions of (other) pure states, i.e., $e^{-isA}\, U_t\psi = U_t\, e^{-isA}\psi$.

(c) *The action on observables (2.90) of the one-parameter group of symmetries generated by A, $\mathbb{R} \ni s \mapsto e^{isA}$ leaves H invariant. That is*

$$e^{-isA}He^{isA} = H\,, \qquad \text{for all } s \in \mathbb{R}\,.$$

Proof. Suppose that (a) holds, by definition $U_t^{-1}AU_t = A$. By (i) in Proposition 2.2.66, we have that $U_t^{-1}e^{-isA}U_t = e^{-isA}$ which is equivalent to (b). If (b) is true, we have that $e^{-isA}e^{-itH}e^{isA} = e^{-itH}$. Here an almost direct application of Stone theorem yields $e^{-isA}He^{isA} = H$. Finally suppose that (c) is valid. Again (i) in Proposition 2.2.66 produces $e^{-isA}U_te^{isA} = U_t$ which can be rearranged into $U_t^{-1}e^{-isA}U_t = e^{-isA}$. Finally Stone theorem leads to $U_t^{-1}AU_t = A$ which is (a), concluding the proof. □

Remark 2.3.68.

(a) *In physics textbooks, the above statements are almost always stated using time derivatives and commutators. This is useless and involves many subtle troubles with domains of the involved operators.*

(b) *The theorem can be extended to observables $A(t)$ parametrically depending on time already in the Schrödinger picture [5]. In this case (a) and (b) are equivalent too. With this more general situation, (2.102) in (b) has to be re-written as*

$$e^{-isA(t)}U_t = U_t e^{-isA(0)} \qquad \text{for all } s, t \in \mathbb{R}$$

and Heisenberg evolution considered in (a) encompasses both time dependences

$$A_t = U_t^{-1}A(t)U_t\,.$$

At this juncture, (c) can similarly be stated but, exactly as it happens in Hamiltonian classical mechanics, it has a more complicated interpretation [5].

An example is the generator of the boost one-parameter subgroup along the axis $\mathbf{n}$ *of transformations of the Galilean group $\mathbb{R}^3 \ni x \mapsto x + tv\mathbf{n} \in \mathbb{R}^3$, where the speed*

$v \in \mathbb{R}$ is the parameter of the group. The generator is [5] the unique selfadjoint extension of

$$K_{\mathbf{n}}(t) = \sum_{j=1}^{3} n_j (mX_j|_D - tP_j|_D)\,, \tag{2.103}$$

the constant $m > 0$ denoting the mass of the system and D being the Gåding or the Nelson domain of the representation of (central extension of the) Galilean group as we will discuss later.

(c) *In QM, there are symmetries described by operators which are simultaneously selfadjoint and unitary, so they are also observables and can be measured. The* **parity** *is one of them: $(\mathcal{P}\psi)(x) := \psi(-x)$ for a particle described in $L^2(\mathbb{R}^3, d^3x)$. These are constants of motion ($U_t^{-1}\mathcal{P}U_t = \mathcal{P}$) if and only if they are dynamical symmetries ($\mathcal{P}U_t = \mathcal{P}U_t$). This phenomenon has no classical correspondence.*

(d) *The* time reversal symmetry, *when described by an antiunitary operator T is supposed to satisfy: $THT^{-1} = H$. However, since it is antilinear, it gives rise to the identity (exercise) $Te^{-itH}T^{-1} = e^{-itTHT^{-1}}$, so that $TU_t = U_{-t}T$ as physically expected. There is no conserved quantity associated with this operator because it is not selfadjoint.*

Exercise 2.3.69.

(1) *Prove that if the Hamiltonian observable does not depend on time, it is a constant of motion.*

Solution. In this case, the time translation is described by $U_t = e^{itH}$ and trivially it commute with U_s. Noether theorem implies the thesis.

(2) *Prove that for the free particle in $\mathbb{R}^3$, the momentum along x_1 is a constant of motion as consequence of translational invariance along that axis. Assume that the unitary group representing translations along x_1 is U_u with $(U_u\psi)(x) = \psi(x - u\mathbf{e}_1)$ if $\psi \in L^2(\mathbb{R}^3, d^3x)$.*

Solution. The Hamiltonian is $H = \frac{1}{2m}\sum_{j=1}^{3} P_j^2$. It commutes with the one-parameter unitary group describing displacements along x_1, because as one can prove, the said groups is generated by P_1 itself: $U_u := e^{-iuP_1}$. Theorem 2.3.67 yields the thesis.

(3) *Prove that if $\sigma(H)$ is bounded below but not above, the time reversal symmetry cannot be unitary.*

Solution. We look for an operator, unitary or antiunitary such that $TU_t = U_{-t}T$ for all $t \in \mathbb{R}$. If the operator is unitary, the said identity easily implies $THT^{-1} = -H$ and therefore, with obvious notation, $\sigma(THT^{-1}) = -\sigma(H)$. (e) in remark 2.2.42 immediately yields $\sigma(H) = -\sigma(H)$ which is false if $\sigma(H)$ is bounded below but not above.

2.3.6.5 *Strongly continuous unitary representations of Lie groups, Nelson theorem*

Topological and Lie groups are intensively used in QM [24]. More precisely they are studied in terms of their strongly continuous unitary representations. The reason to consider strongly continuous representations is that they immediately induce continuous representations of the group in terms of quantum symmetries (Def. 2.3.54). In the rest of the section, we consider only the case of a *real Lie group*, G, whose Lie algebra is indicated by $\mathfrak{g}$ endowed with the Lie bracket or commutator $\{\ ,\ \}$.

Definition 2.3.70. *If G is a Lie group, a* **strongly continuous unitary representation** *of G over the complex Hilbert space $\mathcal{H}$ is a group homomorphism $G \ni g \mapsto U_g \in \mathfrak{B}(\mathcal{H})$ such that every U_g is unitary and $U_g \to U_{g_0}$, in the strong operator topology, if $g \to g_0$.*

We leave to the reader the elementary proof that strong continuity is equivalent to strong continuity at the unit element of the group and in turn, this is equivalent to weak continuity at the unit element of the group.

A fundamental technical fact is that the said unitary representations are associated with representations of the Lie algebra of the group in terms of (anti)selfadjoint operators. These operators are often physically interpreted as constants of motion (generally parametrically depending on time) when the Hamiltonian of the system belongs to the representation of the Lie algebra. We want to study this relation between the representation of the group on the one hand and the representation of the Lie algebra on the other hand. First of all we define the said operators representing the Lie algebra.

Definition 2.3.71. *Let G be a real Lie group and consider a strongly continuous unitary representation U of G over the complex Hilbert space* H.

If $\mathsf{A} \in \mathfrak{g}$ let $\mathbb{R} \ni t \mapsto \exp(t\mathsf{A}) \in G$ be the generated one-parameter Lie subgroup. The **selfadjoint generator associated with** A

$$A : D(A) \to \mathcal{H}$$

is the generator of the strongly continuous one-parameter unitary group

$$\mathbb{R} \ni t \mapsto U_{\exp\{t\mathsf{A}\}} = e^{-isA}$$

in the sense of Theorem 2.3.62.

The expected result is that these generators (with a factor $-i$) define a representation of the Lie algebra of the group. The utmost reason is that they are associated with the unitary one-parameter subgroups exactly as the elements of the Lie algebra are associated to the Lie one-parameter subgroups. In particular, we expect that the Lie parenthesis correspond to the commutator of operators. The technical problem is that the generators A may have different domains. Thus we look for a common invariant (because the commutator must be defined thereon) domain, where all them can be defined. This domain should embody all the amount of information about the operators A themselves, disregarding the fact that they are defined in larger domains. In other words we would like that the domain be a *core* ((3) in Def. 2.2.20) for each generator. There are several candidates for this space, one of the most appealing is the so called *Gårding space*.

Definition 2.3.72. *Let G be a (finite dimensional real) Lie group and consider a strongly continuous unitary representation U of G over the complex Hilbert space $\mathcal{H}$. If $f \in C_0^\infty(G;\mathbb{C})$ and $x \in \mathcal{H}$, define*

$$x[f] := \int_G f(g) U_g x \, dg \tag{2.104}$$

where dg denotes the Haar measure over G and the integration is defined in a weak sense exploiting Riesz' lemma: Since the map $\mathcal{H} \ni x \mapsto \int_G f(g)\langle y|U_g x\rangle dg$ is continuous (the proof being elementary), $x[f]$ is the unique vector in $\mathcal{H}$ such that

$$\langle y|x[f]\rangle = \int_G f(g)\langle y, U_g x\rangle dg \,, \quad \forall y \in \mathcal{H} \,.$$

The complex span of all vectors $x[f] \in \mathcal{H}$ with $f \in C_0^\infty(G;\mathbb{C})$ and $x \in \mathsf{H}$ is called **Gårding space** *of the representation and is denoted by $D_G^{(U)}$.*

The subspace $D_G^{(U)}$ enjoys very remarkable properties we state in the next theorem. In the following $L_g : C_0^\infty(G;\mathbb{C}) \to C_0^\infty(G;\mathbb{C})$ denotes the standard left-action of $g \in G$ on complex valued smooth compactly supported functions defined on G:

$$(L_g f)(h) := f(g^{-1}h) \quad \forall h \in G \,, \tag{2.105}$$

and, if $\mathsf{A} \in \mathfrak{g}$, $X_{\mathsf{A}} : C_0^\infty(G;\mathbb{C}) \to C_0^\infty(G;\mathbb{C})$ is the smooth vector field over G (a smooth differential operator) defined as:

$$(X_{\mathsf{A}}(f))\,(g) := \lim_{t\to 0} \frac{f\left(\exp\{-t\mathsf{A}\}g\right) - f(g)}{t} \quad \forall g \in G\,. \tag{2.106}$$

so that that map

$$\mathfrak{g} \ni \mathsf{A} \mapsto X_{\mathsf{A}} \tag{2.107}$$

defines a representation of $\mathfrak{g}$ in terms of vector fields (differential operators) on $C_0^\infty(G;\mathbb{C})$. We conclude with the following theorem [24], establishing that the Gårding space has all the expected properties.

Theorem 2.3.73. *Referring to Definitions 2.3.71 and 2.3.72, the Gårding space $D_G^{(U)}$ satisfies the following properties.*

(a) *$D_G^{(U)}$ is dense in $\mathcal{H}$.*

(b) *If $g \in G$, then $U_g(D_G^{(U)}) \subset D_G^{(U)}$. More precisely, if $f \in C_0^\infty(G)$, $x \in \mathcal{H}$, $g \in G$, it holds*

$$U_g x[f] = x[L_g f]\,. \tag{2.108}$$

(c) *If $\mathsf{A} \in \mathfrak{g}$, then $D_G^{(U)} \subset D(A)$ and furthermore $A(D_G^{(U)}) \subset D_G^{(U)}$. More precisely*

$$-iAx[f] = x[X_{\mathsf{A}}(f)]\,. \tag{2.109}$$

(d) *The map*

$$\mathfrak{g} \ni \mathsf{A} \mapsto -iA|_{D_G^{(U)}} =: U(\mathsf{A}) \tag{2.110}$$

is a Lie algebra representation in terms of antisymmetric operators defined on the common dense invariant domain $D_G^{(U)}$. In particular if $\{\ ,\ \}$ is the Lie commutator of $\mathfrak{g}$ we have:

$$[U(\mathsf{A}), U(\mathsf{A}')] = U(\{\mathsf{A}, \mathsf{A}'\}) \quad \text{if } \mathsf{A}, \mathsf{A}' \in \mathfrak{g}.$$

(e) *$D_G^{(U)}$ is a core for every selfadjoint generator A with $\mathsf{A} \in \mathfrak{g}$, that is*

$$A = \overline{A|_{D_G^{(U)}}}\,, \quad \forall \mathsf{A} \in \mathfrak{g}\,. \tag{2.111}$$

Now we tackle the inverse problem: We suppose a certain representation of a Lie algebra $\mathfrak{g}$ in terms of symmetric operators defined in common invariant domain of a complex Hilbert space $\mathcal{H}$. We are interested in lifting this representation to a whole strongly continuous representation of the unique simply connected Lie group G admitting $\mathfrak{g}$ as Lie algebra. This is a much more difficult problem solved by Nelson.

Given a strongly continuous representation U of a (real) Lie group G, there is another space $D_N^{(U)}$ with similar features to $D_G^{(U)}$. Introduced by Nelson [24], it turns out to be more useful than the Gårding space to *recover* the representation U by exponentiating the Lie algebra representation.

By definition $D_N^{(U)}$ consists of vectors $\psi \in \mathcal{H}$ such that $G \ni g \mapsto U_g\psi$ is *analytic* in g, i.e. expansible in power series in (real) analytic coordinates around every point of G. The elements of $D_N^{(U)}$ are called **analytic vectors of the representation** U and $D_N^{(U)}$ is the **space of analytic vectors of the representation** U. It turns out that $D_N^{(U)}$ is invariant for every U_g, $g \in G$.

A remarkable relationship exists between analytic vectors in $D_N^{(U)}$ and analytic vectors according to Definition 2.2.36. Nelson proved the following important result [24], which implies that $D_N^{(U)}$ is dense in $\mathcal{H}$, as we said, because analytic vectors for a selfadjoint operator are dense (exercise 2.2.73). An operator is introduced, called *Nelson operator*, that sometimes has to do with the *Casimir operators* [24] of the represented group.

Proposition 2.3.74. *Let G be a (finite dimensional real) Lie group and $G \ni g \mapsto U_g$ a strongly continuous unitary representation on the Hilbert space $\mathcal{H}$. Take $\mathsf{A}_1, \ldots, \mathsf{A}_n \in \mathfrak{g}$ a basis and define* **Nelson's operator** *on $D_G^{(U)}$ by*

$$\Delta := \sum_{k=1}^{n} U(\mathsf{A}_k)^2 \, ,$$

where the $U(\mathsf{A}_k)$ are, as before, the selfadjoint generators A_k restricted to the Gårding domain $D_G^{(U)}$. Then:

(a) *Δ is essentially selfadjoint on $D_G^{(U)}$.*
(b) *Every analytic vector of the selfadjoint operator $\overline{\Delta}$ is analytic an element of $D_N^{(U)}$, in particular $D_N^{(U)}$ is dense.*
(c) *Every vector in $D_N^{(U)}$ is analytic for every selfadjoint operator $\overline{U(\mathsf{A}_k)}$, which is thus essentially selfadjoint in $D_N^{(U)}$ by Nelson's criterion.*

We finally state the well-known theorem of Nelson that enables to associate representations of the only simply connected Lie group with a given Lie algebra to representations of that Lie algebra.

Theorem 2.3.75 (Nelson theorem). *Consider a real n-dimensional Lie algebra V of operators $-iS$ – with each S symmetric on the Hilbert space $\mathcal{H}$, defined on* *a common invariant subspace $\mathcal{D}$ dense in $\mathcal{H}$ and V-invariant – with the usual commutator of operators as Lie bracket.*

Let $-iS_1, \cdots, -iS_n \in V$ be a basis of V and define Nelson's operator with domain $\mathcal{D}$:

$$\Delta := \sum_{k=1}^{n} S_k^2 .$$

If Δ is essentially selfadjoint, there exists a strongly continuous unitary representation

$$G_V \ni g \mapsto U_g$$

on $\mathcal{H}$, of the unique simply connected Lie group G_V with Lie algebra V.

U is completely determined by the fact that the closures $\overline{S}$, for every $-iS \in V$, are the selfadjoint generators of the representation of the one-parameter subgroups of G_V in the sense of Def. 2.3.71.

In particular, the symmetric operators S are essentially selfadjoint on $\mathcal{D}$, their closure being selfadjoint.

Exercise 2.3.76. *Let A, B be selfadjoint operators in the complex Hilbert space $\mathcal{H}$ with a common invariant dense domain D where they are symmetric and commute. Prove that if $A^2 + B^2$ is essentially selfadjoint on D, then the spectral measures of A and B commute.*

Solution. Apply Nelson's theorem observing that A, B define the Lie algebra of the additive Abelian Lie group $\mathbb{R}^2$ and that D is a core for A and B, because they are essentially selfadjoint therein again by Nelson theorem.

Example 2.3.77.

(1) Exploiting spherical polar coordinates, the Hilbert space $L^2(\mathbb{R}^3, d^3x)$ can be factorised as $L^2([0, +\infty), r^2dr) \otimes L^2(\overline{S}^2, d\Omega)$, where $d\Omega$ is the natural rotationally invariant Borel measure on the sphere $\overline{S}^2$ with unit radius in $\mathbb{R}^3$, with $\int_{\overline{S}^2} 1 d\Omega = 4\pi$. In particular, a Hilbertian basis of $L^2(\mathbb{R}^3, d^3x)$ is therefore made of the products $\psi_n(r)Y_m^l(\theta, \phi)$ where $\{\psi_n\}_{n\in\mathbb{N}}$ is any Hilbertian basis in $L^2([0, +\infty), r^2dr)$ and $\{Y_m^l \mid l = 0, 1, 2, \ldots, m = 0, \pm 1, \pm 2, \ldots \pm l\}$ is the standard Hilbertian basis of *spherical harmonics* of $L^2(\overline{S}^2, d\Omega)$ [24]. Since the function Y_m^l are smooth on $\overline{S}^2$, it is possible to arrange the basis of ψ_n made of compactly supported smooth functions whose derivatives in 0 vanish at every order, in order that $\mathbb{R}^3 \ni x \mapsto (\psi_n \cdot Y_m^l)(x)$ are elements of $\mathbb{C}^\infty(\mathbb{R}^n; \mathbb{C})$ (and therefore also of $\mathcal{S}(\mathbb{R}^3)$). Now consider the three symmetric operators defined on the common dense invariant domain $\mathcal{S}(\mathbb{R}^3)$

$$\mathcal{L}_k = \sum_{i,j=1}^{3} \epsilon_{kij} X_i P_j|_{\mathcal{S}(\mathbb{R}^3)}$$

where ϵ_{ijk} is completely antisymmetric in ijk and $\epsilon_{123} = 1$. By direct inspection, one sees that

$$[-i\mathcal{L}_k, -i\mathcal{L}_h] = \sum_{r=1}^{3} \epsilon_{khr}(-i\mathcal{L}_r)$$

so that the finite real span of the operators $i\mathcal{L}_k$ is a representation of the Lie algebra of the simply connected real Lie group $SU(2)$ (the universal covering of $SO(3)$). Define the Nelson operator $\mathcal{L}^2 := -\sum_{k=1}^3 \mathcal{L}_k^2$ on $\mathcal{S}(\mathbb{R}^3)$. Obviously this is a symmetric operator. A well-known computation proves that

$$\mathcal{L}^2\, \psi_n(r) Y_m^l = l(l+1)\, \psi_n(r) Y_m^l\,.$$

We conclude that $\mathcal{L}^2$ admits a Hilbertian basis of eigenvectors. Corollary 2.2.38 implies that $\mathcal{L}^2$ is essentially selfadjoint. Therefore we can apply Theorem 2.3.75 concluding that there exists a strongly continuous unitary representation $SU(2) \ni M \mapsto U_M$ of $SU(2)$ (actually it can be proven to be also of $SO(3)$). The three selfadjoint operators $L_k := \overline{\mathcal{L}_k}$ are the generators of the one-parameter of rotations around the corresponding three orthogonal Cartesian axes x_k, $k = 1, 2, 3$. The one-parameter subgroup of rotations around the generic unit vector $\mathbf{n}$, with components n_k, admits the selfadjoint generator $L_{\mathbf{n}} = \overline{\sum_{k=1}^3 n_k \mathcal{L}_k}$. The observable $L_{\mathbf{n}}$ has the physical meaning of the $\mathbf{n}$-*component of the angular momentum* of the particle described in $L^2(\mathbb{R}^3, d^3x)$. It turns out that, for $\psi \in L^2(\mathbb{R}^3, d^3x)$,

$$(U_M\psi)(x) = \psi(\pi(M)^{-1}x)\,, \quad M \in SU(2)\,, \quad x \in \mathbb{R}^3 \tag{2.112}$$

where $\pi : SU(2) \to SO(3)$ is the standard covering map. (2.112) is the action of the rotation group on pure states in terms of quantum symmetries. This representation is, in fact, a subrepresentation of the unitary representation of $IO(3)$ already found in (1) of example 2.3.50.

(2) Given a quantum system, a quite general situation is the one where the quantum symmetries of the systems are described by a strongly continuous representation $V : G \ni g \mapsto V_g$ on the Hilbert space $\mathcal{H}$ of the system, and the time evolution is the representation of a one-parameter Lie subgroup with generator $\mathsf{H} \in \mathfrak{g}$. So that

$$V_{\exp(t\mathsf{H})} = e^{-itH} =: U_t\,.$$

This is the case, for instance, of relativistic quantum particles, where G is the *special orthochronous Lorentz group*, $SO(1,3)_+$, (or its universal covering $SL(2,\mathbb{C})$). Describing non-relativistic quantum particles, the relevant group G is an $U(1)$ central extension of the universal covering of the (connected orthochronous) Galilean group.

In this situation, every element of $\mathfrak{g}$ determines a constant of motion. Actually there are two cases.

(i) If $\mathsf{A} \in \mathfrak{g}$ and $\{\mathsf{H},\mathsf{A}\} = 0$, then the Lie subgroups $\exp(t\mathsf{H})$ and $\exp(s\mathsf{A})$ commute as, for example, follows from *Baker-Campbell-Hausdorff* formula (see [24; 5], for instance). Consequently A is a constant of motion because $V_{\exp(t\mathsf{H})} = e^{-itH}$ and $V_{\exp(s\mathsf{A})} = e^{-isA}$ commute as well and Theorem 2.3.67 is valid. In this case e^{-isA} defines a dynamical symmetry in accordance with the aforementioned theorem. This picture applies in particular, referring to a free particle, to $A = J_{\mathbf{n}}$, the observable describing total angular momentum along the unit vector $\mathbf{n}$ computed in an inertial reference frame.

(ii) A bit more complicated is the case of $\mathsf{A} \in \mathfrak{g}$ with $\{\mathsf{H},\mathsf{A}\} \neq 0$. However, even in this case A defines a constant of motion in terms of selfadjont operators (observables) belonging to the representation of the Lie algebra of G. The difference with respect to the previous case is that, now, the constant of motion *parametrically depend on time.* We therefore have a class of observables $\{A(t)\}_{t\in\mathbb{R}}$ in the Schrödinger picture, in accordance with (b) in remark 2.3.69, such that $A_t := U_t^{-1}A(t)U_t$ are the corresponding observables in the Heisenber picture. The equation stating that we have a constant of motion is therefore $A_t = A_0$.

Exploiting the natural action of the Lie one-parameters subgroups on $\mathfrak{g}$, let us define the time parametrized class of elements of the Lie algebra

$$\mathsf{A}(t) := \exp(t\mathsf{H})\mathsf{A}\exp(-t\mathsf{H}) \in \mathfrak{g}\,, \quad t \in \mathbb{R}\,.$$

If $\{\mathsf{A}_k\}_{k=1,\ldots,n}$ is a basis of $\mathfrak{g}$, it must consequently hold

$$\mathsf{A}(t) = \sum_{k=1}^{n} a_k(t)\mathsf{A}_k \tag{2.113}$$

for some real-valued smooth functions $a_k = a_k(t)$. By construction, the corresponding class of selfadjoint generators $A(t)$, $t \in \mathbb{R}$, define a parametrically time dependent constant of motion. Indeed, since (exercise)

$$\exp(s\exp(t\mathsf{H})\mathsf{A}\exp(-t\mathsf{H})) = \exp(t\mathsf{H})\exp(s\mathsf{A})\exp(-t\mathsf{H})\,,$$

we have

$$-iA(t) = \frac{d}{ds}\Big|_{s=0} V_{\exp(t\mathsf{H})\mathsf{A}\exp(-t\mathsf{H})} = \frac{d}{ds}\Big|_{s=0} V_{\exp(t\mathsf{H})\exp(s\mathsf{A})\exp(-t\mathsf{H})}$$
$$= \frac{d}{ds}\Big|_{s=0} V_{\exp(t\mathsf{H})} V_{\exp(s\mathsf{A})} V_{\exp(-t\mathsf{H})} = -iU_t A U_t^{-1}$$

Therefore

$$A_t = U_t^{-1} A(t) U_t = U_t^{-1} U_t A U_t^{-1} U_t = A = A_0 \,.$$

In view of Theorem 2.3.73, as the map $\mathfrak{g} \ni A \mapsto A|_{D_G^{(V)}}$ is a Lie algebra isomorphism, we can recast (2.113) for selfadjoint generators

$$A(t)|_{D_G^{(V)}} = \sum_{k=1}^{n} a_k(t) A_k|_{D_G^{(V)}} \tag{2.114}$$

(where $D_G^{(V)}$ may be replaced by $D_N^{(V)}$ as the reader can easily establish, taking advantage of Proposition 2.3.74 and Theorem 2.3.75). Since $D_G^{(V)}$ (resp. $D_N^{(V)}$) is a core for $A(t)$, it also holds

$$A(t) = \overline{\sum_{k=1}^{n} a_k(t) A_k|_{D_G^{(V)}}} \,, \tag{2.115}$$

the bar denoting the closure of an operator as usual. (The same is true replacing $D_G^{(V)}$ for $D_N^{(V)}$.) An important case, both for the non-relativistic and the relativistic case is the selfadjoint generator $K_{\mathbf{n}}(t)$ associated with the boost transformation along the unit vector $\mathbf{n} \in \mathbb{R}^3$, the rest space of the inertial reference frame where the boost transformation is viewed as an active transformation. In fact, referring to the Lie generators of (a $U(1)$ central extension of the universal covering of the connected orthochronous) Galilean group, we have $\{h, k_{\mathbf{n}}\} = -p_{\mathbf{n}} \neq 0$, where $p_{\mathbf{n}}$ is the generator of spatial translations along $\mathbf{n}$, corresponding to the observable momentum along the same axis when passing to selfadjoint generators. The non-relativistic expression of $K_{\mathbf{n}}(t)$, for a single particle, appears in (2.103). For a more extended discussion on the non-relativistic case see [5]. A pretty complete discussion including the relativistic case is contained in [24].

2.3.6.6 *Selfadjoint version of Stone-von Neumann-Mackey theorem*

A remarkable consequence of Nelson's theorem is a selfadjoint operator version of Stone-von Neumann theorem usually formulated in terms of unitary operators

[5], proving that the CCRs always give rise to the standard representation in $L^2(\mathbb{R}^n, d^n x)$. We state and prove this version of the theorem, adding a last statement which is the selfadjoint version of Mackey completion to Stone von Neumann statement [5; 6].

Theorem 2.3.78 (Stone-von Neumann-Mackey theorem). *Let $\mathcal{H}$ be a complex Hilbert space and suppose that there are $2n$ selfadjoint operators in $\mathcal{H}$ we indicate with $Q_1, \ldots, Q_n$ and $M_1, \ldots, M_n$ such the following requirements are valid where $h, k = 1, \ldots, n$.*

(1) *There is a common dense invariant subspace $D \subset \mathcal{H}$ where the CCRs hold*

$$[Q_h, M_k]\psi = i\hbar\delta_{hk}\psi\,, \quad [Q_h, Q_k]\psi = 0\,, \quad [M_h, M_k]\psi = 0 \quad \psi \in D\,. \tag{2.116}$$

(2) *The representation is irreducible, in the sense that there is no closed subspace $\mathcal{K} \subset \mathcal{H}$ such that $Q_k(\mathcal{K} \cap D(Q_k)) \subset \mathcal{K}$ and $M_k(\mathcal{K} \cap D(M_k)) \subset \mathcal{K}$ such that Q_k and M_k are selfadjoint as operators over $\mathcal{K}$.*

(3) *The operator $\sum_{k=1}^n Q_k^2|_D + M_k^2|_D$ is essentially selfadjoint.*

Under these conditions, there is a Hilbert space isomorphism, that is a surjective isometric map, $U : \mathcal{H} \to L^2(\mathbb{R}^n, d^n x)$ such that

$$UQ_kU^{-1} = X_k \quad \textit{and} \quad UM_kU^{-1} = P_k \quad k = 1, \ldots, n \tag{2.117}$$

where X_k and P_k respectively are the standard position (2.35) and momentum (2.36) selfadjoint operators in $L^2(\mathbb{R}^n, d^n x)$. In particular $\mathcal{H}$ results to be separable.

If (1), (2) and (3) are valid with the exception that the representation is not reducible, then $\mathcal{H}$ decomposes into an orthogonal Hilbertian sum $\mathcal{H} = \oplus_{r \in R}\mathcal{H}_k$ where R is finite or countable if $\mathcal{H}$ is separable, the $\mathcal{H}_r \subset \mathcal{H}$ are closed subspaces with

$$Q_k(\mathcal{H}_r \cap D(Q_k)) \subset \mathcal{H}_r \quad \textit{and} \quad M_k(\mathcal{H}_r \cap D(M_k)) \subset \mathcal{H}_r$$

for all $r \in R$, $k = 1, \ldots, n$ and the restrictions of all the Q_k and M_k to each $\mathcal{H}_r$ satisfy (2.117) for suitable surjective isometric maps $U_r : \mathcal{H}_r \to L^2(\mathbb{R}^n, d^n x)$.

Proof. If (1) holds, the restrictions to D of the selfadjoint operators Q_k, M_k define symmetric operators (since they are selfadjoint and D is dense and included in their domains), also their powers are symmetric since D is invariant. If also (2) is valid, in view of Nelson theorem (since evidently the symmetric operator $I|_D^2 + \sum_{k=1}^n Q_k^2|_D + M_k^2|_D$ is essentially selfadjoint if $\sum_{k=1}^n Q_k^2|_D + M_k^2|_D$ is), there is a strongly continuous unitary representation $W \ni g \mapsto V_g \in \mathfrak{B}(\mathcal{H})$ of the simply connected $2n+1$-dimensional Lie group W whose Lie algebra is defined by (2.116) (correspondingly re-stated for the operators $-iI, -iQ_k, -iM_k$) together with $[-iQ_h, -iI] = [-iM_k, -iI] = 0$, where the operator $-iI$ restricted to D is the remaining Lie generator. W is the *Heisenberg-Weyl* group [5]. The selfadjoint generators of this representation are just the operators Q_k and P_k (and I), since they coincide with the closure of their restrictions to D, because they are selfadjoint (so they admit unique selfadjoint extensions) and D is a core. If furthermore the Lie algebra representation is irreducible, the unitary representation is irreducible, too: If $\mathcal{K}$ were an invariant subspace for the unitary operators, Stone theorem would imply that $\mathcal{K}$ be also invariant under the selfadjoint generators of the one parameter Lie subgroups associated to each Q_k and P_k. This is impossible if the Lie algebra representation is irreducible as we are assuming. The standard version of Stone-von Neumann theorem [5] implies that there is isometric surgective operator $U : \mathcal{H} \to L^2(\mathbb{R}^n, d^nx)$ such that $W \ni g \mapsto UV_gU^{-1} \in \mathfrak{B}(L^2(\mathbb{R}^n, d^nx))$ is the standard unitary representation of the group W in $L^2(\mathbb{R}^n, d^nx)$ genernated by X_k and P_k (and I) [5]. Again, Stone theorem immediately yields (2.117). The last statement easily follows from the standard form of Mackey's theorem completing Stone-von Neumann result [5]. □

Remark 2.3.79.

(a) *The result* a posteriori *gives, in particular, a strong justification of the requirement that the Hilbert space of an elementary quantum system, like a particle, must be separable.*

(b) *Physical Hamiltonian operators have spectrum bounded from below to avoid thermodynamical instability. This fact prevents the definition of a "time operator" canonically conjugated with H following the standard way. This result is sometime quoted as* **Pauli theorem**. *As a consequence, the meaning of Heisenberg relations $\Delta E \Delta T \geq \hbar/2$ is different from the meaning of the analogous relations for position and momentum. It is however possible to define a sort of time osservable just extending the notion of PVM to the notion of POVM i.e.,* positive valued operator measure *(see the first part and [5]). POVMs are exploited to describe concrete physical phenomena related to measurement procedures, especially in quantum information theory [30; 31].*

Corollary 2.3.80. *If the Hamiltonian operator $\sigma(H)$ of a quantum system is bounded below, there is no selfadjoint operator (time operator) T satisfying the standard CCR with H and the hypotheses (1), (2), (3) of Theorem 2.3.78.*

Proof. The couple H, T should be mapped to a corresponding couple X, P in $L^2(\mathbb{R}, dx)$, or a direct sum of such spaces, by means of a Hilbert space isomorphism. In all cases the spectrum of H should therefore be identical to the one of X, namely is $\mathbb{R}$. This fact is false by hypotheses. □

2.4 Just Few Words about the Algebraic Approach

The fundamental theorem 2.3.6.6 of Stone-von Neumann and Mackey is stated in the jargon of theoretical physics as follows:

"*all irreducible representations of the CCRs with a finite, and fixed, number of degrees of freedom are unitarily equivalent*".

The expression *unitarily equivalent* refers to the existence of the Hilbert-space isomorphism U, and the finite number of degrees of freedom is the dimension of the Lie algebra spanned by the generators I, X_k, P_k.

What happens then in infinite dimensions? This is the case when dealing with *quantum fields*, where the $2n+1$ generators I, X_k, P_k $(k = 1, 2, \ldots, n)$, are replaced by a *continuum* of generators, the so-called *field operators at fixed time* and the **conjugated momentum at fixed time**: $I, \Phi(f), \Pi(g)$ which are smeared by arbitrary functions $f, g \in C_0^\infty(\mathbb{R}^3)$. Here $\mathbb{R}^3$ is the rest space of a given reference frame in the spacetime. Those field operators satisfy commutation relations similar to the ones of X_k and P_k (e.g., see [26; 27; 28]). Then the Stone–von Neumann theorem no longer holds. In this case, theoretical physicists would say that

"*there exist irreducible non-equivalent CCR representations with an infinite number of degrees of freedom*".

What happens in this situation, in practice, is that one finds two *isomorphic* *-algebras of field operators, the one generated by $\Phi(f), \Pi(g)$ in the Hilbert space $\mathcal{H}$ and the other generated by $\Phi'(f), \Pi'(g)$ in the Hilbert space $\mathcal{H}'$ that admit *no Hilbert space* isomorphism $U : \mathsf{H}' \to \mathsf{H}$ satisfying:

$$U\Phi'(f)\,U^{-1} = \Phi(f)\,, \quad U\Pi'(g)\,U^{-1} = \Pi(g) \quad \text{for any pair } f, g \in C_0^\infty(\mathbb{R}^3)\,.$$

Pairs of this kind are called *(unitarily) non-equivalent.* Jumping from the finite-dimensional case to the infinite-dimensional one corresponds to passing from Quantum Mechanics to Quantum Field Theory (possibly relativistic, and on curved spacetime [28]). The presence of non-equivalent representations of one single physical system shows that a formulation in a fixed Hilbert space is fully inadequate, at least because it insists on a fixed Hilbert space, whereas the physical system is characterized by a more abstract object: An algebra of observables which may be represented in different Hilbert spaces in terms of operators. These representations are not unitarily equivalent and none can be considered more fundamental than the remaining ones. We must abandon the structure of Hilbert space in order to lay the foundations of quantum theories in broader generality.

This program has been widely developed (see e.g., [13; 25; 26; 27]), starting from the pioneering work of von Neumann himself, and is nowadays called *algebraic formulation of quantum (field) theories.* Within this framework it was possible to formalise, for example, field theories in curves spacetime in relationship to the quantum phenomenology of black-hole thermodynamics.

2.4.1 *Algebraic formulation*

The algebraic formulation prescinds, anyway, from the nature of the quantum system and may be stated for systems with finitely many degrees of freedom as well [25]. The new viewpoint relies upon two assumptions [26; 27; 25; 29; 5].

(AA1) A physical system S is described by its **observables**, viewed now as selfadjoint elements in a certain C^*-algebra $\mathfrak{A}$ with unit $\mathbb{1}$ associated to S.

(AA2) An **algebraic state** on $\mathfrak{A}_S$ is a linear functional $\omega : \mathfrak{A}_S \to \mathbb{C}$ such that:

$$\omega(a^*a) \geq 0 \quad \forall a \in \mathfrak{A}_S, \qquad \omega(\mathbb{1}) = 1 ,$$

that is, *positive* and *normalized to* 1.

We have to stress that $\mathfrak{A}$ is not seen as a concrete C^*-algebra of operators (a von Neumann algebra for instance) on a given Hilbert space, but remains an abstract C^*-algebra. Physically, $\omega(a)$ is the *expectation value* of the observable $a \in \mathfrak{A}$ in state ω.

Remark 2.4.1.

(a) $\mathfrak{A}$ *is usually called* the algebra of observables of S *though, properly speaking, the observables are the selfadjoint elements of* $\mathfrak{A}$ *only.*

(b) *Differently form the Hilbert space formulation, the algebraic approach can be adopted to describe* both classical and quantum systems. *The two cases are distinguished on the base of commutativity of the algebra of observables* $\mathfrak{A}_S$*: A commutative algebra is assumed to describe a classical system whereas a non-commutative one is supposed to be associated with a quantum systems.*

(c) *The notion of* **spectrum** *of an element* a *of a* C^**-algebra* $\mathfrak{A}$*, with unit element* $\mathbb{1}$*, is defined analogously to the operatorial case [5].* $\sigma(a) := \mathbb{C} \setminus \rho(a)$ *where we have introduced the* **resolvent** *set:*

$$\rho(a) := \{\lambda \in \mathbb{C} \mid \exists (a - \lambda\mathbb{1})^{-1} \in \mathfrak{A}\} .$$

When applied to the elements of $\mathfrak{B}(\mathcal{H})$*, this definition coincides with the one provi-ously discussed for operators in view of (2) in exercise 2.2.43. It turns out that if* $a^*a = aa^*$*, namely* $a \in \mathfrak{A}$ *is normal, then*

$$||a|| = \sup_{\lambda \in \sigma(a)} |\lambda| .$$

The right-hand side of the above identity is called **spectral radius** *of* a. *If* a *is not normal,* a^*a *is selfadjoint and thus normal in any cases. Therefore the* C^**-property of the norm* $||a||^2 = ||a^*a||$ *permits us to write down* $||a||$ *in terms of the spectrum of* a^*a. *As the spectrum is a completely algebraic property, we conclude that it is impossible to change the norm of a* C^**-algebra preserving the* C^**-algebra property of the new norm. A unital* **-algebra admits at most one* C^**-norm.*

(d) *Unital* C^**-algebras are very rigid structures. In particular, every* $*$*-homomorphism* $\pi : \mathfrak{A} \to \mathfrak{B}$ *(which is a pure algebraic notion) between two unital* C^**-algebras is necessarily [5] norm decreasing* ($||\pi(a)|| \leq ||a||$) *thus continuous. Its image,* $\pi(\mathfrak{A})$, *is a* C^**-subalgebra of* $\mathfrak{B}$. *Finally* π *is injective if and ony if it is isometric. The spectra satisfy a certain permanence property [5], with obvious meaning of the symbols*

$$\sigma_{\mathfrak{B}}(\pi(a)) = \sigma_{\pi(\mathfrak{A})}(a) \subset \sigma_{\mathfrak{A}}(a)\,, \quad \forall a \in \mathfrak{A}\,,$$

where the last inclusion becomes and equality if π *is injective.*

The most evident *a posteriori* justification of the algebraic approach lies in its powerfulness [26]. However there have been a host of attempts to account for assumptions **(AA1)** and **(AA2)** and their physical meaning in full generality (see the study of [32], [27] and [25; 29] and especially the work of I. E. Segal [33] based on so-called *Jordan algebras*). Yet none seems to be definitive [34].

An evident difference with respect to the standard QM, where states are measures on the lattice of elementary propositions, is that we have now a complete identification of the notion of state with that of expectation value. This identification would be natural within the Hilbert space formulation, where the class of observables includes the elementary ones, represented by orthogonal projectors, and corresponding to "Yes-No" statements. The expectation value of such an observable coincides with the probability that the outcome of the measurement is "Yes". The set of all those probabilities defines, in fact, a quantum state of the system as we know. However, the analogues of these elementary propositions generally do not belong to the C^*-algebra of observables in the algebraic formulation. Nevertheless, this is not an insurmountable obstruction. Referring to a completely general physical system and following [27], the most general notion of state, ω, is the assignment of all probabilities, $w_\omega^{(A)}(a)$, that the outcome of the measurement of the observable A is a, for all observables A and all of values a. On the other hand, it is known [25] that all experimental information on the measurement of an observable A in the state ω – the probabilities $w_\omega^{(A)}(a)$ in particular – is recorded in the expectation values of the polynomials of A. Here, we should think of $p(A)$ as the observable whose values are the values $p(a)$ for all values a of A. This characterization of an observable is theoretically supported by the various solutions to the *moment problem* in probability measure theory. To adopt this paradigm we have thus to assume that the set of observables must include at least all real polynomials $p(A)$ whenever it contains the observable A. This is in agreement with the much stronger requirement **(AA1)**.

2.4.1.1 *The GNS reconstruction theorem*

The set of algebraic states on $\mathfrak{A}_S$ is a convex subset in the dual $\mathfrak{A}'_S$ of $\mathfrak{A}_S$: if ω_1 and ω_2 are positive and normalized linear functionals, $\omega = \lambda\omega_1 + (1-\lambda)\omega_2$ is clearly still the same for any $\lambda \in [0,1]$.

Hence, just as we saw for the standard formulation, we can define *pure algebraic states* as extreme elements of the convex body.

Definition 2.4.2. *An algebraic state $\omega : \mathfrak{A} \to \mathbb{C}$ on the C^*-algebra with unit $\mathfrak{A}$ is called a* **pure algebraic state** *if it is extreme in the set of algebraic states. An algebraic state that is not pure is called* **mixed**.

Surprisingly, most of the entire abstract apparatus introduced, given by a C^*-algebra and a set of states, admits elementary Hilbert space representations when a reference algebraic state is fixed. This is by virtue of a famous procedure that Gelfand, Najmark and Segal came up with, and that we prepare to present [26; 27; 25; 5; 6].

Theorem 2.4.3 (GNS reconstruction theorem). *Let $\mathfrak{A}$ be a C^*-algebra with unit $\mathbb{1}$ and $\omega : \mathfrak{A} \to \mathbb{C}$ a positive linear functional with $\omega(\mathbb{1}) = 1$. Then the following holds.*

(a) *There exist a triple $(\mathcal{H}_\omega, \pi_\omega, \Psi_\omega)$, where $\mathcal{H}_\omega$ is a Hilbert space, the map $\pi_\omega : \mathfrak{A} \to \mathfrak{B}(\mathcal{H}_\omega)$ a $\mathfrak{A}$-representation over $\mathcal{H}_\omega$ and $\Psi_\omega \in \mathcal{H}_\omega$, such that:*

  (i) *Ψ_ω is cyclic for π_ω. In other words, $\pi_\omega(\mathfrak{A})\Psi_\omega$ is dense in $\mathcal{H}_\omega$;*
  (ii) *$\langle \Psi_\omega, \pi_\omega(a)\Psi_\omega \rangle = \omega(a)$ for every $a \in \mathfrak{A}$.*

(b) *If $(\mathcal{H}, \pi, \Psi)$ satisfies (i) and (ii), there exists a unitary operator $U : \mathcal{H}_\omega \to \mathcal{H}$ such that $\Psi = U\Psi_\omega$ and $\pi(a) = U\pi_\omega(a)U^{-1}$ for any $a \in \mathfrak{A}$.*

Remark 2.4.4. *The GNS representation $\pi_\omega : \mathfrak{A} \to \mathfrak{B}(\mathcal{H}_\omega)$ is a *-homomorphism and thus (c) in remark 2.4.1 applies. In particular π_ω is norm decreasing and continuous. Moreover, again referring to the same remark, if π_ω is faithful – i.e., injective – it is isometric and preserves the spactra of the elements. If $a \in \mathfrak{A}$ is selfadjoint $\pi_\omega(a)$ is a selfadjoint* operator *and its spectrum has the well-known quantum meaning. This meaning, in view of the property of permanence of the* *spectrum, can be directly attributed to the spectrum of $a \in \mathfrak{A}$: If $a \in \mathfrak{A}$ represents an abstract observable, $\sigma(a)$ is the set of the possible values attained by a.*

As we initially said, it turns out that different algebraic states ω, ω' generally give rise to unitarily inequivalent GNS representations $(\mathcal{H}_\omega, \pi_\omega, \Psi_\omega)$ and $(\mathcal{H}_{\omega'}, \pi_{\omega'}, \Psi_{\omega'})$: There is no isometric surjective operator $U : \mathcal{H}_{\omega'} \to \mathcal{H}_\omega$ such that

$$U\pi_{\omega'}(a)U^{-1} = \pi_\omega(a) \quad \forall a \in \mathsf{A}\,.$$

The fact that one may simultaneously deal with all these inequivalent representations is a representation of the power of the algebraic approach with respect to the Hilbert space framework.

However one may also focus on states referred to as a fixed GNS representation. If ω is an algebraic state on $\mathfrak{A}$, every statistical operator on the Hilbert space of a GNS representation of ω – i.e. every positive, trace-class operator with unit trace $T \in \mathfrak{B}_1(\mathcal{H}_\omega)$ – determines an algebraic state

$$\mathfrak{A} \ni a \mapsto tr\,(T\pi_\omega(a))\,,$$

evidently. This is true, in particular, for $\Phi \in \mathcal{H}_\omega$ with $||\Phi||_\omega = 1$, in which case the above definition reduces to

$$\mathfrak{A} \ni a \mapsto \langle \Phi, \pi_\omega(a)\Phi\rangle_\omega\,.$$

Definition 2.4.5. *If ω is an algebraic state on the C^*-algebra with unit $\mathfrak{A}$, every algebraic state on $\mathfrak{A}$ obtained either from a density operator or a unit vector, in a GNS representation of ω, is called* **normal state** *of ω. Their set $Fol(\omega)$ is the* **folium** *of the algebraic state ω.*

Note that in order to determine $Fol(\omega)$ one can use a fixed GNS representation of ω. In fact, as the GNS representation of ω varies, normal states do not change, as implied by part (b) of the GNS theorem.

2.4.1.2 *Pure states and irreducible representations*

To conclude we would like to explain how pure states are characterised in the algebraic framework. To this end we have the following simple result (e.g., see [26; 27; 25; 5].

Theorem 2.4.6 (Characterization of pure algebraic states). *Let ω be an algebraic state on the C^*-algebra with unit $\mathfrak{A}$ and $(\mathcal{H}_\omega, \pi_\omega, \Psi_\omega)$ a corresponding GNS triple. Then ω is pure if and only if π_ω is irreducible.*

The algebraic notion of pure state is in nice agreement with the Hilbert space formulation result where pure states are represented by unit vectors (in the absence of superselection rules). Indeed we have the following proposition which make a comparison between the two notions.

Proposition 2.4.7. *Let ω be a pure state on the C^*-algebra with unit $\mathfrak{A}$ and $\Phi \in \mathsf{H}_\omega$ a unit vector. Then:*

(a) *the functional*

$$\mathfrak{A} \ni a \mapsto \langle \Phi, \pi_\omega(a)\Phi \rangle_\omega ,$$

defines a pure algebraic state and $(\mathsf{H}_\omega, \pi_\omega, \Phi)$ is a GNS triple for it. In that case, GNS representations of algebraic states given by non-zero vectors in $\mathcal{H}_\omega$ are all unitarily equivalent;

(b) *unit vectors $\Phi, \Phi' \in \mathsf{H}_\omega$ give the same (pure) algebraic state if and only if $\Phi = c\Phi'$ for some $c \in \mathbb{C}$, $|c| = 1$, i.e. if and only if Φ and Φ' belong to the same ray.*

The correspondence pure (algebraic) states vs. state vectors, automatic in the standard formulation, holds in Hilbert spaces of GNS representations of pure algebraic states, but in general not for mixed algebraic states. The following exercise focusses on this apparent problem.

Exercise 2.4.8. *Consider, in the standard (not algebraic) formulation, a physical system described on the Hilbert space $\mathcal{H}$ and a mixed state $\rho \in \mathfrak{S}(\mathcal{H})$. The map $\omega_\rho : \mathfrak{B}(\mathcal{H}) \ni A \mapsto tr(\rho A)$ defines an algebraic state on the C^*-algebra $\mathfrak{B}(\mathcal{H})$. By the GNS theorem, there exist another Hilbert space $\mathcal{H}_\rho$, a representation $\pi_\rho : \mathfrak{B}(\mathcal{H}) \to \mathfrak{B}(\mathcal{H}_\rho)$ an unit vector $\Psi_\rho \in \mathsf{H}_\rho$ such that*

$$tr(\rho A) = \langle \Psi_\rho, \pi_\rho(A)\Psi_\rho \rangle$$

for $A \in \mathfrak{B}(\mathcal{H})$. Thus it seems that the initial mixed state has been transformed into a pure state! How is this fact explained?

Solution. There is no transformtion from mixed to pure state because the mixed state is represented by a vector, Ψ_ρ, in a different Hilbert space, $\mathcal{H}_\rho$. Moreover, there is no Hilbert space isomorphism $U : \mathcal{H} \to \mathcal{H}_\rho$ with $UAU^{-1} = \pi_\rho(A)$, so that $U^{-1}\Psi_\rho \in \mathcal{H}$. In fact, the representation $\mathfrak{B}(\mathcal{H}) \ni A \mapsto A \in \mathfrak{B}(\mathcal{H})$ is irreducible, whereas π_ρ cannot be irreducible (as it would be if U existed), because the state ρ is not an extreme point in the space of non-algebraic states, and so it cannot be extreme in the larger space of algebraic states.

Bibliography

[1] G. Ghirardi, *Sneaking a Look at God's Cards: Unraveling the Mysteries of Quantum Mechanics* Princeton University Press; Revised edition (March 25, 2007)
[2] Edward N. Zalta (ed.) *The Stanford Encyclopedia of Philosophy* http://plato.stanford.edu/
[3] P.A.M. Dirac, *The principles of Quantum Mechanics.* Oxford University Press, Oxford (1930)
[4] I. M. Gelfand and N. J. Vilenkin, *Generalized Functions*, vol. 4: Some Applications of Harmonic Analysis. Rigged Hilbert Spaces. Academic Press, New York (1964).
[5] V. Moretti, *Spectral Theory and Quantum Mechanics: Mathematical Foundations of Quantum Theories, Symmetries and Introduction to the Algebraic Formulation*, Springer, 2018
[6] V. Moretti, *Fundamental Mathematica Structures of Quantum Theories.* Springer, Berlin (2019)
[7] J. von Neumann, *Mathematische Grundlagen der Quantenmechanik.* Springer-Verlag, Berlin (1932)
[8] W. Rudin, *Functional Analysis* 2nd edition, McGraw Hill (1991)
[9] K. Schmüdgen, *Unbounded Self-adjoint Operators on Hilbert Space*, Springer (2012)
[10] G. K. Pedersen, *Analysis Now*, Graduate Texts in Mathematics, Vol. 118 (Springer-Verlag, New York, 1989)
[11] V.S. Varadarajan, *Geometry of Quantum Theory*, Second Edition, Springer, Berlin (2007)
[12] Ph. Blanchard, D. Giulini D., E. Joos, C. Kiefer, I.-O. Stamatescu (Eds.): Decoherence: Theoretical, Experimental, and Conceptual Problems. Lecture Notes in Physics. Springer-Verlag, Berlin (2000)
[13] O. Bratteli, D.W. Robinson, *Operator Algebras and Quantum Statistical Mechanics* (Vol I and II, Second Edition). Springer, Berlin (2002)
[14] G. Birkhoff and J. von Neumann, *The logic of quantum mechanics*, Ann. of Math. (2) 37(4) (1936) 823–843
[15] E.G. Beltrametti, G. Cassinelli, *The logic of quantum mechanics.* Encyclopedia of Mathematics and its Applications, vol. 15, Addison-Wesley, Reading, Mass. (1981)
[16] K. Engesser, D. M. Gabbay, D. Lehmann (Eds), *Handbook of Quantum Logic and Quantum Structures.* Elsevier, Amsterdam (2009)
[17] G. Mackey, *The Mathematical Foundations of Quantum Mechanics.* Benjamin, New York (1963)
[18] C. Piron, *Axiomatique Quantique* Helv. Phys. Acta **37** 439-468 (1964)
[19] J.M., Jauch and C. Piron, *On the structure of quantal proposition system* Helv. Phys. Acta **42**, 842 (1969)
[20]J.M., Jauch, *Foundations of Quantum Mechanics* Addison-Wesley Publishing Company, Reading USA (1978)
[21] A. Dvurecenskij, *Gleasons theorem and its applications.* Kluwer academic publishers, Dordrecht (1992)
[22] A. S. Wightman, *Superselection rules; old and new* Nuovo Cimento B 110, 751-769 (1995)

[23] B. Simon, *Quantum dynamics: From automorphism to Hamiltonian.* Studies in Mathematical Physics, Essays in Honor of Valentine Bargmann (ed. E.H. Lieb, B. Simon and A.S. Wightman), Princeton University Press, Princeton, 327-349 (1976)

[24] A.O. Barut, R. Raczka, *Theory of group representations and applications*, World Scientific (1984)

[25] F. Strocchi, *An Introduction To The Mathematical Structure Of Quantum Mechanics: A Short Course For Mathematicians*, World Scientific, Singapore (2005)

[26] R. Haag, *Local Quantum Physics* (Second Revised and Enlarged Edition). Springer Berlin (1996)

[27] H. Araki,*Mathematical Theory of Quantum Fields.* Oxford University Press, Oxford (2009)

[28] I. Khavkine, V. Moretti, *Algebraic QFT in Curved Spacetime and quasifree Hadamard states: an introduction.* Advances in Algebraic Quantum Field Theory by Springer 2015 (Eds R. Brunetti, C. Dappiaggi, K. Fredenhagen, and J. Yngvason)

[29] F. Strocchi, *The Physical Principles of Quantum Mechanics.* European Physics Journal Plus **127**, 12 (2012)

[30] P. Busch, *Quantum states and generalized observables: a simple proof of Gleason's theorem.* Physical Review Letters **91**, 120403 (2003)

[31] P. Busch, M. Grabowski, P.J. Lahti, *Operational Quantum Physics.* Springer, Berlin (1995)

[32] G.G, Emch, *Algebraic Methods in Statistical Mechanics and Quantum Field Theory.* Wiley-Interscience, New York (1972)

[33] I. Segal, *Postulates for general quantum mechanics*, Annals of Mathematics (2), **48**, 930-948 (1947)

[34] R.F., Streater, *Lost Causes in and beyond Physics*, Springer-Verlag, Berlin (2007)

Chapter 3

A Concise Introduction to Quantum Field Theory

Manuel Asorey

Departamento de Física Teórica, Universidad de Zaragoza and Centro de Astopartículas y Física de Altas Energías
email: asorey@unizar.es

We review the basic principles of Quantum Field Theory in a brief but comprehensive introduction to the foundations of Quantum Field Theory. The principles of Quantum Field Theory are introduced in canonical and covariant formalisms. The problem of ultraviolet divergences and its renormalization is analyzed in the canonical formalism. As an application, we review the roots of Casimir effect. For simplicity, we focus on the scalar field theory but the generalization for fermion fields is straightforward. However, the quantization of gauge fields requires extra techniques which are beyond the scope of these lectures. The special cases of free field theories and conformal invariant theories in lower space-time dimensions illustrate the relevance of the foundations of the theory. Finally, a short introduction to functional integrals and perturbation theory in the Euclidean formalism is included in the last section.

3.1 Introduction

Quantum Field Theory (QFT) is the current paradigm of Fundamental Physics. It emerges from the convolution of Quantum Physics and Relativity, the two major theoretical revolutions of the 20th century physics. The search for a theory of quantum fields started right after the discovery of Quantum Mechanics, but the appearance of ultraviolet divergences postponed the formulation of a consistent theory for two decades. The main problem was solved in perturbation theory by the renormalization of all physical quantities: vacuum energy, fundamental fields, particle masses and charges, and other couplings constants. One further step in the formulation of the theory beyond perturbation theory was achieved by Wilson's renormalization group approach, a quarter of a century later.

It is usually considered that there were *only* two conceptual revolutions in 20th century physics: the theory of relativity and the quantum theory. It is not quite true, the formulation of QFT required also a radical deep conceptual change in the relations between theory and observations that might be considered as a third major revolution of physics. The need of renormalization of ultraviolet (UV) divergences required a *dramatic* solution (*a la Planck*): the parameters which appear in the Lagrangian do not necessary coincide with those associated with observations. Moreover the parameters of the Lagrangian of interacting field theories become divergent when the UV cutoff is removed, while the physical parameters remain finite in that limit.

The first attempts to quantize free field theories started just one year after the discovery of quantum mechanics by Heisenberg, Born, and Jordan. One year later in 1927 Dirac introduced quantum electrodynamics (QED), the first quantum theory of interacting fields. The general program of quantization of relativistic field theories was initiated by Jordan and Pauli one year later.

However, the development of the theory was suddenly stopped by the appearance of ultraviolet divergences. The situation was so desperate that Heisenberg noted in 1938 that the revolutions of special relativity and of quantum mechanics were associated with fundamental dimensional parameters: the speed of light, c, and Planck's constant, h. These constants outlined the boundaries of classical physics. He proposed that the next revolution could be associated with the introduction of a new fundamental unit of length, that would outline the boundaries of the domain where the concept of fields and local interactions would be applicable.

Dirac was even more pessimistic. He wrote in the last paragraph of the fourth edition of his book on the Principles of Quantum Mechanics [10]: *It would seem that we have followed as far as possible the path of the logical development of the ideas of quantum mechanics as they are at present understood. The difficulties being of a profound character can be removed only by some drastic change in the foundations of the theory, probably a change as drastic as the passage from Bohr's orbit theory to the present quantum mechanics.*

The resolution of the renormalization problem required two decades to be solved by Bethe, Feynman, Schwinger and Tomonaga.

After the resolution of this problem, QED became a powerful predictive theory for atomic physics and provided results which matched the experimental values with the major accuracy ever found in physics. The QED prediction for the magnetic dipole moment of the muon

$$(g_\mu - 2)_{\text{theor.}} = 233\ 169\ 436 \times 10^{-11}, \tag{3.1}$$

fits impressively well with the experimentally measured value

$$(g_\mu - 2)_{\text{exp.}} = 233\ 184\ 182 \times 10^{-11}. \tag{3.2}$$

If we include all corrections due to the Standard Model of Particle Physics

$$(g_\mu - 2)_{\text{theor.}} = 233\ 183\ 606 \times 10^{-11}, \tag{3.3}$$

the theoretical agreement is even more impressive.

However, in the late 1950s it was remarked that the renormalization of the theory generates another UV catastrophe due to the appearance of a singular pole in the effective electric charge of the electron. The phenomenon know as Landau pole motivated that the majority of particle physicists considered that quantum field theory was not a suitable theory for weak and strong interactions of the newly discovered *elementary* particles.

However, with the discovery of the theory of gauge fields and the formulation of the Standard Model of Particle Physics the perspective radically changed and today, quantum field theory is the basic framework of Fundamental Physics.

(1) **What are the essential features of quantum field theory which make it so special?**

A first answer is that QFT provides a unified framework where the quantum theory and the theory of relativity are consistently integrated.

Sometimes field theory is identified as the theory of particle physics. This is not completely correct. Field theory is a framework which goes beyond particle physics. In fact there are field theories where there is no particle interpretation of the states of the theory.

But it is also true that most of the successful field theories admit a particle interpretation. That means that there are states which can be correctly interpreted as particle states and in those cases, field theory provides a causal framework for particle interactions where action at a distance is replaced by local field interactions. Although this picture also holds for classical field theories, the difference between the classical and quantum theories lies on the fact that in the quantum theory, the interaction between the particles can be interpreted as a process of creation and destruction of messenger particles. The association of forces and interactions with particle exchange is one of the most interesting features of QFT.

The particles that appear in field theory are very special: they are all identical. This means that the electrons in the earth are the same as the electrons in Alpha Centauri because all of them are excitations of the same electron field in QED.

Another essential characteristic of relativistic field theories is that when the field theory admits particle states they are accompanied by antiparticle states, i.e. the theory requires the existence of antiparticles. This interesting property is also a source of the ultraviolet problems of the theory.

(2) **What are the mathematical tools of quantum field theory?**

As Gamow remarked, the first two revolutions had at their disposal the required mathematical tools [11]: *In their efforts to solve the riddles of Nature, physicists often looked for the help of pure mathematics, and in many cases obtained it. When Einstein wanted to interpret gravity as the curvature of four-dimensional, continuum space-time, he found waiting for him Riemann's theory of curved multi-dimensional space. When Heisenberg looked for some unusual kind of mathematics to describe the motion of electrons inside an atom, noncommutative algebra was ready for him.* However, the revolution of QFT was lacking an appropriate mathematical tool. The required mathematical theory to deal with UV singularities in a rigorous way, the theory of distributions, was only formulated in the late 1940s by L. Schwartz.

The fact that the quantum fields involve distributions is behind the existence of UV divergences which in the quantum field theory requires a renormalization program.

The goal of these lectures is to summarize the foundations of QFT and provide some physical and mathematical insights to a reader with a solid mathematical background. However, due to the space limitations, the level of mathematical rigor will be softened. We will follow a path between the standards levels fixed by von Neumann and Dirac in their approaches to quantum mechanics[1].

In Section 2, we summarize the basic elements of quantum mechanics and special relativity. The principles of QFT in the canonical approach are introduced in Section 3. The first problems of the canonical quantization approach appear with the ultraviolet divergences of the vacuum energy. The renormalization of the vacuum energy and the quantum Hamiltonian and their implications in the Casimir effect are analyzed in Section 4. The relations between quantum fields and particles are analyzed in the context of free field theories in Section 5, whereas the introduction of interactions is postponed to Section 6. The covariant formulation of QFT is analyzed in Section 7 and the functional integral approach to the quantization of classical field theories in Section 8. We conclude with a brief outlook of the main topics not covered by this review. Finally in three appendices we ad-

[1]There are excellent textbooks in QFT, e.g. [1]–[13]. Our approach will be close that of Simon and Glimm-Jaffe books [23], [13].

dress the underpinnings of the Casimir effect calculations, the theory of functional Gaussian measures and an introduction to the fairly unknown Peierls approach to the canonical formulation of classical field theories.

3.2 Quantum Mechanics and Relativity

3.2.1 *Quantum mechanics*

A quantum theory is defined by a space of states which are projective rays of vectors $|\psi\rangle$ of a Hilbert space $\mathcal{H}$[2]. The physical observables are Hermitian operators in this Hilbert space (see Section 2.2 of this volume). In any quantum system, there is a special observable, the Hamiltonian $H(t)$, which governs the time evolution of the quantum states by the first order differential equation

$$i\partial_t|\psi(t)\rangle = H(t)|\psi(t)\rangle\,.$$

The symmetries of a quantum system are unitary operators U which commute with the Hamiltonian of the system. In the particular case that the Hamiltonian $H(t)$ is time independent the unitary group defined by

$$U(t) = e^{itH}$$

is a symmetry group, i.e. $[U(t), H] = 0$, and defines the dynamics of the quantum system,

$$|\psi(t)\rangle = U(t)|\psi(t)\rangle\,.$$

Some interesting cases of quantum systems are those who arise from the canonical quantization of classical mechanical systems. The archetype of those systems is the harmonic oscillator. Let us analyze this case in some detail because it will be useful to understand its generalization to field theory.

Classical Harmonic Oscillator

The Lagrangian of a harmonic oscillator is

$$L = \frac{1}{2}m\dot{x}^2 - \frac{1}{2}m\omega^2 x^2\,.$$

The Euler-Lagrange equations give rise to the classical Newton's equation of motion

$$\ddot{x} = -\omega^2 x\,. \tag{3.4}$$

[2]The foundations of quantum mechanics are summarized in the first and second parts of this volume.

The solution of Eq. (3.4) in terms of the initial Cauchy conditions $(x(0), \dot{x}(0))$ is

$$x(t) = x(0)\cos\omega t + \frac{\dot{x}(0)}{\omega}\sin\omega t$$
$$= \frac{1}{2}\left[x(0) + i\frac{\dot{x}(0)}{\omega}\right]e^{-i\omega t} + \frac{1}{2}\left[x(0) - i\frac{\dot{x}(0)}{\omega}\right]e^{i\omega t}. \tag{3.5}$$

Upon Legendre transformation

$$p = m\dot{x}\,,$$

and the Poisson bracket structure,

$$\{x, x\} = \{p, p\} = 0\,; \quad \{x, p\} = 1 \tag{3.6}$$

one obtains the Hamiltonian of the harmonic oscillator

$$H = \frac{1}{2m}p^2 + \frac{1}{2}m\omega^2 x^2\,,$$

and the corresponding Hamilton equations of motion

$$\dot{x} = \{x, H\} = \frac{p}{m}, \qquad \dot{p} = \{p, H\} = -m\omega^2 x\,,$$

that are equivalent to Newton's equations Eq. (3.4).

In field theory, it is very convenient to use the coherent variables

$$a = \sqrt{\frac{m\omega}{2}}x + \frac{i}{\sqrt{2m\omega}}p\,; \qquad a^* = \sqrt{\frac{m\omega}{2}}x - \frac{i}{\sqrt{2m\omega}}p\,,$$

in terms of which time evolution becomes

$$a(t) = a(0)\,e^{-i\omega t}\,,$$

which gives Eq. (3.5),

$$x(t) = \frac{1}{\sqrt{2m\omega}}\,(a(t) + a^*(t)) = \frac{1}{\sqrt{2m\omega}}\left(a(0)\,e^{-i\omega t} + a^*(0)\,e^{i\omega t}\right).$$

The Quantum Harmonic Oscillator

The canonical quantization prescription proceeds by mapping the classical observables, position x and momentum p into selfadjoint operators $\hat{x}$ and $\hat{p}$ of a Hilbert space $\mathcal{H}$ and replacing the Poisson bracket $\{\cdot,\cdot\}$ by the operators commutator $-i[\cdot,\cdot]$, i.e.

$$\{x, p\} = 1 \Rightarrow [\hat{x}, \hat{p}] = i\,\mathbb{I}\,, \tag{3.7}$$

where we assume that the Planck constant $\hbar = 1$.

The Hamiltonian is given by

$$\hat{H} = \frac{1}{2m}\left(\hat{p}^2 + m^2\omega^2\hat{x}^2\right),$$

which in terms of creation and annihilation operators (see Example 1.2.4 of the first part)

$$a = \sqrt{\frac{m\omega}{2}}\hat{x} + \frac{i}{\sqrt{2m\omega}}\hat{p}; \qquad a^\dagger = \sqrt{\frac{m\omega}{2}}\hat{x} - \frac{i}{\sqrt{2m\omega}}\hat{p}$$

reads as

$$\hat{H} = \omega\left(a^\dagger a + \frac{1}{2}\right).$$

Now, because of the commutation relations

$$[\hat{H}, a^\dagger] = \omega a^\dagger \qquad [\hat{H}, a] = -\omega a \qquad [a, a^\dagger] = 1,$$

we have that we can associate to any eigenstate $|E\rangle$ of the Hamiltonian $\hat{H}$, $\hat{H}|E\rangle = E|E\rangle$, two more eigenstates $a^\dagger|E\rangle$ and $a|E\rangle$ with eigenvalues $E+\omega$ and $E-\omega$, respectively. Indeed,

$$\hat{H}a^\dagger|E\rangle = a^\dagger\hat{H}|E\rangle + \omega a^\dagger|E\rangle = (E+\omega)a^\dagger|E\rangle,$$

$$\hat{H}\,a|E\rangle = a\,\hat{H}\,|E\rangle - \omega\,a\,|E\rangle = (E-\omega)a\,|E\rangle.$$

The quantum Hamiltonian $\hat{H}$ is a positive operator since for any physical state $|\psi\rangle \in \mathcal{H}$, $\langle\psi|\hat{H}|\psi\rangle = \omega\langle\psi|a^\dagger a + \frac{1}{2}\mathbb{I}|\psi\rangle = \omega(||a|\psi\rangle||^2 + \frac{1}{2}|||\psi\rangle||^2) \geq 0$. However, the ladder of quantum states $a^n|E\rangle$ generated by any eigenvalue $|E\rangle$ of $\hat{H}$ will contain negative energy states unless before reaching negative eigenvalues the state of the ladder vanishes, i.e. $a^{n_0}|E\rangle = 0$. Thus, the only possibility which is compatible with the positivity of the Hamiltonian $\hat{H}$ is the existence of a final ground state such that $a|E_0\rangle = 0$. But then, the energy of this ground state $H|E_0\rangle = \frac{1}{2}\omega|E_0\rangle$ is non-vanishing, unlike the energy of the classical vacuum configuration which vanishes. The non-trivial value of the quantum vacuum energy has remarkable consequences for the physics of the vacuum in the quantum field theory.

We shall denote from now on the ground state $|E_0\rangle$ by $|0\rangle$. Higher energy states are obtained by applying the creation operator $a^\dagger$ to the ground state $|0\rangle$:

$$|n\rangle = \frac{1}{\sqrt{(n+1)!}}(a^\dagger)^n|0\rangle; \qquad H|n\rangle = \left(n + \frac{1}{2}\right)\omega|n\rangle \tag{3.8}$$

for any positive integer $n \in \mathbb{N}$. The state $|n\rangle$ has unit norm, i.e. $\|\,|n\rangle\|^2 = 1$ and satisfies that

$$a|n\rangle = \sqrt{n}|n-1\rangle.$$

In the harmonic oscillator, the fundamental observables are the position x and the momentum p. Since by the quantization prescription $\hat{x}$ and $\hat{p}$ do not commute, the space of states has to be infinite dimensional, because if $\dim \mathcal{H} = n < \infty$, $\operatorname{tr}[x, p] = 0 \neq i \operatorname{tr} \mathbb{I} = in$. Moreover, any other operator O that commutes with both fundamental operators $[\hat{p}, O] = [\hat{x}, O] = 0$ has to be proportional to the identity $O = cI$. Now, since the projector $\mathbb{P}$ to the subspace spanned by the stationary states $|n\rangle$ commutes with $\hat{x}$ and $\hat{p}$ ($[\hat{p}, \mathbb{P}] = [\hat{x}, \mathbb{P}] = 0$), it follows that $\mathbb{P} = cI$. This implies that the subspace spanned by the vectors $|n\rangle$ is complete, i.e. does coincide with the whole Hilbert space $\mathcal{H}$.

Although the principles of quantum mechanics are identical for all quantum system and all separable Hilbert spaces are isomorphic, different systems can be distinguished by their algebra of observables. In the particular case of the harmonic oscillator, the position $\hat{x}$ and momentum $\hat{p}$ operators belong to the algebra of observables, which does not only imply that the Hilbert space is infinite dimensional, but also that it can be identified with the space of square integrable functions $L^2(\mathbb{R})$ of the position (Schrödinger representation) or momentum (Heisenberg representation).

In the Schrödinger representation the ground state reads

$$\langle x|0\rangle = \sqrt{\frac{\omega}{\pi}} e^{-\frac{1}{2}\omega x^2} .$$

The position and momentum operator are given by

$$\hat{x}\psi(x) = x\psi(x) ; \qquad \hat{p}\psi(x) = -i\partial_x \psi(x) .$$

The Hamiltonian reads

$$\hat{H} = -\frac{1}{2m}\frac{d^2}{dx^2} + \frac{1}{2} m\omega^2 x^2 ,$$

and the creation and destruction operators are

$$a = \frac{1}{\sqrt{2m\omega}}\left(\frac{d}{dx} + m\omega x\right); \qquad a^\dagger = \frac{1}{\sqrt{2m\omega}}\left(-\frac{d}{dx} + m\omega x\right).$$

It is easy to check that the excited states Eq. (3.8) are given by

$$\langle x|n\rangle = H_n(x) e^{-\frac{1}{2}\omega x^2},$$

in terms of the Hermite polynomials

$$H_n(x) = e^{\frac{1}{2}\omega x^2} \frac{1}{\sqrt{(n+1)!}} (a^\dagger)^n e^{-\frac{1}{2}\omega x^2} .$$

The generalization for multidimensional harmonic oscillators is straightforward. If we have n-harmonic oscillators of frequencies ω_i and masses

m_i; $i = 1, 2, \cdots, n$. The position and momentum operators are given in the Schrödinger representation by

$$\hat{x}_i \psi(\mathbf{x}) = x_i \psi(\mathbf{x}) ; \qquad \hat{p}_i \psi(\mathbf{x}) = -i\partial_i \psi(\mathbf{x}) ,$$

where $\mathbf{x} = (x_1, x_2, \ldots, x_n) \in \mathbb{R}^n$. They satisfy the canonical commutation relations

$$[\hat{x}_i, \hat{p}_j] = i\delta_{ij} .$$

The Hamiltonian

$$\hat{H} = -\sum_{i=1}^{n} \frac{1}{2m_i} \left(\partial_i^2 + m_i^2 \omega_i^2 x_i^2\right) ,$$

in terms of the creation and destruction operators

$$a_i = \frac{1}{\sqrt{2m_i\omega_i}} \left(\partial_i + m_i\omega_i x_i\right) ; \qquad a^\dagger = \frac{1}{\sqrt{2m_i\omega_i}} \left(-\partial_i + m_i\omega_i x_i\right) ,$$

reads

$$\hat{H} = \sum_{i=1}^{n} \omega_i \left(a_i^\dagger a_i + \frac{1}{2}\right) .$$

The ground state

$$\langle \mathbf{x} | 0 \rangle = \left(\prod_{i=1}^{n} \sqrt{\frac{\omega_i}{\pi}} \right) e^{-\frac{1}{2} \sum_{i=1}^{n} \omega_i x_i^2} ,$$

has an energy given by the sum of the ground state energies of the different harmonic modes

$$E_0 = \frac{1}{2} \sum_{i=1}^{n} \omega_i .$$

In a similar way, it is easy to check that the excited states Eq. (3.8) are given by

$$\langle \mathbf{x} | n_1, n_2, \ldots, n_n \rangle = \left(\prod_{i=1}^{n} H_{n_i}(x_i) \right) e^{-\frac{1}{2} \sum_{i=1}^{n} \omega_i x_i^2} ,$$

in terms of the Hermite polynomials $H_{n_i}, i = 1, 2, \ldots, n$.

3.2.2 *Relativity and the Poincaré group*

The Einstein theory of Relativity is based on the unification of space and time into a four-dimensional space-time $\mathbb{R}^4$ equipped with the Minkowski metric

$$dx^2 = dx_0^2 - dx_1^2 - dx_2^2 - dx_3^2 = \sum_{\mu=0}^{3} \eta_{\mu\nu} dx_\mu dx_\nu ,$$

where $x_0 = ct$ denotes the time-like coordinate and

$$\eta_{\mu\nu} = \begin{pmatrix} 1 & 0 & 0 & 0 \\ 0 & -1 & 0 & 0 \\ 0 & 0 & -1 & 0 \\ 0 & 0 & 0 & -1 \end{pmatrix}.$$

From now on we shall assume that the speed of light c is normalized to unit.

The geodesic which connects two points x, y of Minkowski space-time is a straight line in $\mathbb{R}^4$ and the Minkowski distance between x and y is

$$d(x, y) = (y_0 - x_0)^2 - (y_1 - x_1)^2 - (y_2 - x_2)^2 - (y_3 - x_3)^2 .$$

When this distance vanishes, the geodesic line represents the trajectory of a light ray connecting the two points.

The Minkowski metric introduces a causal structure in the space-time. The line connecting two points x, y is time-like if $d(x, y) > 0$; space-like if $d(x, y) < 0$; light-like if $d(x, y) = 0$; Two points x, y are causally separated iff $d(x, y) \geq 0$; and spatially separated iff $d(x, y) < 0$; A causal line connecting x, y is future oriented if $y_0 - x_0 > 0$; past oriented if $y_0 - x_0 < 0$ (See Figure 3.1).

The space-time symmetries of a relativistic theory are space-time translations,

$$x' = x + a \tag{3.9}$$

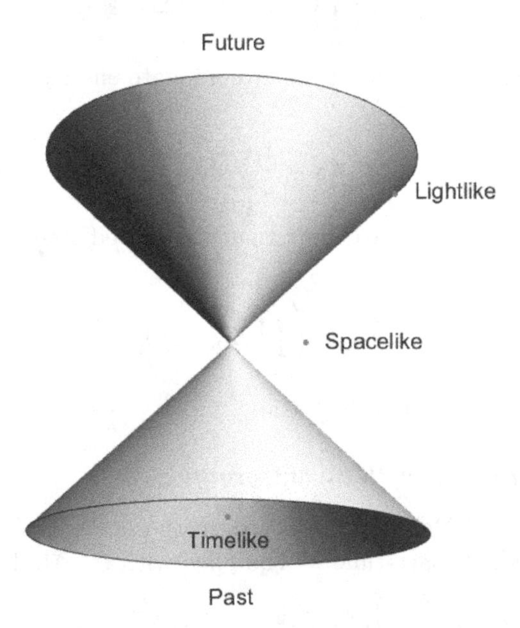

Fig. 3.1 Light cone and causal structure of Minkowski space-time.

space rotations, e.g.

$$x'^{\mu} = \Lambda^{\mu}_{\nu} x^{\nu}, \qquad \Lambda^{\mu}_{\nu} = \begin{pmatrix} 1 & 0 & 0 & 0 \\ 0 & \cos\theta & \sin\theta & 0 \\ 0 & -\sin\theta & \cos\theta & 0 \\ 0 & 0 & 0 & 1 \end{pmatrix}, \quad \text{with } 0 \leq \theta \leq 2\pi \tag{3.10}$$

and Lorentz transformations, e.g.

$$x'^{\mu} = \Lambda^{\mu}_{\nu} x^{\nu}, \qquad \Lambda^{\mu}_{\nu} = \begin{pmatrix} \gamma & -\gamma v & 0 & 0 \\ -\gamma v & \gamma & 0 & 0 \\ 0 & 0 & 1 & 0 \\ 0 & 0 & 0 & 1 \end{pmatrix}, \quad \text{with } v \in \mathbb{R} \text{ and } \gamma = \sqrt{1 - v^2} \tag{3.11}$$

which are linear transformations that leave the Minkowski metric invariant

$$\eta_{\mu\nu} = \Lambda^{\sigma}_{\mu} \eta_{\sigma\tau} \Lambda^{\tau}_{\nu}.$$

There are some extra discrete symmetries which play an important role in field theory. They are generated by time reversal,

$$T : (\mathbf{x}, t) \to (\mathbf{x}, -t)$$

and parity

$$P : (\mathbf{x}, t) \to (-\mathbf{x}, t).$$

The whole group of space-time symmetries is the Poincaré group

$$\mathcal{P} = ISO(3,1) = \{(\Lambda, a); \Lambda \in O(3,1), a \in \mathbb{R}^4\},$$

which contains all these continuous and discrete symmetries and has four connected components.

The first attempts to make the quantum theory compatible with the theory of relativity where based on covariant equations of motion like the Maxwell equations of classical electrodynamics. This leads to the discovery of Klein-Gordon equation for scalar fields and the Dirac equation for spinorial fields. Wigner introduced a completely different approach. If the quantum theories are defined in a Hilbert space and the relativity is based on Poincaré invariance, he conjectured that elementary quantum particles as the most elementary quantum systems must arise from irreducible projective representations of the Poincaré group. This program was achieved by Wigner and Mackey and their results show that the simplest irreducible representations are characterized by two numbers which represent the mass $m \in \mathbb{R}^+$ and the spin $s \in \mathbb{N}/2$ of the particle. The case of spin zero correspond to a scalar field satisfying the Klein-Gordon equation. The case of spin $\frac{1}{2}$ corresponds to spinor fields satisfying the Dirac equation, and the case of massless helicity 1 to electromagnetic fields satisfying Maxwell equations. To be concise we shall not analyze these higher spin field theories in these lectures and we shall concentrate on scalar fields.

3.3 Quantum Field Theory

The most interesting case of field theory is that which concerns relativistic fields. The compatibility of quantum theories with the theory of relativity is not immediate. The first attempts to formulate a quantum dynamics compatible with the theory of relativity lead to puzzling theories full of paradoxes, like the Klein paradox, which arises in the dynamics defined by Klein-Gordon or Dirac equations. The solution to those puzzles comes from the quantization of classical field theories.

A quantum field theory is a quantum theory which is relativistic invariant and where there is a special type of quantum operators which are associated with classical fields.

In the case of a real scalar field ϕ a consistent theory should satisfy the following principles.

- P1 *Quantum principle:* The space of quantum states is the space of rays in a separable Hilbert space $\mathcal{H}$.
- P2 *Unitarity:* There is a antiunitary representation $U(\Lambda, a)$ of the Poincaré group in $\mathcal{H}$, where time reversal T is represented as an antiunitary operator $U(T)$.
- P3 *Spectral condition:* The spectrum of generators of space-time translations P_μ is contained in the forward like cone

$$\bar{V}_+ = \{p_\mu; p^2 \geq 0, p_0 \geq 0\}\,.$$

- P4 *Vacuum state:* There is a unique state $|0\rangle \in \mathcal{H}$, satisfying that $P_\mu|0\rangle \;=\; 0$.
- P5 *Field theory (real boson):* For any classical field f in the space $\mathcal{S}(\mathbb{R}^3)$ of *fast* decreasing smooth $C^\infty(\mathbb{R}^3)$ functions[3] there is a field operator $\phi(f)$ in $\mathcal{H}$ which satisfies $\phi(f) = \phi(f)^\dagger$. The field operator can be considered as the smearing by f of a fundamental field operator $\phi(x)$

$$\phi(f) = \int d^3\mathbf{x} f(\mathbf{x})\phi(x)\,.$$

[3]In the case of massless fields the classical field test functions f must have compact support, i.e. $f \in \mathcal{D}(\mathbb{R}^3) = C_0^\infty(\mathbb{R}^3)$.

The subspace spanned by the vectors $\phi(f_1)\phi(f_2)\cdots\phi(f_n)|0\rangle$ for arbitrary test functions $f_1, f_2, \cdots f_n \in \mathcal{S}(\mathbb{R}^3)$ is a dense subspace of $\mathcal{H}$.

- P6 *Poincaré covariance:* Let $\tilde{f} \in \mathcal{S}(\mathbb{R}^4)$ be a test function defined in Minkowski space-time and

$$\phi(\tilde{f}) = \int_{\mathbb{R}^4} d^4x\, \phi(x)\tilde{f}(x)\,,$$

where $\phi(x) = \phi(\mathbf{x}, t) = e^{itP_0}\phi(\mathbf{x})e^{-itP_0}$. Then[4],

$$U(\Lambda, a)\phi(\tilde{f})U(\Lambda, a)^\dagger = \phi(\tilde{f}_{(\Lambda,a)})\,,$$

with

$$\tilde{f}_{(\Lambda,a)}(x) = \tilde{f}(\Lambda^{-1}(x - a))\,.$$

- P7 *(Bosonic) local causality:* For any $f, g \in \mathcal{S}(\mathbb{R}^3)$ the corresponding field operators $\phi(f), \phi(g)$ commute[5],

$$[\phi(f), \phi(g)] = 0\,. \tag{3.12}$$

3.3.1 *Canonical quantization*

As in the case of quantum mechanics, there are special cases where the quantum field theory arises from the quantization of a classical field theory[6].

Let us consider a scalar real field ϕ in Minkowski space-time $\mathbb{R}^4$. The classical field theory is defined according to the variational principle by stationary field configurations $\phi(x)$ of the classical action functional

$$S[\phi(x)] \equiv \int d^4x\, \mathcal{L}(\phi, \partial_\mu\phi) = \int d^4x \left(\tfrac{1}{2}\partial_\mu\phi\, \partial^\mu\phi - V(\phi)\right). \tag{3.13}$$

The equations of motion are obtained, thus, from the Euler-Lagrange equations

$$\partial_\mu \left[\frac{\delta\mathcal{L}}{\delta(\partial_\mu\phi)}\right] - \frac{\delta\mathcal{L}}{\delta\phi} = 0 \qquad \Longrightarrow \qquad \Box\phi + \frac{\delta V}{\delta\phi} = 0\,, \tag{3.14}$$

where $\Box = \partial_\mu\partial^\mu$. Notice that the Poincaré invariance of the action implies the Poincaré invariance of the equations of motion.

---

[4]For higher spin fields, the Poincaré representation satisfies the covariant transfomation law $U(\Lambda, a)\phi(f)U(\Lambda, a)^\dagger = S(\Lambda)^{-1}\phi(f_{(\Lambda,a)})$, where S is a linear n-dimensional representation of Lorentz group and ϕ is a field with n-components.

[5]In the fermionic case, the commutator $[\cdot,\cdot]$ is replaced by an anticommutator $\{\cdot,\cdot\}$.

[6]See section 1.3.

The quantization is usually formulated in the Hamiltonian formalism. Thus, it is necessary to start from the classical canonical formalism. Let $\mathcal{M}$ be the configuration space of square integrable classical fields at any fixed time (e.g. $t = 0$),

$$\mathcal{M} = \left\{ \phi(\mathbf{x}) = \phi(\mathbf{x}, 0); \| \phi \|^2 = \int d^3\mathbf{x} |\phi(\mathbf{x}, 0)|^2 < \infty \right\}. \tag{3.15}$$

The Legendre transformation maps the tangent space $T\mathcal{M}$ into the cotangent space $T^*\mathcal{M}$, fixing the value of the canonical momentum

$$\pi = \frac{\delta \mathcal{L}}{\delta \dot{\phi}} = \dot{\phi},$$

from the Lagrangian

$$L = \int d^3\mathbf{x} \left(\frac{1}{2} \dot{\phi}^2 - \frac{1}{2} (\nabla \phi)^2 - V(\phi) \right),$$

where $\dot{\phi} = \partial_t \phi$. The corresponding Hamiltonian is given by

$$H = \int d^3\mathbf{x} \left(\frac{1}{2} \pi^2 + \frac{1}{2} (\nabla \phi)^2 + V(\phi) \right). \tag{3.16}$$

In the case of a free massive theory with mass m, $V(\phi) = \frac{1}{2} m^2 \phi^2$ and the Hamiltonian reads

$$H = \frac{1}{2} (\| \pi \|^2 + \| \nabla \phi \|^2 + m^2 \| \phi \|^2), \tag{3.17}$$

where we have used the $L^2(\mathbb{R}^3)$ norm introduced in Eq. (3.15).

In this case the classical vacuum solution is unique $\phi = 0$. However, in the massless case $m = 0$ the vacuum is degenerated, because any constant configuration $\phi = \text{cte}$ is a solution with finite energy, although such configurations in the massive case have infinite energy.

The symplectic structure of $T^*\mathcal{M}$

$$\omega = \int d^3\, \mathbf{x}\; \delta\pi \wedge \delta\phi$$

induces a Poisson structure in the space of functionals of $T^*\mathcal{M}$. Given two local functionals $\mathcal{F}(\phi, \pi)$, $\mathcal{G}(\phi, \pi)$ of the canonical variables of the form

$$\mathcal{F}(\phi, \pi) = \int d^3\mathbf{x}\; F(\phi, \pi), \qquad \mathcal{G}(\phi, \pi) = \int d^3\mathbf{x}\; G(\phi, \pi).$$

Their Poisson bracket is defined by

$$\{\mathcal{F}, \mathcal{G}\} \equiv \int d^3\mathbf{x} \left[\frac{\delta F}{\delta \phi} \frac{\delta G}{\delta \pi} - \frac{\delta F}{\delta \pi} \frac{\delta G}{\delta \phi} \right],$$

where the functional derivative $\frac{\delta}{\delta\phi}$ is given by

$$\frac{\delta F}{\delta\phi} = \frac{\delta\mathcal{F}}{\delta\phi} - \partial_\mu\left[\frac{\delta\mathcal{F}}{\delta(\partial_\mu\phi)}\right].$$

The Poisson brackets of fundamental fields are

$$\{\phi(\mathbf{x}_1), \phi(\mathbf{x}_2)\} = \{\pi(\mathbf{x}_1), \pi(\mathbf{x}_2)\} = 0\,,$$
$$\{\phi(\mathbf{x}_1), \pi(\mathbf{x}_2)\} = \delta^3(\mathbf{x}_1 - \mathbf{x}_2)\,,$$

because of the basic rules of functional derivation

$$\frac{\delta\phi(\mathbf{x}_1)}{\delta\phi(\mathbf{x}_2)} = \delta^3(\mathbf{x}_1 - \mathbf{x}_2)\,; \qquad \frac{\delta\pi(\mathbf{x}_1)}{\delta\pi(\mathbf{x}_2)} = \delta^3(\mathbf{x}_1 - \mathbf{x}_2)\,.$$

The appearance of delta functions in the Poisson structure of the fields reflects the fact that a mathematically sound analysis of field theory requires the use of distributions. This will be even more necessary for the quantum fields. Thus, it is convenient to consider smeared field functionals. To a classical field test function f which might be more regular that $L^2(\mathbb{R}^3)$ fields (e.g. $f \in \mathcal{S}(\mathbb{R}^3)$ for massive fields, or $f \in \mathcal{D}(\mathbb{R}^3)$ for massless fields) we can associate a smeared field defined by the image of the linear functional

$$\phi(f) = \int d^3\mathbf{x} f(\mathbf{x})\phi(x)\,; \qquad \pi(f) = \int d^3\mathbf{x} f(\mathbf{x})\pi(x)$$

in $L^2(\mathbb{R}^3)$.

The Poisson structure can be expressed in terms of smeared fields $\phi(f)$ as

$$\{\phi(f_1), \phi(f_2)\} = \{\pi(f_1)\,, \pi(f_2)\} = 0,$$
$$\{\phi(f_1), \pi(f_2)\} = (f_1, f_2)$$

where $(\cdot,\cdot)$ denotes the canonical Hilbert product of $L^2(\mathbb{R}^3)$.

By choosing an orthonormal Hilbert basis of test functions f_n in $L^2(\mathbb{R}^3)$ we can get a discrete representation of the Poisson structure,

$$\{\phi_n, \phi_m\} = \{\phi_n, \phi_m\} = 0\,; \quad \{\phi_n, \pi_m\} = \delta_{mn}$$

where $\phi_n = \phi(f_n)$ and $\pi_n = \pi(f_n)$.

In that representation, the Hamiltonian operator Eq. (3.17) becomes

$$H = \frac{1}{2}\sum_{n=0}^{\infty}{\pi_n}^2 - \frac{1}{2}\sum_{n,m=0}^{\infty}\Delta_{mn}\phi_m\phi_n + \frac{1}{2}m^2\sum_{n=0}^{\infty}{\phi_n}^2\,, \tag{3.18}$$

where

$$\Delta_{mn} = (f_m, \Delta f_n) = (f_m, \boldsymbol{\nabla}^2 f_n)\,. \tag{3.19}$$

In this representation, it is clear that the system describes an infinite series of coupled harmonic oscillators. The way of disentangling the coupling is to find the normal modes, i.e. to choose a basis of test functions f_n where the interaction operator Δ is diagonal. The normal modes are plane waves which do not belong to $L^2(\mathbb{R}^3)$. For such a reason it is convenient to introduce an infrared regulator, i.e. to consider the system in a finite volume. There are many physical reasons why this method is necessary. In the quantum case there are two types of divergences:

(i) ultraviolet (UV) divergences, which are due to short range singularities associated to the local products of distributions; and
(ii) infrared (IR) divergences which are due to the infinite volume of space. Both need to be regularized and renormalized as we will see later on.

From this perspective, the introduction of a finite volume can be considered as a regulator of IR divergences. Poincaré invariance will be recovered in the limit of infinite volume at the very end.

We shall consider mostly the torus T^3 compactification of $\mathbb{R}^3$, that is, a box with periodic boundary conditions. The normal modes in this case are normalizable plane waves,

$$f^+_{\mathbf{n}}(\mathbf{x}) = \frac{1}{2}(f_{\mathbf{n}}(\mathbf{x}) + f_{\mathbf{n}}(\mathbf{x})^*)\,; \quad f^-_{\mathbf{n}}(\mathbf{x}) = -\frac{i}{2}(f_{\mathbf{n}}(\mathbf{x}) - f_{\mathbf{n}}(\mathbf{x})^*) \qquad \mathbf{n} \in \mathbb{Z}^3_+\,,$$

with

$$f_{\mathbf{n}}(x) = \frac{1}{\sqrt{L^3}} e^{i2\pi \mathbf{n}\cdot x/L}\,, \tag{3.20}$$

where $\mathbf{n} \in \mathbb{Z}^3$, $\mathbf{n}\cdot\mathbf{x} = n_1x_1 + n_2x_2 + n_3x_3$ and L is the length of each side of the torus, i.e. $\mathbf{x} \in [0, L]^3$. The normal modes diagonalize the Hamiltonian because

$$\Delta f^{\pm}_{\mathbf{n}}(x) = -\left(\frac{2\pi}{L}\right)^2 (\mathbf{n}\cdot\mathbf{n}) f^{\pm}_{\mathbf{n}}(x)\,. \tag{3.21}$$

However, it is more convenient to use the complex modes f_n, provided that in the mode expansion the fields

$$\phi(\mathbf{x}) = \sum_{\mathbf{n}\in\mathbb{Z}^3} \phi_n f_{\mathbf{n}}(\mathbf{x})\,,$$

the coefficients $\phi_n = \phi(f)$ satisfy the reality conditions $\phi^*_{\mathbf{n}} = \phi_{-\mathbf{n}}$ in order to guarantee the reality of the fields $\phi^* = \phi$.

In terms of the complex modes the Hamiltonian is diagonal

$$H = \frac{1}{2}\sum_{\mathbf{n}\in\mathbb{Z}^3}\left(|\pi_{\mathbf{n}}|^2 + \left(\left|\frac{2\pi\mathbf{n}}{L}\right|^2 + m^2\right)|\phi_{\mathbf{n}}|^2\right), \tag{3.22}$$

and it is evident that the system Eq. (3.22) describes an infinite series of harmonic oscillators with frequencies

$$\omega_{\mathbf{n}} = \sqrt{\left|\frac{2\pi \mathbf{n}}{L}\right|^2 + m^2}\,. \tag{3.23}$$

Canonical quantization maps classical fields into operators in a Hilbert space $\mathcal{H}$ satisfying the commutation relations obtained by replacing Poisson brackets by operator commutators

$$\{\cdot,\cdot\} \Longrightarrow -i\,[\cdot,\cdot]\,, \tag{3.24}$$

which can be realized in the Schrödinger representation on space of functionals of $\mathcal{M}$ by

$$\hat{\pi}(\mathbf{x}) = -i\,\frac{\delta}{\delta\phi(\mathbf{x})}\,; \qquad \hat{\phi}(\mathbf{x}) = \phi(\mathbf{x})\,. \tag{3.25}$$

The corresponding quantum Hamiltonian is

$$\hat{H} = \frac{1}{2}(\parallel \hat{\pi} \parallel^2 + \parallel \boldsymbol{\nabla}\phi \parallel^2 + m^2 \parallel \phi \parallel^2)\,, \tag{3.26}$$

In terms of smeared functions, the Schrödinger representation of the momentum operator

$$\hat{\pi}(f) = -i\int d^3\mathbf{x}\, f(\mathbf{x})\frac{\delta}{\delta\phi(\mathbf{x})}\,,$$

becomes just a Gateaux derivative operator

$$\hat{\pi}(f)\mathcal{F}(\phi) = -i\,\lim_{s\to 0}\frac{1}{s}\left(\mathcal{F}(\phi + sf) - \mathcal{F}(\phi)\right). \tag{3.27}$$

For any orthonormal basis of test functions f_n in $L^2(\mathbb{R}^3)$ we have the expansion of classical fields $\phi \in L^2(\mathbb{R}^3)$

$$\phi(\mathbf{x}) = \sum_{n=0}^{\infty}\phi_n f_n(\mathbf{x})\,,$$

where $\phi_n = \phi(f_n)$. Moreover, by linearity $\phi(f) = \sum_{n=0}^{\infty}\phi_n(f_n, f)$

$$\hat{\pi}(\mathbf{x})\mathcal{F}(\phi) = -i\,\frac{\delta}{\delta\phi(\mathbf{x})}\mathcal{F}\left(\sum_{n=0}^{\infty}\phi_n f_n(\mathbf{x})\right) \tag{3.28}$$

and from Eq. (3.27) it follows that

$$\hat{\pi}(f_n) = -i\,\frac{\delta}{\delta\phi_n}\,.$$

In the plane wave basis, the quantum Hamiltonian

$$\hat{H} = \frac{1}{2}\sum_{\mathbf{n}\in\mathbb{Z}^3}\left(\frac{\delta}{\delta\phi_{-\mathbf{n}}}\frac{\delta}{\delta\phi_{\mathbf{n}}} + \omega_{\mathbf{n}}^2\,|\phi_{\mathbf{n}}|^2\right), \tag{3.29}$$

again corresponds to an infinite series of harmonic oscillators with frequencies $\omega_{\mathbf{n}}$.

3.4 The Quantum Vacuum

The advantage of the diagonal structure of the quantum Hamiltonian in the plane wave basis is that it facilitates the analysis of its spectrum.

In particular the ground state, known in QFT as vacuum state, is given by

$$\Psi_0 = \langle \phi | 0 \rangle = \prod_{\mathbf{n}\in\mathbb{Z}^3} \sqrt{\frac{\omega_n}{\pi}} e^{-\frac{1}{2}\omega_{\mathbf{n}}|\phi_{\mathbf{n}}|^2} = \exp\left\{ -\frac{1}{2} \sum_{\mathbf{n}\in\mathbb{Z}^3} \left(\omega_{\mathbf{n}} |\phi_{\mathbf{n}}|^2 + \log\frac{\pi}{\omega_n} \right) \right\}. \tag{3.30}$$

Indeed,

$$\hat{H}\Psi_0 = E_0 \Psi_0 \,, \tag{3.31}$$

where the vacuum energy

$$E_0 = \frac{1}{2} \sum_{\mathbf{n}\in\mathbb{Z}^3} \omega_{\mathbf{n}} \tag{3.32}$$

is half the divergent sum of all normal modes frequencies. This appearance of this divergence is a genuine quantum effect which is induced by the fact that the lowest energy (*zero-point energy*) of each quantum oscillator is non-vanishing. The divergence is generated by the large momentum $\mathbf{n}$ (*ultraviolet*) modes. The vacuum energy is the simplest quantity of the quantum theory which presents UV divergences.

3.4.1 *Renormalization of vacuum energy*

The appearance of UV divergences postponed the formulation of quantum field theories for two decades. The solution of the UV puzzle came from the renormalization program. The main idea behind the renormalization program is the disassociation of the fundamental observables like the quantum Hamiltonian and the observed quantities.

To implement the renormalization program, we need a previous step which is known as regularization. This requires to introduce a modification of all fundamental (*bare*) operators depending on a UV scale parameter Λ in a way that they become well defined operators with a finite spectrum, e.g. by cutting the infinite sum in Eq. (3.32) to a finite sum. The modification must disappear in the limit $\Lambda \to \infty$ to recover the original divergent expressions. The second step consists of a *physical* modification of the fundamental observables by absorbing the sources of divergences into the physical parameters like particle masses and charges, coupling constants or energy scales of the theory in a way that the they remain finite in the $\Lambda \to \infty$.

To illustrate the implementation of the renormalization mechanism, let us consider the case of the Hamiltonian operator of the free field theory Eq. (3.26).

The regularization can be introduced in several different ways. Let us consider two different methods:

(i) Sharp momentum cut-off

$$\hat{H}_\Lambda = \frac{1}{2}\sum_{\mathbf{n}\in\mathbb{Z}^3}^{\omega_{\mathbf{n}}<\Lambda}\left(\frac{\delta}{\delta\phi_{-\mathbf{n}}}\frac{\delta}{\delta\phi_{\mathbf{n}}} + \omega_{\mathbf{n}}^2\,|\phi_{\mathbf{n}}|^2\right), \tag{3.33}$$

(ii) Heat kernel regularization

$$\hat{H}_\epsilon = \frac{1}{2}\sum_{\mathbf{n}\in\mathbb{Z}^3}\left(\frac{\delta}{\delta\phi_{-\mathbf{n}}}\frac{\delta}{\delta\phi_{\mathbf{n}}} + \omega_{\mathbf{n}}^2\,e^{-\epsilon\omega_{\mathbf{n}}^2}\,|\phi_{\mathbf{n}}|^2\right), \tag{3.34}$$

which can be related by choosing $\epsilon = \frac{\sqrt{2}}{\Lambda^2}$.

There are other methods which include higher derivative terms or lattice discretization of the continuum space, but for simplicity we shall not discuss them in this course.

The renormalization of the fundamental Hamiltonian is obtained by subtracting an *unobservable* constant quantity E_0 in such a way that the observable (*renormalized*) quantum Hamiltonian

$$\hat{H}_{\text{ren}} = \lim_{\Lambda\to\infty}\left(\hat{H}_\Lambda - E_0(\Lambda)\right), \tag{3.35}$$

is a well defined quantum operator with a finite energy spectrum.

Even if the renormalization of the Hamiltonian solves the divergence problem, one might wonder about its physical meaning. First, let us analyze the structure of the divergences of vacuum energy.

In the large L limit the vacuum energy E_0 becomes a good approximation to the Riemann integral

$$E_0(\Lambda, L) = \frac{1}{2}L^3\int_{|k|\leq\Lambda}\frac{d^3k}{(2\pi)^3}\sqrt{k^2+m^2} + \mathcal{O}(L\Lambda).$$

Now it becomes clear that the infrared divergence is just due to the infinite volume of the system and translation invariance. However the vacuum energy density

$$\mathcal{E}_0(\Lambda) = \lim_{L\to\infty}\frac{E_0(\Lambda, L)}{L^3} = \frac{1}{2}\int_{|k|\leq\Lambda}\frac{d^3k}{(2\pi)^3}\sqrt{k^2+m^2} \tag{3.36}$$

is free of IR divergences. However, the integral Eq. (3.36) is UV divergent. In the sharp momentum cut-off regularization

$$\mathcal{E}_0(\Lambda) = \frac{\Lambda^4}{16\pi^2} + \frac{m^2\Lambda^2}{16\pi^2} + \frac{m^4}{64\pi^2}\log\frac{m^2}{\Lambda^2} + \frac{m^4(1-\log 16)}{128\pi^2} + \mathcal{O}\left(\frac{1}{\Lambda^2}\right)$$

in the large Λ limit. Whereas in the heat kernel regularization

$$\mathcal{E}_0(\Lambda) = \frac{1}{8\pi^2\epsilon^2} - \frac{m^2}{16\pi^2\epsilon} + \frac{m^4}{64\pi^2}\left(2+\gamma+\log\frac{\epsilon m^2}{4}\right) + \mathcal{O}(\epsilon)$$

for small values of $\epsilon = \frac{\sqrt{2}}{\Lambda^2}$. The leading quartic and logarithmic divergent terms are the same in both regularizations whereas the quadratically divergent term is different.

The source of divergence is of ultraviolet origin because it comes from the integration of $\omega(\mathbf{k}) = \sqrt{\mathbf{k}^2+m^2}$ at large values of the momentum. The quantum field theory of free scalar fields is an infinite set of harmonic oscillators, each one labelled by $\mathbf{k}$. Each of these oscillators contribute to the vacuum energy with their zero-point energy $\frac{1}{2}\omega(\mathbf{k})$. This total contribution of zero-points energies to the vacuum energy density is divergent, since there are modes with arbitrary high momentum. This is why this divergence has ultraviolet origin. It appears in any quantum field theory and not only in the free scalar quantum field. It is something intrinsic to the theory of quantum fields.

3.4.2 *Momentum operator*

The generator of space translations is the momentum operator $\hat{P}_i$. In the free scalar field theory it is given by

$$\hat{P}_i = \int d^3\mathbf{x}\,(\hat{\pi}\partial_i\phi), \qquad i = 1,2,3. \tag{3.37}$$

Since the vacuum state is translation invariant,

$$\begin{aligned}\hat{P}_i\Psi_0 &= i\int d^3\mathbf{x}\;\phi\sqrt{\boldsymbol{\nabla}^2+m^2}\,\partial_i\phi\Psi_0\\ &= \frac{i}{2}\int d^3\mathbf{x}\,\partial_i\left(\phi\sqrt{\boldsymbol{\nabla}^2+m^2}\,\phi\right)\Psi_0 = 0,\end{aligned} \tag{3.38}$$

which apparently does not require renormalization as the vacuum energy. However, the above analysis is too naive. If we write Eq. (3.38) in terms of the Fourier modes of the field,

$$\hat{P}_i\Psi_0 = \frac{2\pi}{L}\sum_{\mathbf{n}\in\mathbb{Z}^3}(n_i\,\omega_{\mathbf{n}})\Psi_0\,, \tag{3.39}$$

we realize that the sum is divergent. However, in the cut-off or heat kernel regularizations, the regularized eigenvalues vanish

$$\hat{P}_i\Psi_0 = \frac{2\pi}{L}\sum_{\mathbf{n}\in\mathbb{Z}^3}^{|\omega_{\mathbf{n}}|<\Lambda}(n_i\,\omega_{\mathbf{n}})\Psi_0 = \frac{2\pi}{L}\sum_{\mathbf{n}\in\mathbb{Z}^3}(n_i\,\omega_{\mathbf{n}}e^{-\epsilon\omega_{\mathbf{n}}^2})\Psi_0 = 0\,, \tag{3.40}$$

and thus, the renormalized value of the vacuum eigenvalue of $\hat{P}_i$ vanish. This is due to the spherical symmetry of both regularizations. What is remarkable is that in the case of the vacuum energy, any choice of the regularization provides a non-vanishing value. In this sense there is a difference between the quantum generators of space and time translations, which seems to be not in agreement with the Lorentz symmetry. This is a genuine characteristic of canonical quantization as will be emphasized later in the course.

3.4.3 *Casimir effect*

The existence of UV divergences in vacuum energy is not a special property of scalar fields. Any quantum field theory faces the same problem, e.g. the electromagnetic field in quantum electrodynamics or the fields of the Standard Model have UV divergent vacuum energies. We have renormalized the divergences by removing the whole contribution of vacuum energy. However, this does not mean that it is an unphysical quantity. The fact that the vacuum energy can have observable consequences was first pointed out by Casimir [6]. He remarked that although we can remove a fixed vacuum energy for the free fields, the variation of the vacuum energy under external conditions could be detected and observed.

Consider a pair of infinite, perfectly conducting plates placed parallel to each other at a distance d (see Figure 2). The conducting character of the plates implies that the electromagnetic forces vanishes at both plate surfaces. The presence of

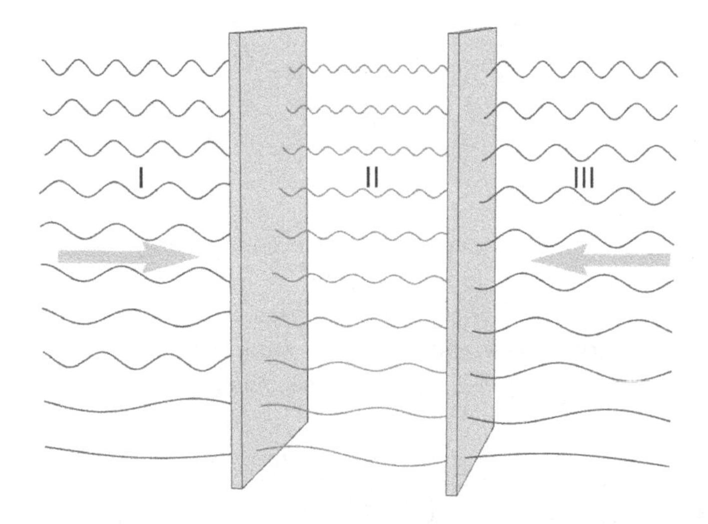

Fig. 3.2 Three different domains of vacuum fluctuations in the Casimir effect.

the plates modifies the vacuum energy in a d-dependent way. If the perturbation increases the vacuum energy with the distance it will induce an attractive force between the plates and this force will be repulsive if the energy decreases with the distance.

The main modification introduced by the plates is the presence of boundary conditions on the classical fields. Since the free electromagnetic field corresponds to photons with two polarizations, the electromagnetic vacuum energy is twice the vacuum energy of a massless scalar field with vanishing boundary conditions on the plates. The physical $\mathbb{R}^3$ space is split into three disjoint domains:

$$\begin{aligned} \Omega_{I} &= \{\mathbf{x} \in \mathbb{R}^3; -\infty < x_3 \leq -\tfrac{d}{2}\}, \\ \Omega_{II} &= \{\mathbf{x} \in \mathbb{R}^3; -\tfrac{d}{2} < x_3 \leq \tfrac{d}{2}\}, \\ \Omega_{III} &= \{\mathbf{x} \in \mathbb{R}^3;\ \tfrac{d}{2} < x_3 \leq \infty\}. \end{aligned}$$

The physical vacuum is the product of the vacua of the different sectors

$$\psi_0(\phi) = \psi_{I}(\phi_{I})\, \psi_{II}(\phi_{II})\, \psi_{III}(\phi_{III})$$

and the vacuum energy the sum of the vacuum energies of the three domains

$$E_0 = E_{I} + E_{II} + E_{III}$$

The calculation of vacuum energy density in the domain between the plates $\mathcal{E}_{II} = E_{II}/V_{II}$ by using the heat kernel regularization gives (see appendix A)

$$\mathcal{E}_{II} = \frac{1}{8\pi^2\epsilon^2} - \frac{1}{16\sqrt{\pi}\epsilon^{\frac{3}{2}}}\frac{1}{d} - \frac{\pi^2}{1440 d^4} + \mathcal{O}(\epsilon^{\frac{1}{2}}). \tag{3.41}$$

In the other two domains, we only get

$$\mathcal{E}_{I} = \frac{1}{8\pi^2\epsilon^2}, \qquad \mathcal{E}_{III} = \frac{1}{8\pi^2\epsilon^2} \tag{3.42}$$

because of their infinite transversal size. The common divergent term corresponds to the vacuum density in infinite volume. Thus, it disappears under vacuum energy renormalization. The $\epsilon^{\frac{3}{2}}$ divergent term between the plates correspond to the selfenergy of the plates and has to be renormalized as well. The remaining renormalized vacuum energy density between the plates

$$\mathcal{E}_{II}^{\rm ren} = -\frac{\pi^2}{1440 d^4} \tag{3.43}$$

is negative inducing an attractive force between the plates, which is the Casimir effect of the vacuum energy on the plates.

3.5 Fields versus Particles

We have assumed that the field operators act on a separable Hilbert space $\mathcal{H}$ without any special properties. In canonical quantization, we assumed that the space of quantum states is given by functionals of the configuration space of square integrable classical fields $\mathcal{M}$ Eq. (3.15). However, such a space has not a canonical translation invariant Hilbert product. The reason being that in quantum mechanics the equivalent space of functions of the configuration space is given by the space of square integrable functions L^2 with the standard L^2-product with respect to the Lebesgue measure $d^n x$ of $\mathbb{R}^n$. However in infinite dimensional Hilbert spaces, the equivalent Lebesgue measure is not well defined because the basic building blocks of hypercubes of size L have an infinite volume if $L > 1$ or zero volume if $L < 1$. Thus, although all operators: the fields $\phi(f)$, the Hamiltonian $\hat{H}$ and the momentum operator $\hat{\mathbf{P}}$ are formally selfadjoint with respect to the naive generalization of Lebesgue measure $\delta\phi$, the definition of the quantum field theory requires a rigorous definition of the Hilbert product and a redefinition of the physical observables.

The key ingredient is that the naive vacuum state Eq. (3.30) defines a good measure in the space of functionals on space of classical fields $\mathcal{M}$. Indeed, the measure defined by

$$\delta\mu(\phi) = \mathcal{N} e^{-(\phi,\sqrt{-\boldsymbol{\nabla}^2+m^2}\phi)}\delta\phi = \prod_{\mathbf{n}\in\mathbb{Z}^3}\sqrt{\frac{\omega_n}{\pi}}e^{-\omega_{\mathbf{n}}\phi_n^2}\delta\phi_n\,, \tag{3.44}$$

is a well-defined probability measure. In Eq. (3.44), $\mathcal{N}$ denotes the normalization factor which guarantees that the volume of the whole configuration space is unit. According to Minlos' theorem [12] (see Appendix B) the Gaussian measure $\delta\mu$ is supported on the space of tempered distributions $\mathcal{S}'(\mathbb{R}^3)$ in the massive case and on the space of generalized distributions $\mathcal{D}'$ in the massless case.

The above definition requires a redefinition of all physical states and operators by a similarity transformation

$$\Psi(\phi) \Rightarrow e^{\frac{1}{2}(\phi,\sqrt{-\boldsymbol{\nabla}^2+m^2}\phi)}\Psi(\phi)\,; \qquad O \Rightarrow e^{\frac{1}{2}(\phi,\sqrt{-\boldsymbol{\nabla}^2+m^2}\phi)}Oe^{-\frac{1}{2}(\phi,\sqrt{-\boldsymbol{\nabla}^2+m^2}\phi)}\,.$$

The field operator $\phi(f)$ remains unchanged whereas the canonical momentum operator $\hat{\pi}(f)$ becomes

$$\hat{\pi}(f) = -i\int d^3\mathbf{x}\, f(\mathbf{x})\left(\frac{\delta}{\delta\phi(\mathbf{x})} - \sqrt{-\boldsymbol{\nabla}^2+m^2}\phi(x)\right),$$

and both are selfadjoint with respect to the Hilbert product

$$(\mathcal{F},\mathcal{G}) = \int_{\mathcal{S}'(\mathbb{R}^3)} \delta\mu(\phi)\,\mathcal{F}(\phi)^*\mathcal{G}(\phi)$$

of $\mathcal{H} = L^2(\mathcal{S}'(\mathbb{R}^3),\delta\mu)$.

The vacuum state becomes trivial

$$\Psi_0 = 1\,,$$

which now is normalizable with respect to the Gaussian measure Eq. (3.44), i.e. $(\Psi_0, \Psi_0) = 1$.

The new renormalized Hamiltonian is

$$\hat{H}_{\rm ren} = -\int d^3\mathbf{x} \left(\frac{\delta}{\delta\phi(\mathbf{x})} - 2\sqrt{-\nabla^2 + m^2}\phi(x) \right) \frac{\delta}{\delta\phi(\mathbf{x})}\,,$$

and the excited states are just field polynomials.

3.5.1 *Fock space*

The space of physical states is generated by polynomials of field operators, e.g.

$$\mathcal{F}_{(f_1, f_2, \cdots, f_n)} = \phi(f_1)\phi(f_2)\cdots\phi(f_n)\,. \tag{3.45}$$

The simplest state is a degree zero polynomial: the vacuum state. The degree one monomials $\phi(f)$ correspond to one-particle states, where f is the quantum wave packet state of the particle. The functional

$$\mathcal{F}_f(\phi) = \phi(f)$$

associated to one-particle states is a linear map in the space of quantum field $\phi \in \mathcal{S}'(\mathbb{R}^3)$. In mathematical terms, one-particle state constitute the dual space of the configuration space of classical fields[7].

Higher order monomials correspond to linear combinations of quantum states with different number of particles. To pick up only states with a defined number of particles, one has to proceed as in the harmonic oscillator case where the eigenstates of the Hamiltonian are given by Hermite polynomials which involve suitable combinations of monomials.

For such a reason to identify physical states in terms of particles, it is convenient to introduce a coherent state basis. This can be achieved in terms of creation and annihilation operators,

$$a(f) = \phi(\sqrt{-\nabla^2 + m^2} f) + i\hat{\pi}(f)\,, \qquad a(f)^\dagger = \phi(\sqrt{-\nabla^2 + m^2}\, f) - i\hat{\pi}(f)^\dagger\,.$$

It is easy to show that

$$[a(f), a(g)] = [a(f)^\dagger, a(g)^\dagger] = 0\,,$$
$$\left[a(f), a(g)^\dagger\right] = 2(f, \sqrt{-\nabla^2 + m^2}\, g)\,.$$

[7]This explains why in the case of gauge fields, where the configuration space of classical gauge fields modulo gauge transformations is a curved manifold, the particle interpretation of quantum states is so difficult.

Using the basis of plane waves Eq. (3.20) we have

$$\hat{H}_{\mathbf{ren}} = \frac{1}{4} \sum_{\mathbf{n} \in \mathbb{Z}^3} (a^\dagger_{\mathbf{n}} a_{\mathbf{n}} + a_{\mathbf{n}} a^\dagger_{\mathbf{n}}) - E_0 = \frac{1}{2} \sum_{\mathbf{n} \in \mathbb{Z}^3} a^\dagger_{\mathbf{n}} a_{\mathbf{n}}$$

where $a_{\mathbf{n}} = a(f_{\mathbf{n}})$,

$$\hat{\mathbf{P}}_{\mathbf{ren}} = \frac{1}{4} \sum_{\mathbf{n} \in \mathbb{Z}^3} \mathbf{n} \big(a^\dagger_{\mathbf{n}} a_{\mathbf{n}} + a_{\mathbf{n}} a^\dagger_{\mathbf{n}} \big) ,$$

and we can define the number operator as

$$\hat{N} = \frac{1}{2} \sum_{\mathbf{n} \in \mathbb{Z}^3} \frac{1}{\omega_{\mathbf{n}}} a^\dagger_{\mathbf{n}} a_{\mathbf{n}} .$$

The main property of the number operator are its commutation relations with the creation and annihilation operators,

$$[\hat{N}, a_{\mathbf{n}}] = -a_{\mathbf{n}} , \quad [\hat{N}, a^\dagger_{\mathbf{n}}] = a^\dagger_{\mathbf{n}} .$$

The creation operators $a(f)$ can generate by iterative actions on the vacuum a basis of physical states. In particular, the state[8]

$$|f\rangle = a(f)^\dagger |0\rangle$$

can be considered as a one-particle state with wave packet f. Indeed, it easy to check that

$$\hat{N}|f\rangle = |f\rangle .$$

The relativistic invariant normalization of the creation operators simplifies the identification of the norm of one-particle states,

$$\langle f|f\rangle_r = \| f \|_r^2 = \frac{1}{2} \int d^3\mathbf{x} \int d^3\mathbf{y}\, f^*(\mathbf{x})(-\Delta + m^2)^{-1}(\mathbf{x}, \mathbf{y}) f(\mathbf{y}) .$$

The completion of the space of one-particle states with this norm is then

$$\mathcal{H} = L^2(\mathbb{R}^3, \mathbb{C}) = \{f, \| f \|_r^2 < \infty\} ,$$

which again can be identified with the dual space of the configuration space of classical fields.

The next step is the identification of two particle states. They are of the form

$$|f_1, f_2\rangle = \frac{1}{\sqrt{2}} a(f_1)^\dagger a(f_2)^\dagger |0\rangle = \frac{1}{\sqrt{2}} a(f_2)^\dagger |f_1\rangle ,$$

[8]We use the Dirac notation, where $|0\rangle$ is the vacuum state, and the ket $|f\rangle$ denotes the state $|f\rangle = a(f)^\dagger |0\rangle$.

and satisfy

$$\hat{N}|f_1, f_2\rangle = 2|f_1, f_2\rangle\,.$$

The n-particle states can be identified with

$$|f_1, f_2, \cdots, f_n\rangle = \frac{1}{\sqrt{n!}} a(f_1)^\dagger a(f_2)^\dagger \cdots a(f_n)^\dagger |0\rangle\,,$$

and the number operator satisfies

$$\hat{N}|f_1, f_2, \cdots, f_n\rangle = n|f_1, f_2, \cdots, f_n\rangle\,.$$

Now because of the commutation properties of the bosonic field operators, the space of states with n-particles is not the tensor product $\mathcal{H}^{\otimes n}$ of n Hilbert spaces of one-particle states $\mathcal{H}$. Instead it can be identified with the subspace of symmetric states involving n particles,

$${}^s\mathcal{H}^{\otimes \mathbf{n}} \subset \mathcal{H} \otimes \mathcal{H} \otimes \overset{\mathbf{n}}{\cdots} \otimes \mathcal{H} = \mathcal{H}^{\otimes n}\,,$$

in the space of quantum states of n distinguishable particles.

In this sense, the bosonic nature of the commutation relations implies the bosonic statistics of the corresponding particles. To some extent this example illustrates the existence of a link between the spin of the fields and the statistics of the corresponding particles. In general, for any field theory the spin-statistics connection follows from fundamental principles (spin-statistics theorem) [25].

The Fock space is the Hilbert space of all multiparticle bosonic states[9],

$$\mathcal{F} = \bigoplus_{n=0}^{\infty} {}^s\mathcal{H}^{\otimes n}\,.$$

$\mathcal{F}$ is the Hilbert space of the (bosonic) quantum field theory. In the free theory the Hamiltonian $\hat{H}$ and the number of particles operator $\hat{N}$ commute. Thus, all energy levels have a definite number of particles. However, in the presence of interactions this is not longer true, e.g. for

$$V(\phi) = \frac{1}{2} m^2 \phi^2 + \frac{\lambda}{4!} \phi^4$$

we have that

$$[\hat{N}, V] \neq 0\,,$$

which means that the number of particles might change by time evolution. This is one of the novel characteristics of quantum field theory. New process like, decaying of particles, pair creation and photon emission in atoms can occur in the theory.

[9]See the first part of this volume.

Quantum Field Theory is a natural way to describe such process in an accurate manner.

Although we have identified the Fock structure of the space of quantum states with the stratification given by the number of particles, the notion of particle is secondary and it is just the quantum field that is fundamental. The explanation of why two particles are identical come from the fact that they are generated by the same field, which permeates the whole Universe. In this way we understand why the particles coming in cosmic rays from far galaxies are identical to the same particles on earth. The link is the quantum field. Moreover, there are field theories where the particle composition is unclear. For example in gauge theories, the fundamental fields are quark and gluon fields, but the physical particles are mesons, baryons and glueballs which are bounded composites of quarks and gluons.

3.5.2 *Wick theorem*

An important consequence of the Gaussian nature of the ground state of a free field theory is the clustering property of the vacuum expectation values of the product of field operators

$$\langle 0|\phi(f_1)\phi(f_1)\cdots\phi(f_n)|0\rangle = \langle f_1 f_1 \cdots f_n\rangle = \int \delta\mu\, \phi(f_1)\phi(f_1)\cdots\phi(f_n)\,.$$

The cluster property is a fundamental characteristic of Gaussian measures which gives rise to the Wick theorem which states that

$$\langle f_1 f_2 \cdots f_n\rangle = \begin{cases} 0 & \text{for } n = 2m+1, \\ \dfrac{1}{2!m!}\displaystyle\sum_{\sigma\in S_n} \langle f_{\sigma(1)} f_{\sigma(2)}\rangle \cdots \langle f_{\sigma(2m-1)} f_{\sigma(2m)}\rangle & \text{for } n = 2m\,. \end{cases}$$

3.6 Fields in Interaction

The free QFT considered in the previous section shows the basic properties of a relativistic quantum field theory, but the challenge is to quantize field theories of interacting fields. The procedure is basically the same. The main difference is that the interacting Hamiltonian is not exactly solvable. For example, let us consider the $\frac{\lambda}{4!}\phi^4$ theory Hamiltonian

$$\hat{H} = \frac{1}{2}\left(\| \hat{\pi} \|^2 + \| \nabla\phi \|^2 + m^2 \| \phi \|^2 + \frac{\lambda}{12} \| \phi^2 \|^2\right).$$

Using the same quantization rules as in Eq. (3.25) we get a formal quantum Hamiltonian $\hat{H}$ which is defined in the space of functionals in the space of classical fields

$\mathcal{M}$. However, as we have seen in the case of free fields, the theory needs a renormalization.

Using the plane wave basis on a finite torus, we have

$$\hat{H} = \frac{1}{2} \sum_{\mathbf{n}\in\mathbb{Z}^3} \left(\frac{\delta}{\delta\phi_{-\mathbf{n}}} \frac{\delta}{\delta\phi_{\mathbf{n}}} + \omega_{\mathbf{n}}^2 |\phi_{\mathbf{n}}|^2 + \frac{\lambda}{12} \sum_{\mathbf{n}_1,\mathbf{n}_2\in\mathbb{Z}^3} \phi_{\mathbf{n}}\phi_{\mathbf{n}_1}\phi_{\mathbf{n}_2}\phi_{-\mathbf{n}-\mathbf{n}_1-\mathbf{n}_2} \right) .$$

Again the regularization of UV divergences requires the introduction of a regularization, e.g.

$$\hat{H}_\Lambda = \frac{1}{2} \sum_{\mathbf{n}\in\mathbb{Z}^3}^{\omega_{\mathbf{n}}<\Lambda} \left(\frac{\delta}{\delta\phi_{-\mathbf{n}}} \frac{\delta}{\delta\phi_{\mathbf{n}}} + \omega_{\mathbf{n}}^2 |\phi_{\mathbf{n}}|^2 + \frac{\lambda}{12} \sum_{\mathbf{n}_1,\mathbf{n}_2\in\mathbb{Z}^3}^{\omega_{\mathbf{n}_1},\omega_{\mathbf{n}_2}<\Lambda} \phi_{\mathbf{n}}\phi_{\mathbf{n}_1}\phi_{\mathbf{n}_2}\phi_{-\mathbf{n}-\mathbf{n}_1-\mathbf{n}_2} \right) . \quad (3.46)$$

The Hamiltonian Eq. (3.46) can be split into two terms

$$\hat{H}_\Lambda = H_0 + \hat{H}_\Lambda^{\text{int}} .$$

The first term

$$\hat{H}_0 = \frac{1}{2} \sum_{\mathbf{n}\in\mathbb{Z}^3}^{\omega_{\mathbf{n}}<\Lambda} \left(\frac{\delta}{\delta\phi_{-\mathbf{n}}} \frac{\delta}{\delta\phi_{\mathbf{n}}} + \omega_{\mathbf{n}}^2 |\phi_{\mathbf{n}}|^2 \right)$$

is just the Hamiltonian of the free bosonic theory, whereas the second term

$$\hat{H}_\Lambda^{\text{int}} = \frac{\lambda}{4!} \sum_{\mathbf{n}_1,\mathbf{n}_2,\mathbf{n}_3\in\mathbb{Z}^3}^{\omega_{\mathbf{n}_1},\omega_{\mathbf{n}_2},\omega_{\mathbf{n}_3}<\Lambda} \phi_{\mathbf{n}_1}\phi_{\mathbf{n}_2}\phi_{\mathbf{n}_3}\phi_{-\mathbf{n}_1-\mathbf{n}_2-\mathbf{n}_3} .$$

contains the interaction terms. The renormalization of H_0 can be performed as in previous section by subtracting the vacuum energy of the free theory,

$$\hat{H}_\Lambda^{\text{ren}} = H_0^{\text{ren}} + \hat{H}_\Lambda^{\text{int}} .$$

But there are new divergences generated by the interacting terms which require an extra renormalization.

The easiest way of dealing with the interacting theory is to consider the interacting term $\hat{H}_\Lambda^{\text{int}}$ as a perturbation. In first order of perturbation theory, the vacuum energy gets an additional contribution

$$\Delta E_0 = \langle 0|\hat{H}_{\text{int}}|0\rangle ,$$

which by Wick's theorem

$$\Delta E_0 = \frac{\lambda}{8} \sum_{\mathbf{n}_1,\mathbf{n}_2\in\mathbb{Z}^3}^{\omega_{\mathbf{n}_1},\omega_{\mathbf{n}_2}<\Lambda} \langle 0|\phi_{-\mathbf{n}_1}\phi_{\mathbf{n}_1}|0\rangle\langle 0|\phi_{-\mathbf{n}_2}\phi_{\mathbf{n}_2}|0\rangle ,$$

gives an extra divergent contribution to the vacuum energy

$$\Delta E_0 = \frac{\lambda}{512\pi^4}\left(\Lambda^4 + \Lambda^2 m^2\left(\log\frac{m^2}{2\Lambda^2}+2\right) + \frac{m^4}{4}\left(\log\frac{m^2}{2\Lambda^2}+2\right)^2\right) + \mathcal{O}\left(\frac{m^2}{\Lambda^2}\right). \tag{3.47}$$

That contribution has to be subtracted from the Hamiltonian to renormalize the vacuum energy to zero.

3.6.1 *Renormalization of excited states*

The vacuum energy is not the only divergent quantity of the theory. The energy of one-particle states gets a perturbative correction which is also UV divergent. The energy of the excited state $|f_{\mathbf{n}}\rangle = a^\dagger_{\mathbf{n}}|0\rangle$ in the free theory is $\omega_{\mathbf{n}}$. The first order correction to this energy is

$$\Delta E_{\mathbf{n}} = \langle f_{\mathbf{n}}|\hat{H}_{\text{int}}|f_{\mathbf{n}}\rangle\,.$$

Using Wick's theorem, a simple calculation shows that

$$\Delta E_{\mathbf{n}} = \Delta E_0 + \frac{\lambda}{8}\sum_{\mathbf{n}_1,\mathbf{n}_2\in\mathbb{Z}^3}^{\omega_{\mathbf{n}_1},\omega_{\mathbf{n}_2}<\Lambda} \langle f_{\mathbf{n}}|\phi_{-\mathbf{n}_1}|0\rangle\langle 0|\phi_{\mathbf{n}_1}|f_{\mathbf{n}}\rangle\langle 0|\phi_{-\mathbf{n}_2}\phi_{\mathbf{n}_2}|0\rangle\,.$$

Both terms are divergent. The first term corresponds to the vacuum energy correction, which is removed by the previous renormalization of vacuum energy Eq. (3.47). The second term gives a new type of UV quadratic divergence. In the sharp momentum cutoff it is given by

$$\frac{\lambda}{64\,\pi^2\omega^2_{\mathbf{n}}}\left(\Lambda^2 + \frac{m^2}{2}\left(\log\frac{m^2}{2\Lambda^2}+2\right)\right).$$

This renormalization of the divergence can be absorbed by a renormalization of the mass of the theory. Indeed if we redefine the Hamiltonian of the theory as

$$\hat{H}^{\text{ren}}_{\text{int}} = \frac{1}{2}\Delta m^2\sum_{\mathbf{n}\in\mathbb{Z}^3}|\phi_{\mathbf{n}}|^2 + \frac{\lambda}{4!}\sum_{\mathbf{n}_1,\mathbf{n}_2,\mathbf{n}_3\in\mathbb{Z}^3}^{\omega_{\mathbf{n}_1},\omega_{\mathbf{n}_2},\omega_{\mathbf{n}_3}<\Lambda} \phi_{\mathbf{n}_1}\phi_{\mathbf{n}_2}\phi_{\mathbf{n}_3}\phi_{-\mathbf{n}_1-\mathbf{n}_2-\mathbf{n}_3}\,,$$

where

$$\Delta m^2 = -\frac{\lambda}{32\pi^2}\left(\Lambda^2 - m^2\left(\log\frac{m^2}{2\Lambda^2}-\frac{1}{2}\right)\right),$$

the first order correction to the energy of all one-particle levels is finite. With the above prescription there is no correction to the free value $\omega_{\mathbf{n}}$, but we could have renormalized the mass of the theory by a different subtraction: $m^2 \to m^2 - \Delta m^2 + a^2$. In that case, the *renormalized* value of the energy one-particle states will be after resummation of the perturbative series,

$$\omega_n^r = \sqrt{\mathbf{n}^2 + m^2 + a^2}\,.$$

With the above renormalizations of the vacuum energy and mass, the theory is finite at first order of perturbation theory. This means that the corrections to the higher energy levels are finite at first order in λ.

We can understand now the physical meaning of the renormalization program. The physical parameters which appear in the classical Lagrangian do not necessarily coincide with the corresponding quantum physical parameters. This includes the constant term which can always be added to the classical Lagrangian without changing the dynamics (but determines the vacuum energy of the quantum theory), the mass of the theory m and the coupling constant λ.

Until now we have only renormalized the mass and the vacuum energy. However, in higher orders of perturbation theory new UV divergences appear. They can be absorbed by new renormalizations of the vacuum energy E_0, the mass m^2, the coupling constant λ and the field operators $\hat{\phi}(f)$.

However, the proof of consistency of the resulting theory is quite involved and required few decades to be completely achieved. One of the main problems is that in the canonical approach, the preservation of the relativistic invariance is not guaranteed. Among other things the use of UV cutoff breaks Lorentz invariance and one has to prove that the renormalization prescriptions do preserve the relativistic symmetries. In general, it is not obvious that the interacting theory satisfies the general quantum field principles of section 3.

For such reasons it is convenient to develop a new approach to quantization based on a covariant formalism, where time and space are treated on the same footing.

3.7 Covariant Approach

If we consider the Heisenberg representation of quantum operators the field operator evolves according to the Heisenberg law

$$\phi(\mathbf{x}, t) = U(t)\phi(\mathbf{x})U(t)^\dagger\,,$$

which smeared with test functions $\tilde{f} \in \mathcal{S}(\mathbb{R}^4)$

$$\phi(\tilde{f}) = \int_{\mathbb{R}^4} d^4x \, \phi(\mathbf{x}, t) \tilde{f}(x, t)$$

satisfy

$$U(t)\phi(\tilde{f})U(t)^\dagger = \phi(\tilde{f}_{(\mathbb{I},t)}),$$

where

$$\tilde{f}_{(\mathbb{I},a)}(x, t) = \tilde{f}(x, t - a).$$

In terms of the new covariant field operators $\phi(\tilde{f})$, the principles of quantum field theory are similar to the ones introduced in section 3. The only changes affect the last three principles which now read:

- PC5 *Field theory (real boson):* For any $\tilde{f} \in \mathcal{S}(\mathbb{R}^4)$ there is field operator $\phi(\tilde{f})$ in $\mathcal{H}$ which satisfies $\phi(\tilde{f}) = \phi(\tilde{f})^*$. The subspace spanned by the vectors $\phi(\tilde{f}_1)\phi(\tilde{f}_2)\cdots\phi(\tilde{f}_n)|0\rangle$ for arbitrary test functions $\tilde{f}_1, \tilde{f}_2, \cdots \tilde{f}_n \in \mathcal{S}(\mathbb{R}^4)$ is a dense subspace of $\mathcal{H}$.
- PC6 *Poincaré covariance:* For any Poincaré transformation (Λ, a) and classical field test function defined in Minkowski space-time $\tilde{f} \in \mathcal{S}(\mathbb{R}^4)$

$$U(\Lambda, a)\phi(\tilde{f})U(\Lambda, a)^\dagger = \phi(\tilde{f}_{(\Lambda,a)}),$$

  where

$$\tilde{f}_{(\Lambda,a)}(x) = \tilde{f}(\Lambda^{-1}(x - a)).$$

- PC7 *(Bosonic) local causality:* For any $\tilde{f}, \tilde{g} \in \mathcal{S}(\mathbb{R}^4)$ whose domains are space-like separated[10] the corresponding field operators $\phi(\tilde{f}), \phi(\tilde{g})$ commute[11]

$$[\phi(\tilde{f}), \phi(\tilde{g})] = 0. \tag{3.48}$$

It is not difficult to show that these principles are satisfied by the free field theory. The only non-trivial test is the calculation of the commutator of free fields $[\phi(f), \phi(g)]$. After some simple algebra, it can be shown that it is an operator proportional to the identity operator times a real function of f and g which can be estimated from the vacuum expectation value of the operator

$$[\phi(\tilde{f}), \phi(\tilde{g})] = \mathbb{I} \int_{\mathbb{R}^4} d^4x \int_{\mathbb{R}^4} d^4y \, \tilde{f}(x) \, \Delta(x - y) \, \tilde{g}(y) = \mathbb{I} \, \langle 0|[\phi(\tilde{f}), \phi(\tilde{g})]|0\rangle ,$$

[10] $\tilde{f}, \tilde{g}$ are space-like separated if for any $x, y \in \mathbb{R}^4$ such that $\tilde{f}(x) \neq 0$ and $\tilde{g}(y) \neq 0$, $d(x, y) < 0$.
[11] In the fermionic case the commutator $[\cdot, \cdot]$ is replaced by an anticommutator $\{\cdot, \cdot\}$.

where

$$\Delta(x-y) = \int \frac{d^3k}{(2\pi)^3} \frac{1}{2\sqrt{\mathbf{k}^2+m^2}} \left(e^{ik\cdot(x-y)} - e^{-ik\cdot(x-y)} \right),$$

and $k \cdot (x-y) = \mathbf{k} \cdot (\mathbf{x}-\mathbf{y}) - \sqrt{\mathbf{k}+m^2}(x_0 - y_0)$. The local causality property Eq. (3.48) follows from the fact that the causal propagator kernel $\Delta(x-y)$ vanishes for equal times $x_0 = y_0$, since the two terms in

$$\Delta(\mathbf{x}-\mathbf{y}, 0) = \int \frac{d^3k}{(2\pi)^3} \frac{1}{2\sqrt{\mathbf{k}^2+m^2}} \left(e^{i\mathbf{k}\cdot(\mathbf{x}-\mathbf{y})} - e^{-i\mathbf{k}\cdot(\mathbf{x}-\mathbf{y})} \right) = 0,$$

give the same contributions, as can be shown by flipping the sign of $\mathbf{k}$ in one of them. Although the expression of causal propagator kernel $\Delta(x-y)$ is relativistic invariant, it seems to be non-covariant. However, it can be written in an explicitly covariant form

$$\Delta(x-y) = D(x-y) - D(y-x),$$

where

$$D(x-y) = \int \frac{d^4k}{(2\pi)^4} \theta(k_0)\delta(k^2+m^2) e^{ik\cdot(x-y)}.$$

This result shows that the covariant quantization approach could also be derived from the Peierls covariant classical approach to field theory [17], by replacing Peierls brackets by commutators (see appendix C).

To check that fundamental principles are satisfied in an interacting theory is more difficult, but it can be shown that in perturbation theory they are satisfied even after renormalization.

3.7.1 *Euclidean approach*

Working with field operators in the Fock space is hard because they are unbounded operators. For such a reason it is more convenient to consider their expectation values on the different states. Since the full Fock space is generated by the completeness principle by the field operators, it is enough to consider the expectation values of the products of field operators on the vacuum state.

These expectation values are known as Wightman functions

$$W(\tilde{f}_1, \tilde{f}_2, \ldots, \tilde{f}_n) = \langle 0|\phi(\tilde{f}_1)\phi(\tilde{f}_2)\ldots\phi(\tilde{f}_n)|0\rangle.$$

However the unbounded character of the field operators $\phi(\tilde{f})$ reflects in the oscillating behavior of the Wightman functions. For such a reason it is much more convenient to introduce the Euclidean time analytic extensions of the quantum fields and the corresponding Wightman functions which after analytic continuation become Schwinger functions. Indeed if one considers a Euclidean time $\tau = it$ the τ-evolution of the field operators becomes

$$\phi_E(\mathbf{x}, \tau) = e^{\tau H} \phi(\mathbf{x}, 0) e^{-\tau H} .$$

The smearing of $\phi_E(\mathbf{x}, t)$ by a test function $\tilde{f}$ defines the Euclidean field operators

$$\phi_E(\tilde{f}) = \int_{\mathbb{R}^4} d^4x \, \phi_E(\mathbf{x}, t) \tilde{f}(\mathbf{x}, t) .$$

Now the vacuum expectation values of products of field operators $\phi_E(\tilde{f})$ is not always well defined because the Euclidean time evolution is given by hermitian operators $U_E(\tau) = U(it)$ which define a semigroup instead of a group unlike the case of real time evolution. The hermitian operators $U_E(\tau)$ are only bounded for positive values of the Euclidean time $\tau < 0$. For such a reason the vacuum expectation values of products of field operators $\phi_E(\tilde{f})$ require some time-ordering of the domains of the test functions. If the support of the family of functions $\{\tilde{f}_i\}_{i=1,2,\cdots,n}$ of $\mathcal{S}(\mathbb{R}^4)$ are time ordered, i.e. $\tau_1 > \tau_2 > \cdots > \tau_n$ for any point $x = (x_1, x_2, \cdots, x_n)$ with $f_i(x_i) \neq 0$ for $i = 1, 2, \cdots, n$, then

$$S_n(\tilde{f}_1, \tilde{f}_2, \ldots, \tilde{f}_n) = \langle 0 | \phi_E(\tilde{f}_1) \phi_E(\tilde{f}_2) \ldots \phi_E(\tilde{f}_n) | 0 \rangle \tag{3.49}$$

is a well defined function and does coincide in that case with the analytic extension of the corresponding Minkowskian vacuum expectation values.

Moreover, it can be extended for multivariable test functions $\tilde{\mathbf{f}}_\mathbf{n} \in \mathcal{S}(\mathbb{R}^{4n})$ defined in $\mathbb{R}^{4n}$ by

$$S_n(\tilde{\mathbf{f}}_\mathbf{n}) = \int_{\mathbb{R}^4} d^4x_1 \int_{\mathbb{R}^4} d^4x_2 \ldots \int_{\mathbb{R}^4} d^4x_n \, \phi_E(x_1) \phi_E(x_2) \ldots \phi_E(x_n) \tilde{f}_n(x_1, x_2, \ldots, x_n) \tag{3.50}$$

when $\tilde{\mathbf{f}}_\mathbf{n}$ has support in a time ordered subset of $\mathbb{R}^{4n}$, i.e. $\tilde{f}_n(x_1, x_2, \ldots, x_n) = 0$ if $x \in \mathbb{R}^{4n}$ does not satisfy any of the inequalities $\tau_1 > \tau_2 > \cdots > \tau_n$. In the particular case of $\tilde{\mathbf{f}}_\mathbf{n} = \tilde{f}_1 \tilde{f}_2 \ldots \tilde{f}_n$ the expectation value Eq. (3.50) reduces to Eq. (3.49). But $S_n(\tilde{\mathbf{f}}_\mathbf{n})$ can be extended to multivariable functions with more general support by analytic extension from the Mikowskian definition.

The interesting point is that the analytic extension also provides a finite value for the case where the supports are not time-ordered. These analytically extended functions are known as Schwinger functions. Although they can only be expressed as vacuum expectation values of products of Euclidean fields Eq. (3.49) when the supports of the test functions are time-ordered, in practice, because of their symmetry under permutations, they can always be calculated in that way.

The relevance of Schwinger functions is that the quantum field theory can be completely formulated in terms of them and the fundamental principles reformulated in the following way.

Let θ be the Euclidean-time reflection symmetry defined by $\theta(\mathbf{x}, \tau) = (\mathbf{x}, -\tau)$. The action of θ on $\mathbb{R}^4$ induces a transformation on the classical fields test functions given by $\theta \tilde{f}(x) = \tilde{f}(\theta x)$ and in the multivariable test functions $\tilde{\mathbf{f}}_{\mathbf{n}} \in \mathcal{S}$ in a similar way $\theta \tilde{\mathbf{f}}_{\mathbf{n}}(x_1, x_2, \ldots, x_n) = \tilde{\mathbf{f}}_{\mathbf{n}}(\theta x_1, \theta x_2, \ldots, \theta x_n)$.

- E1 *(Regularity):* The Schwinger functions S_n are tempered distributions in $\mathcal{S}$, satisfying the reflection reality condition
$$S_n(\tilde{\mathbf{f}}_{\mathbf{n}})^* = S_n(\theta \tilde{\mathbf{f}}_{\mathbf{n}}^*)\,.$$
- E2 *(Permutation symmetry):* The Schwinger functions are symmetric under permutations, i.e.
$$S_n(\tilde{f}_{\sigma(1)}, \tilde{f}_{\sigma(2)}, \ldots, \tilde{f}_{\sigma(n)}) = S_n(\tilde{f}_1, \tilde{f}_2, \ldots, \tilde{f}_n)$$
for any permutation $\sigma \in \mathcal{S}_n$.
- E3 *(Euclidean invariance):* The Schwinger functions are covariant under Euclidean transformations, i.e.
$$S_n(\tilde{\mathbf{f}}_{\mathbf{n}(\mathbf{\Lambda},\mathbf{a})}) = S_n(\tilde{\mathbf{f}}_{\mathbf{n}})$$
for any Euclidean transformation $(\Lambda, a) \in E_4 = ISO(4)$.
- E4 *(Reflection positivity):* For any family of multivariable functions, test functions $\tilde{\mathbf{f}}_{n_i} \in S(R_+^{4n_i}), i = 0, 1, 2, \ldots, n$ the following inequality[12]
$$\sum_{i,j=0}^{n} S_{n_i+n_j}(\theta \tilde{\mathbf{f}}_{\mathbf{n_i}}^* . \tilde{\mathbf{f}}_{\mathbf{n_j}}) \geq 0 \tag{3.51}$$
holds.
- E5 *(Cluster property).* For any pair of multivariable functions test functions $\tilde{\mathbf{f}}_n \in S(R^{4n}), \tilde{\mathbf{f}}_m \in S(R^{4m})$, we have that
$$\lim_{\sigma \to \infty} S_{n+m}(\tilde{\mathbf{f}}_{\mathbf{n}} . \tilde{\mathbf{f}}_{\mathbf{m}(\mathbb{I},\sigma)}) = S_n(\tilde{\mathbf{f}}_{\mathbf{n}}) S_m(\tilde{\mathbf{f}}_{\mathbf{m}})\,,$$
where $(\mathbb{I}, \sigma)$ is the Euclidean time translation $(\mathbb{I}, \sigma)(\mathbf{x}, \tau) = (\mathbf{x}, \tau + \sigma)$.

These Euclidean principles follow from the field theory principles introduced in section 3. The Euclidean principles E1-E3 are a straightforward consequence of the Minkowskian principles. The permutation symmetry of Euclidean fields (principle E2) follows from the fact that

$$[\phi_E(f_1), \phi_E(f_2)] = 0\,,$$

[12] The space $R_+^4 = \{(x, \tau) \in \mathbb{R}^4, \tau \geq 0\}$ is the half of the n-dimensional Euclidean space $\mathbb{R}^4$ with positive Euclidean time.

for any pair of functions f_1, f_2 which follows from the commutation property of Minkowskian fields for test functions with space-like separated supports (local causality property P7) and Euclidean rotation invariance. The third Euclidean principle E3 follows from the positivity of the norm of the state

$$\sum_{i=0}^{n} \phi(\tilde{\mathbf{f}}_{n_i})|0\rangle \,.$$

Finally the cluster property is a consequence of the uniqueness of the vacuum assumed in the fourth Minkowskian principle P4.

What is more interesting is that the relativistic QFT can be fully reconstructed from the Euclidean principles. The proof was achieved by Osterwalder and Schrader in the early seventies. We will not elaborate in the proof that can be found in the Simon [23] and Glimm-Jaffe [13] books.

3.8 Conformal Invariant Theories

There is a subfamily of field theories which besides the above principles are invariant under a larger symmetry group: the conformal group. Conformal transformations are space-time transformations which leave the Minkowski metric invariant up to a scale factor

$$x' = c(x) \,; \quad \eta'_{\mu,\nu}(x') = \Omega_c^2(x)\, \eta_{\mu,\nu}(x) \,. \tag{3.52}$$

The group of conformal transformations is an extension of the Poincaré symmetry. Besides the translations (3.9), rotations (3.10) and Lorentz transformations (3.11) two new type of transformations preserve the Minkowski metric up to a constant scale factor (3.52):

(i) *dilations*

$$x'^{\mu} = e^{\sigma} x^{\mu},$$

defined by any real number σ; and

(ii) *special conformal transformations*

$$x'^{\mu} = \frac{x^{\mu} - a^{\mu} x^2}{1 - 2a \cdot x + a^2 x^2} \,,$$

defined by any vector a of Minkowski space-time.

In conformal invariant theories, the 2-point Schwinger function is of the form

$$S_2(f_1, f_2) = (f_1, \Delta^{-k} f_2) \,,$$

where k is a positive exponent. For instance, for a free massless bosonic scalar $k = 2$ and the 2-point kernel is

$$\Delta^{-1}(x-y) = \frac{a^2}{|x-y|^2} .$$

The power like behavior of the 2-point kernel is a consequence of conformal invariance. The Schwinger functions of a conformal invariant theory cannot contain any physical dimensionful parameter up to a unique universal space-time scale a.

Conformal invariant theories are associated to second order phase transitions. In 2+1 dimensional space times the free massless bosonic scalar and the 2-point kernel is

$$\Delta^{-1}(x-y) = \frac{a}{|x-y|} .$$

However, in 1+1 space times the corresponding 2-point kernel is not scale covariant,

$$\Delta^{-1}(x-y) = -\frac{1}{2\pi} \log \frac{|x-y|}{a} .$$

Such an anomalous behavior means that the theory is pathological. In fact, the Schwinger function $S_2(f_1, f_2)$ does not satisfies the Osterwalder-Schrader positivity condition (3.51) because

$$\Delta^{-1}(\theta x - x) = -\frac{1}{2\pi} \log \frac{2x_0}{a} ,$$

which is not positive for $x_0 > \frac{a}{2}$.

In fact what happens is simply that the theory has not a normalizable vacuum state. Indeed, the vacuum state measure (3.44) does not define a good Gaussian measure in this case due to infrared problems.

The result agrees with the Coleman-Mermin-Wagner theorem which states that a continuous global symmetry cannot be spontaneously broken in 1+1 dimensions. According to Goldstone theorem such a breaking would imply the existence of massless scalar fields which as we have pointed out are pathological.

On the other hand in 1+1 dimensions, the conformal group is also different. In fact it is an infinite dimensional group of space transformations.

Thus, in 1+1 dimensions the conformal symmetry provides much more constraints than in higher dimensions. However, there is an infinity of consistent field theories which are conformal invariant [2; 9].

The simplest case is the $O(2)$ sigma model. It can be obtained (in the spin wave regime) from the massless scalar by the following field transformation

$$\Phi(x) = e^{\phi(x)} . \tag{3.53}$$

In this case the Schwinger functions can be derived from (3.53). For any two functions $\tilde{f}_1, \tilde{f}_1 \in \mathcal{S}^{\mathbb{R}^4}$ with ordered supports, i.e. for any x_1, x_2 with $\tilde{f}_1(x_1) = \tilde{f}_2(x_2) = 0, \tau_1 > \tau_2$

$$S_2(\tilde{f}_1, \tilde{f}_2) = \langle 0|\Phi_E^*(\tilde{f}_1)\Phi_E(\tilde{f}_2)|0\rangle = \int dx_1 dx_2 \tilde{f}_1(x_1) \frac{a^{2\pi}}{|x_1 - x_2|^{2\pi}} \tilde{f}_2(x_2) .$$

This model corresponds to the spin wave regime of the $O(2)$ sigma model defined by the classical action

$$S[\Phi(x)] = \frac{1}{2} \int d^2x\, \partial_\mu \Phi^*\, \partial^\mu \Phi = \frac{1}{2} \int d^2x\, \partial_\mu \phi\, \partial^\mu \phi .$$

which in that regime does coincide with (3.13) in 1+1 dimensional theories.

3.8.1 *Functional integral approach*

The major advantage of the Euclidean approach is that the Schwinger functions are better behaved than the corresponding Wightman distributions and what is more important they can be derived in most of the cases from functional integration with respect to a probability measure. This also allows by introducing a suitable regularization with a systematic numerical approach.

The result which was first suggested by Symanzik [26] and Nelson [16] is that formally speaking the Schwinger functions can be considered as the momentum operators of a functional measure defined in the space of distributions $\mathcal{S}(R^4)$ by the exponential of the Euclidean classical action S_E, i.e.

$$S_n(\tilde{f}_1, \tilde{f}_2, \ldots, \tilde{f}_n) = \int_{\mathcal{S}'(R^4)} \delta\phi e^{-S_E(\phi)}\, \phi(\tilde{f}_1)\phi(\tilde{f}_2)\ldots\phi(\tilde{f}_n) .$$

In the case of free field theory

$$S_E(\phi) = \frac{1}{2} \parallel \nabla\phi \parallel^2 + \frac{m^2}{2} \parallel \phi \parallel^2 = \frac{1}{2}(\phi, (-\nabla^2 + m^2)\phi)$$

and we have that

$$S_2(\tilde{f}, \tilde{g}) = \int_{\mathcal{S}'(R^4)} \delta\phi\; e^{-S_E(\phi)} \phi(\tilde{f})\phi(\tilde{g}) = \int_{\mathcal{S}'(R^4)} \delta\mu_m \phi(\tilde{f})\phi(\tilde{g})$$

$$= \frac{1}{2}(\tilde{f}, (-\nabla^2 + m^2)^{-1}\tilde{g}) \qquad (3.54)$$

where $\delta\mu_m$ is the Gaussian measure defined on $\mathcal{S}'(R^4)$ with vanishing mean and covariance operator $(-\nabla^2 + m^2)^{-1}$.

It is obvious that Schwinger function (3.54) does coincide with the analytic extension of the Wightman function of the free theory. In fact, from the functional integral formulation it is easy to check that the Schwinger functions satisfy the regularity, symmetry and Euclidean covariance principles. Concerning the reflection positivity property, it is not so evident. Let us check that this is the case to illustrate the subtleties of the very special Osterwalder-Schrader's property. Let us consider the case of an one-particle state with a complex function $\tilde{f} \in \mathcal{S}(R^4_+, \mathbb{C})$

$$S_2(\theta \tilde{f}^*, \tilde{f}) = \int_{\mathcal{S}'(R^4)} \delta\phi \, e^{-S_E(\phi)} \theta\phi(\tilde{f})^* \phi(\tilde{f}) = \frac{1}{2}(\theta \tilde{f}, (-\nabla^2 + m^2)^{-1} \tilde{f}) \, .$$

Let us define $\varphi = (-\nabla^2 + m^2)^{-1} \tilde{f}$. Since θ commutes with $(-\nabla^2 + m^2)$ we have

$$(\theta \tilde{f}, (-\nabla^2 + m^2)^{-1} \tilde{f}) = (\theta \tilde{f}, \varphi) = ((-\nabla^2 + m^2)\theta\varphi, \varphi) \, , \tag{3.55}$$

and since the support of $\tilde{f}$ is contained in $\mathcal{S}(\mathbb{R}^4_+)$, that of $\theta \tilde{f}$ is in $\mathcal{S}(\mathbb{R}^4_-)$[13], thus, we can restrict the integral in Eq. (3.55) to $\mathcal{S}(\mathbb{R}^4_-)$,

$$(\theta \tilde{f}, \varphi) = (\theta \tilde{f}, \varphi)_- = ((-\nabla^2 + m^2)\theta\varphi, \varphi)_- \, .$$

By the same reason, $(\theta\varphi, (-\nabla^2 + m^2)\varphi)_- = (\theta\varphi, \tilde{f})_- = 0$ and

$$(\theta \tilde{f}, \varphi) = ((-\nabla^2 + m^2)\theta\varphi, \varphi)_- \, .$$

Integrating by parts and using Stokes theorem one gets [3]

$$(\theta \tilde{f}, \varphi) = (\theta\varphi, (-\nabla^2 + m^2)\varphi)_- - \int_{\mathbb{R}^3} d^3\mathbf{x} \, (\partial_n \theta\varphi)^* \, \varphi + \int_{\mathbb{R}^3} d^3\mathbf{x} \, \theta\varphi^* \, \partial_n \varphi \, ,$$

where $\partial_n \varphi$ denotes the normal derivative of φ at the boundary $\partial\mathbb{R}^4_- = \mathbb{R}^3$ at Euclidean time $\tau = 0$ of $\mathbb{R}^4_-$. Now, at the boundary $\tau = 0$, $\theta\varphi = \varphi$ and $\partial_n \theta\varphi = -\partial_n \varphi$, thus,

$$(\theta \tilde{f}, \varphi) = 2\mathrm{Re} \int_{\mathbb{R}^3} d^3\mathbf{x} \, \theta\varphi^* \, \partial_n \varphi \, .$$

Finally, by integrating by parts back we get

$$\mathrm{Re} \int_{\mathbb{R}^3} d^3\mathbf{x} \, \theta\varphi^* \partial_n \varphi = (\nabla\varphi, \nabla\varphi)_- + (\varphi, \nabla^2\varphi)_- \, ,$$

and since $\nabla^2 \varphi = \nabla^2(-\nabla^2 + m^2)^{-1} \tilde{f} = -\tilde{f} + m^2 \varphi$,

$$(\theta \tilde{f}, \varphi) = 2\|\nabla\varphi\|^2_- + 2m^2\|\varphi\|^2_- \geq 0 \, .$$

[13]The space $R^4_- = \{(x, \tau) \in \mathbb{R}^4, \tau \leq 0\}$ is the half of the n-dimensional Euclidean space $\mathbb{R}^4$ with negative Euclidean time.

The proof of reflection positivity for higher order Schwinger functions follows from Wick theorem in a similar way.

The cluster property of the two-point Schwinger formula follows from the fact that the kernel $(-\nabla^2 + m^2)^{-1}(x, y)$ vanishes in the limit $\| x - y \| \to \infty$.

In this formalism the functional integral of the interacting theory can be understood as a Riemann-Stieltjes measure with respect to the Gaussian measure of the free theory $\delta\mu_m$, i.e.

$$S_n(\tilde{f}_1, \tilde{f}_2, \ldots, \tilde{f}_n) = \int_{\mathcal{S}'(R^4)} \delta\mu_m(\phi)\; e^{-V(\phi)}\, \phi(\tilde{f}_1)\phi(\tilde{f}_2), \ldots, \phi(\tilde{f}_n)\,.$$

Perturbation theory is defined just by the Taylor expansion of $e^{-V(\phi)}$ in power series and the formal commutation of the Gaussian integration with the Taylor sum. In the $\lambda\phi^4$ case the perturbation theory is defined by

$$S_n(\tilde{f}_1, \tilde{f}_2, \ldots, \tilde{f}_n) = \sum_{n=0}^{\infty} \frac{1}{n!} \frac{\lambda^n}{4!^n} \int_{\mathcal{S}'(R^4)} \delta\mu_m(\phi)\; \|\phi^2\|^{2n} \phi(\tilde{f}_1)\phi(\tilde{f}_2), \ldots, \phi(\tilde{f}_n)\,. \tag{3.56}$$

In this formalism UV divergences appear when computing the different terms of Eq. (3.56) by using Wick's theorem. But the main advantage of the covariant formalism is that in it the preservation of Poincaré symmetries under renormalization is more transparent.

3.9 What is Beyond?

From the Euclidean formulation it follows that the functional integral approach provides us with a constructive method of quantizing a field theory. The perturbative expansion Eq. (3.56) gives us a very explicit way of computing Schwinger functions. The ultraviolet divergences that arise there, in some cases can be renormalized by absorbing them in the bare parameters of the theory.

From this point of view the quantum field theories are classified in two classes:

(i) Theories where a finite set of parameters in the Lagrangian is enough to absorb all UV divergences; and
(ii) Theories that require an infinite set of independent parameters.

Theories of the first family are called renormalizable whereas those of the second class are unrenormalizable. Of course, only theories of the first type are sensible since from a finite number of parameters they can predict the behavior of all quantum states.

To distinguish between both cases one has to work out the perturbation theory and to find a good prescription to renormalize all UV divergences. The best renormalization scheme is the BPHZ method, developed by Bogoliubov, Parasiuk, Hepp and Zimmerman to provide a rigorous proof to the perturbative renormalization program [14].

In a heuristic way, one can distinguish the renormalizable theory just by a power counting algorithm. It proceeds by assigning a physical dimension to the fields according to their scale transformations in a such a way that the kinetic term of the action holds scale invariant (i.e. dimensionless). In the scalar theory this means that the scalar field ϕ has dimension $d_\phi = 1$, like the space-time derivative ∂_x operators. In that way the kinetic term of the action

$$\frac{1}{2}\|\nabla\phi\|^2$$

becomes dimensionless. The theory is renormalizable by power counting if all the terms of the action have non-positive dimensions. This constraint only allows Poincaré invariant terms like

$$\frac{m^2}{2} \parallel \phi \parallel^2 ,$$

which has dimension $d = -2$,

$$\frac{\sigma}{3!}\int_{\mathbb{R}^4} d^4x\ \phi(x)^3 ,$$

which has dimension $d = -1$, or

$$\frac{\lambda}{4!}\int_{\mathbb{R}^4} d^4x\ \phi(x)^4 ,$$

which is dimensionless. No other selfinteracting terms give rise to a renormalizable theory. This limitation became very important in model building because it introduces very stringent constraints on renormalizable models. What is highly remarkable is that Nature has chosen renormalizable models to build the theory of fundamental interactions.

The only fundamental theory which does not satisfy the renormalizability criterium is Einstein theory of Gravitation. In that theory due to diffeomorphism invariance the Einstein term contains an infinity of terms with positive dimensions, which generate an infinite number of new counterterms by renormalization.

One of the advantages of the covariant approach is that it does not require the existence of a classical Lagrangian. It is enough to have a complete set of Schwinger functions satisfying the fundamental properties E1-E5 of a QFT. This opens the possibility of quantum systems which are not defined by quantization of a classical system. There are few examples of that sort. But also it opens the possibility of having different field theories with the same Schwinger functions. In that case they are quantum-mechanically equivalent although their classical theories are completely different.

In two space-time dimensions, theories which in addition to the fundamental properties E1-E5 are conformally invariant, have been analyzed and classified without any reference to the corresponding classical systems [9]. In three dimensions there has been a recent breakthrough which opens the possibility of having similar results [20][22]. However, in four space-time dimensions the problem is far from its solution.

In the early seventies Wilson developed an interpretation of the renormalization method as a nonlinear representation of the one-dimensional group of dilations [29]. In the Euclidean formalism using a space-time lattice regularization Wilson mapped the quantum field theory system into a statistical mechanics model. Then, using the properties of second order phase transitions he interpreted the renormalization of a field theory as a limit process near a critical point of the renormalization group associated to the second order phase transition. The Wilson method provided a new non-perturbative approach to quantum field theory which allows a numerical treatment and has been intensively used in quantum chromodynamics.

However, with Wilson's approach it was also born the possibility of considering QFT not as the ultimate theory of Nature. It can be considered just as a successful approximation to the intimate structure of Nature. This approach, known as effective field theory, considers that the range of validity of the quantum field theory has an energy upper bound beyond which the theory does not hold. The limit scale is sometimes associated to the Planck energy scale, but for some theories might be smaller.

In the last three decades to solve the problem of quantizing gravity there have been many attempts searching for theories which are beyond QFT. From one way or another all these attempts consider the possibility of non-local interactions. The most popular approach is superstring theory. The connection of all non-local approaches with fundamental aspects of Nature has not yet been confirmed by experiments.

Acknowledgments

Lecture notes of a minicourse given at the XXV International Fall Workshop on Geometry and Physics, CSIC, Madrid (2016). I thank the organizers for their invitation and hospitality during the Workshop. I also thank A. P. Balachandran, F. Falceto, G. Marmo, G. Vidal and A. Wipf for many enlightening discussions. This work has been partially supported by the Spanish MINECO/FEDER grant FPA2015-65745-P and DGA-FSE grant 2015-E24/2 and the COST Action MP1405 QSPACE, supported by COST (European Cooperation in Science and Technology).

3.10 Appendix 1. Casimir effect

In the domain Ω_{II} between two parallel plates, the normal modes with Dirichlet boundary conditions $\phi(x_1, x_2, -\frac{d}{2}) = \phi(x_1, x_2, \frac{d}{2}) = 0$ of the scalar field ϕ are wave functions

$$\phi_{k_1,k_2,n}(\mathbf{x}) = \frac{1}{4\pi^2 d} e^{ik_1x_1+ik_2x_2} \sin\left(\frac{n\pi x_3}{d} + \frac{n\pi}{2}\right). \tag{3.57}$$

with arbitrary real values of the transverse momenta k_1 and k_2 and discrete values for the longitudinal modes $k_3 = \frac{n\pi}{d}$ with $n = 1, 2, \cdots$.

For the other two domains outside the plates Ω_I, Ω_{III} the Fourier modes of the classical field ϕ are the same (3.57) but with continuous values of the transverse modes $n \in \mathbb{R}_+$.

The vacuum energy density in the domain Ω_{II} between the plates is given by

$$\mathcal{E}_{II} = \frac{1}{2d} \sum_{n=1}^{\infty} \int_{-\infty}^{\infty} \int_{-\infty}^{\infty} \frac{dk_1 dk_2}{(2\pi)^2} \sqrt{k_1^2 + k_2^2 + \left(\frac{n\pi}{d}\right)^2}.$$

In the heat kernel regularization, the vacuum energy density between the plates is

$$\mathcal{E}^{\epsilon}_{II} = \frac{1}{2d} \sum_{n=1}^{\infty} \int_{-\infty}^{\infty} \int_{-\infty}^{\infty} \frac{dk_1 dk_2}{(2\pi)^2} e^{-\epsilon(k_1^2+k_2^2+(\frac{n\pi}{d})^2)} \sqrt{k_1^2 + k_2^2 + \left(\frac{n\pi}{d}\right)^2}.$$

To calculate this energy density, let us define the function

$$\begin{aligned} f(z) &= \frac{1}{2\pi d} \int_0^{\infty} k\, dk\, e^{-\epsilon\sqrt{k^2+(\frac{z\pi}{d})^2}} \sqrt{k^2 + \left(\frac{z\pi}{d}\right)^2} \\ &= \frac{1}{4\pi d} \int_{(\frac{z\pi}{d})^2}^{\infty} d\kappa\, e^{-\epsilon\sqrt{\kappa}} \sqrt{\kappa} \end{aligned} \tag{3.58}$$

and to use the Euler-MacLaurin formula

$$\begin{aligned} \sum_{n=1}^{\infty} f(n) = \int_0^{\infty} dz\, f(z) + \frac{1}{2}[f(\infty) - f(0)] + \frac{1}{12}[f'(\infty) - f'(0)] \\ - \frac{1}{720}[f'''(\infty) - f'''(0)] + \ldots \end{aligned}$$

Now, from the last three terms only two, $f(0)$ and $f'''(0)$ do not vanish because the function Eq. (3.58) satisfies that $f(\infty) = f'(\infty) = f'''(\infty) = 0$ and $f'(0) = 0$. Thus, we get that

$$\begin{aligned}\mathcal{E}^{\epsilon}_{II} &= \frac{1}{2}\int \frac{d^3\mathbf{k}}{(2\pi)^3} e^{-\epsilon\mathbf{k}^2}\sqrt{\mathbf{k}^2} \\ &- \frac{1}{2d}\int_{-\infty}^{\infty}\int_{-\infty}^{\infty} \frac{dk_1 dk_2}{(2\pi)^2} e^{-\epsilon(k_1^2+k_2^2)}\sqrt{k_1^2+k_2^2} \\ &+ \frac{1}{1440} f'''(0) + \mathcal{O}(\epsilon)\,. \end{aligned} \tag{3.59}$$

The first term in Eq. (3.59) is the vacuum energy of the free field and gives a divergent contribution

$$\mathcal{E}^{(1)} = \frac{1}{8\pi^2\epsilon^2}\,. \tag{3.60}$$

The second term corresponds to the selfenergy of the plates and gives another divergent contribution

$$\mathcal{E}^{(2)} = \frac{1}{16\sqrt{\pi}\epsilon^{\frac{3}{2}}}\frac{1}{d}\,. \tag{3.61}$$

On the contrary the contribution of the third term is finite

$$\mathcal{E}^{(3)}_{II} = -\frac{\pi^2}{1440\, d^4}\,. \tag{3.62}$$

The calculation for the other two domains outside the plates Ω_I and Ω_{III} can be performed in a similar way, but the results can be derived from (3.60), (3.61) and (3.62) just by taking the limit $d \to \infty$. The results are

$$\mathcal{E}^{\epsilon}_{I} = \mathcal{E}^{\epsilon}_{III} = \frac{1}{8\pi^2\epsilon^2}\,. \tag{3.63}$$

3.11 Appendix 2. Gaussian measures

Let us consider the following Gaussian probability measure of zero mean

$$d\mu_c = \frac{e^{-\frac{x^2}{2c}}}{\sqrt{2\pi c}}dx\,,$$

in the real line $\mathbb{R}$. The average of any $L^1(\mathbb{R})$ function f

$$\langle f\rangle = \int_{\mathbb{R}} d\mu_c(x)\, f(x)\,,$$

gets its main contribution from the interval $(-c, c)$. The main properties of the Gaussian measure are given by the average of its momenta,

- $\langle 1 \rangle_c = 1$;
- $\langle x^{2m+1} \rangle_c = 0$;
- $\langle x^{2m} \rangle_c = (2m-1)!! \langle x^2 \rangle^m = (2m-1)!! \, c^{2m}$.

The last formula is known as Wick's theorem.

All characteristics of Gaussian measures can be derived from the average of a single special function

$$g_c(y) = \langle g_c(y) \rangle_c = e^{-\frac{c}{2} y^2} .$$

In particular, all momenta of the measure can be obtained from the derivatives of g_c at the origin

$$\langle x^m \rangle = (-i)^m \frac{d^m}{dy^m} g_c \Big|_{y=0} .$$

The multidimensional generalization is straightforward. Let C be a positive, symmetric matrix, i.e.

$$(x, Cy) = (Cx, y) , \qquad (x, Cx) > 0) .$$

Positivity implies the non-degenerate character of C, $\det C \neq 0$, which guarantees the existence of the inverse matrix C^{-1}.

The Gaussian probability measure is defined by

$$d\mu_C = \frac{e^{-\frac{1}{2} \, (\mathbf{x}, C^{-1} \mathbf{x})}}{\sqrt{2\pi \det C}} d^n \mathbf{x} .$$

The momenta of the multidimensional Gaussian measure are obtained in terms of the covariance matrix C,

- $\langle 1 \rangle_C = 1$;
- $\langle x^{i_1} x^{i_2} \dots x^{i_{2m-1}} \rangle_C = 0$;
- $\langle x^{i_1} x^{i_2} \dots x^{i_{2m}} \rangle_C = \dfrac{1}{2^m m!} \displaystyle\sum_{\sigma \in S_{2m}} C_{\sigma(i_1)\sigma(i_2)} C_{\sigma(i_3)\sigma(i_4)} \cdots C_{\sigma(i_{2m-1}\sigma(i_{2m}))}$.

The last formula is the multidimensional Wick's theorem. The generating function of $d\mu_C$ is

$$g_C(\mathbf{y}) = \langle e^{i(\mathbf{x}, \mathbf{y})} \rangle_C = e^{-\frac{1}{2} (\mathbf{y}, C\mathbf{y})} .$$

All the momenta of the Gaussian measure, can be obtained from the derivatives of g_C at the origin

$$\langle x^{i_1} x^{i_2} \dots x^{i_m} \rangle_C = (-i)^m \frac{\partial^m g_C}{\partial_{x^{i_1}} \partial_{x^{i_2}} \dots \partial_{x^{i_m}}} \Big|_{\mathbf{y}=0} .$$

Gaussian Measures in Hilbert Spaces

Gaussian probability measures can be also defined in infinite dimensional topological vector spaces. The simplest case is the Hilbert space case. Let us consider a positive, selfadjoint, trace class operator C defined in a Hilbert space $\mathcal{H}$:

- $(x, Cy) = (Cx, y)$ for any $x, y \in \mathcal{H}$;
- $(x, Cx) > 0$ for any $x \in \mathcal{H}$;
- $\operatorname{tr} C < \infty$.

Positivity implies the non-degenerate character of C, which guarantees the existence of the inverse operator C^{-1}, that again is positive. Let us assume for concreteness that $\mathcal{H} = L^2(\mathbb{R}^n, \mathbb{C})$ and

$$C_s = (-\Delta + m^2)^{-s}$$

with $s > \frac{n}{2}$. It easy to check that C_s is positive, selfadjoint, trace class operator in $L^2(\mathbb{R}^n, \mathbb{C})$.

The Minlos theorem establishes that the measure defined by C_s is a Borelian probability measure in $\mathcal{H} = L^2(\mathbb{R}^n, \mathbb{C})$ [12]. The momenta of this Gaussian measure are again obtained in terms of the covariance matrix C,

- $\langle 1 \rangle_C = 1$;
- $\langle (g, f_1)(g, f_2) \dots (g, f_{2m-1}) \rangle_C = 0$;
- $\langle (g, f_1)(g, f_2) \dots (g, f_{2m-1} \rangle_C$
$$= \frac{1}{2^m m!} \sum_{\sigma \in S_{2m}} (f_{\sigma(1)}, C f_{\sigma(2)})(f_{\sigma(3)}, C f_{\sigma(4)}) \dots (f_{\sigma(2m-1)}, C f_{\sigma(2m)}),$$

where now we denote by $f_i \in L^2(\mathbb{R}^n, \mathbb{C})$ instead of x the vectors of the Hilbert space $\mathcal{H} = L^2(\mathbb{R}^n, \mathbb{C})$.

The last formula is the infinite dimensional version of Wick's theorem. And again the functional derivatives of the generating functional

$$G_C(f) = \langle e^{i(f,g)} \rangle_C = \int_{L^2(\mathbb{R}^n, \mathbb{C})} d\mu_C(g) e^{i(f,g)} = e^{-\frac{1}{2}(f, Cf)},$$

generate all momenta of the measure,

$$\langle (g, f_1)(g, f_2) \dots (g, f_m) \rangle_C = (-i)^m \frac{\partial^m G_C}{\partial_{f_1} \partial_{f_2} \dots \partial_{f_m}} \bigg|_{f=0}.$$

However, in field theory the natural covariance is not in general of trace class, thus, one needs to make appeal to another version of Minlos theorem which applies to covariances which are not of trace class [12]. In this case the space of test functions is the topological *dual* of the space of distributions where the measure is supported.

Let $\mathcal{S}(\mathbb{R}^n)$ be the space of fast decreasing smooth $C^\infty(\mathbb{R}^n)$ functions and C a positive, symmetric, bounded operator in $\mathcal{S}(\mathbb{R}^n)$. The Minlos theorem states that there is a unique Borelian Gaussian measure with C covariance in the space of tempered distributions $\mathcal{S}'(\mathbb{R}^n)$, which is the topological dual of $\mathcal{S}(\mathbb{R}^n)$. The same holds for the space of smooth functions of compact support $\mathcal{D}(\mathbb{R}^n)$ and its dual $\mathcal{D}'(\mathbb{R}^n)$, the space of generalized distributions.

In the first case the generating function

$$G_C(f) = \langle e^{i(f,g)} \rangle_C = \int_{\mathcal{S}'(\mathbb{R}^n,\mathbb{C})} d\mu_C(g) e^{i(f,g)} = e^{-\frac{1}{2}(f,Cf)},$$

is defined only for test functions $f \in \mathcal{S}(\mathbb{R}^n)$, whereas in the second case it is only defined for functions of compact support $\mathcal{D}(\mathbb{R}^n)$.

There are two special Gaussian measures which arise in quantum field theory. One is defined by the covariance operator

$$C_0 = (-\nabla^2 + m^2)^{-\frac{1}{2}}$$

in $\mathbb{R}^3$ that corresponds to the measure defined by the ground state of a free bosonic field theory. The second one is defined by the covariance operators

$$C_E = (-\Delta + m^2)^{-1}$$

in $\mathbb{R}^4$, which corresponds to the measure given by the Euclidean functional integral of a free bosonic theory. In physical terms C_E is known as the Euclidean propagator of a scalar field.

3.12 Appendix 3. Peierls brackets

There is an alternative canonical approach to classical dynamics developed by R. Peierls which provides a relativistic covariant description of classical field theory.

The standard approach uses the Poisson structure based on equal time commutators Eq. (3.6) Eq. (3.18). However, in relativistic theories the simultaneity of space-like separated points is not a relativistic invariant notion. For such a reason R. Peierls introduced an equivalent dynamical approach which explicitly preserves relativistic covariance.

The phase space in classical mechanics T^*M contains all Cauchy data $(x,p) \in T^*M$ and a canonical symplectic structure

$$\omega_0 = \sum_{i=1}^{n} dx^i \wedge dp^i$$

that determines the time evolution of the system for any kind of Hamiltonian function $H(p,q)$ by means of the Hamilton motion equations

$$\dot{x} = \{x, H\}\,, \qquad \dot{p} = \{p, H\}\,,$$

where $\{\cdot,\cdot\}$ is the Poisson bracket defined by the symplectic form ω_0.

Peierls remarks that the phase space can be identified with the space of trajectories of the system in M induced by any given Lagrangian L_0. In this sense, the Cauchy data are not fixed at a given initial time but by the trajectories themselves, which illustrates why it is the suitable framework for a covariant formulation. The standard canonical approach corresponds to the choice of the singular Lagrangian $L_0 = 0$.

Next, Peierls introduced a Poisson structure in the space of trajectories in the following way. Given time-dependent functions A in $TM \times \mathbb{R}$ we can consider a new dynamical system with Lagrangian $L_A = L_0 + \lambda A$. The trajectories of the new dynamics $x_\lambda(s)$ differ from those of that governed by L_0. If we compare the deviation of the trajectories with the same asymptotic values at $t = -\infty$ in the limit $\lambda \to 0$ we can associate to any other function B defined in $TM \times \mathbb{R}$ a new function in the space of trajectories given by

$$\mathrm{D}_A B(t) = \lim_{\lambda\to 0} \frac{1}{\lambda}\left[\int_{-\infty}^{t} ds\, B(x_\lambda(s), s) - \int_{-\infty}^{t} ds\, B(x_0(s), s)\right].$$

In a similar way, one can associate another function by comparing the deviation of the trajectories with the same asymptotic values at $t = +\infty$

$$\reflectbox{D}_A B(t) = \lim_{\lambda\to 0} \frac{1}{\lambda}\left[\int_{t}^{\infty} ds\, B(x_\lambda(s), s) - \int_{t}^{\infty} ds\, B(x_0(s), s)\right]. \tag{3.64}$$

The Peierls bracket $\{\cdot,\cdot\}_{L_0}$ is defined by

$$\{A, B\}_{L_0} = \mathrm{D}_A B - \reflectbox{D}_A B\,.$$

It has been proven by Peierls in 1952 that, when the Lagrangian L_0 is regular it defines a Hamiltonian system, and the above bracket satisfies the following properties:

$$\begin{aligned}
\{A, B + C\}_{L_0} &= \{A, B\}_{L_0} + \{A, C\}_{L_0} && \text{Distributive,}\\
\{A, BC\}_{L_0} &= \{A, B\}_{L_0} C + B\{A, C\}_{L_0} && \text{Leibnitz rule,}\\
\{A, B\}_{L_0} &= -\{B, A\}_{L_0} && \text{Antisymmetry,}\\
\{A, \{B, C\}_{L_0}\}_{L_0} &= \{\{C, A\}_{L_0}, B\}_{L_0} + \{\{A, B\}_{L_0}, C\}_{L_0} && \text{Jacobi identity,}
\end{aligned}$$

and thus, defines a Poisson structure in the space of classical trajectories.

In the case of the one-dimensional harmonic oscillator, the Poisson bracket of position operator is

$$\{x(t_1), x(t_2)\} = \frac{m}{\omega} \sin \omega(t_2 - t_1)\,.$$

The construction of Peierls brackets depends on the Lagrangian L_0 of the theory. In this sense it is not as universal as the Poisson brackets induced by the symplectic structure of T^*M.

The generalization to field theory is straightforward and the result for a free scalar theory is the following. In the case of the scalar field, a general solution of the field equations

$$(\partial_\mu \partial^\mu + m^2)\phi(x) = 0$$

can be obtained via its Fourier transform

$$(-k^2 + m^2)\widetilde{\phi}(k) = 0\,,$$

whose general solution can be written as $\widetilde{\phi}(p) = 2\pi\hat{\alpha}(k)\delta(k^2 - m^2)$, where $\hat{\alpha}(k)$ is an arbitrary function of k^μ. The solution in position space obtained by inverse Fourier transform is

$$\begin{aligned}\phi(x) &= \int \frac{d^4k}{(2\pi)^4}(2\pi)\delta(k^2 - m^2)\theta(k^0)\left[\alpha(k)e^{-ik\cdot x} + \alpha(k)^* e^{ik\cdot x}\right] \\ &= \int \frac{d^3k}{(2\pi)^3}\frac{1}{2\omega_k}\left[\alpha(\mathbf{k})e^{-i\omega_k t + \mathbf{k}\cdot\mathbf{x}} + \alpha(\mathbf{k})^* e^{i\omega_k t - \mathbf{k}\cdot\mathbf{x}}\right].\end{aligned}$$

The corresponding Peierls bracket is given by

$$\{\phi(x), \phi(y)\} = \Delta(x - y)\,,$$

where $\Delta(x - y)$ is the causal propagator 3.49. At equal times $x_0 = y_0$ the Peierls bracket reduce to the Poisson bracket, and vanishes.

In this framework the covariant quantization rule is a generalization of the canonical one: to replace the Peierls brackets by operator commutators.

Bibliography

[1] L. Alvarez-Gaumé and M. A. Vázquez-Mozo, *An Invitation to Quantum Field Theory*, Springer (2011)

[2] M. Asorey, *Conformal Invariance in Quantum Field Theory and Statistical Mechanics*, Forts. Phys. **40** (1992) 92

[3] M. Asorey, A. Ibort and G. Marmo, *Global theory of quantum boundary conditions and topology change*, Int. J. Mod. Phys. **A 20**, (2005) 1001

[4] T. Banks, *Modern Quantum Field Theory*, Cambridge U. Press (2008)

[5] J. D. Bjorken and S. D. Drell, *Relativistic Quantum Fields*, McGraw-Hill (1965)

[6] H.B.G. Casimir, *On the attraction between two perfectly conducting plates*, Proc. Kon. Ned. Akad. Wetensch. **B51** (1948) 793
[7] S. Coleman, *Notes from Sidney Coleman's Physics 253a*, arXiv:1110.5013 (1987)
[8] B. S. DeWitt, *The Global Approach to Quantum Field Theory*, Vols. 1 & 2, Oxford (2003)
[9] P. Di Francesco, P. Mathieu and D. Senegal, *Conformal Field Theory*, Springer (1997)
[10] P. A. M. Dirac, *The Principles of Quantum Mechanics*, Oxford U. Press (1958)
[11] G. Gamow, *Biography of Physics*, Dover (1961)
[12] I. Gelfand and N. Vilenkin, *Generalized Functions*, Vol. 4 (English Translation), Academic Press, New York (1964)
[13] J. Glimm, A. Jaffe, *Quantum Physics*, Springer (2012)
[14] C. Itzykson and J.-B. Zuber, *Quantum Field Theory*, McGraw-Hill (1980)
[15] V. P. Nair, *Quantum Field Theory. A Modern Perspective*, Springer (2005)
[16] E. Nelson, *A quartic interaction in two dimensions*, In: *Mathematics Theory of Elementary Particles*, R. Goodman and I. Segal, eds., Cambridge: MIT Press (1966)
[17] R. Peierls, *The commutation laws of relativistic field theory*, Proc. Roy. Soc. Lond. **A 214** (1952) 143
[18] M. E. Peskin and D. V. Schroeder, *An Introduction to Quantum Field Theory*, Addison Wesley (1995)
[19] P. Ramond, *Field Theory: A Modern Primer*, Addison-Wesley (1990)
[20] S. Rychkov, *EPFL Lectures on Conformal Field Theory in $D \geq 3$ Dimensions* (2016), arXiv:1601.05000
[21] L. Ryder, *Quantum Field Theory*, Cambridge (1996)
[22] D. Simmons-Duffin, *TASI Lectures on the Conformal Bootstrap* (2016) Preprint arXiv:1602.07982
[23] B. Simon, *$P(\phi)_2$ Euclidean (Quantum) Field Theory*, Princeton U. Press (1974)
[24] M. Srednicky, *Quantum Field Theory*, Cambridge (2007)
[25] R. Streater and A. Wightman, *PCT, Spin and Statistics, and All That*, Benjamin, New York (1964)
[26] K. Symanzik, *Euclidean Quantum Field Theory. I. Equations for a Scalar Model*, J. Math. Phys. **7** (1966) 510
[27] D. Tong, *Quantum Field Theory*, http://www.damtp.cam.ac.uk/user/tong/qft.html (2007)
[28] S. Weinberg, *The Quantum Theory of Fields*, Vol. 1, Cambridge (1995)
[29] K. G. Wilson, *The renormalization group: Critical phenomena and the Kondo problem*, Rev. Mod. Phys. **47** (1975) 773
[30] A. Zee, *Quantum Field Theory in a Nutshell*, Princeton (2003)

Index

CPSIA information can be obtained
at www.ICGtesting.com
Printed in the USA
JSHW021724160120
3607JS00001B/2